Industrial and Applied Mathematics

The **Industrial and Applied Mathematics** series publishes high-quality research-level monographs, lecture notes, textbooks, contributed volumes, focusing on areas where mathematics is used in a fundamental way, such as industrial mathematics, bio-mathematics, financial mathematics, applied statistics, operations research and computer science.

Anurag Jayswal · Preeti · Savin Treanţă

Multi-dimensional Control Problems

Robust Approach

Springer

Anurag Jayswal
Department of Mathematics and Computing
Indian Institute of Technology (ISM)
Dhanbad
Dhanbad, Jharkhand, India

Preeti
Department of Mathematics and Computing
Indian Institute of Technology (ISM)
Dhanbad
Dhanbad, Jharkhand, India

Savin Treanţă
Department of Applied Mathematics
University Politehnica of Bucharest
Bucharest, Romania

ISSN 2364-6837 ISSN 2364-6845 (electronic)
Industrial and Applied Mathematics
ISBN 978-981-19-6560-9 ISBN 978-981-19-6561-6 (eBook)
https://doi.org/10.1007/978-981-19-6561-6

Mathematics Subject Classification: 26A51, 35F15, 49K20, 68T37, 90C25, 93C35

This Springer imprint is published by the registered company Springer Nature Singapore Pte Ltd.
The registered company address is: 152 Beach Road, #21-01/04 Gateway East, Singapore 189721, Singapore

Preface

In real life, everyone faces various decision-making problems, where it requires optimizing several non-comparable objectives simultaneously, subject to different criteria. In this context, optimization refers to the process of selecting the optimal option from a set of limited or infinite possibilities. In essence, optimization provides quantitative models and techniques, unlike qualitative assessment of choices in daily life in the disciplines of applied research, electronics, industry, banking and finance, and many more. Mathematically, optimization is a structured methodology for optimizing a chosen quantity or function subject to such constraints. But, the most difficult and important task in such an optimization problem is the conflict between several goals, and these problems are known as multi-objective optimization problems.

Since data uncertainty is essential in modeling and studying the various optimization problems in recent years, the uncertain optimization problems are such types of problems in which the involved data is not exact. No one can ignore that the information inaccuracy and instability in data of the problems are a disadvantage. In this case, a robust approach is well-applicable to find the solution to such uncertain problems in an efficient way.

This book deals with different types of uncertain multi-dimensional multi-objective variational control problems. It contains eight chapters based on the penalty approach.

Chapter 1 is introductory and includes basic information on the variational control problem and its solution techniques. A short historical overview of the control problem involving multiple integral functionals is presented. Additionally, we formulate the structure of three types of uncertain optimization problems. Further, this chapter involves a brief overview of optimality conditions, dual problems and solution strategies to understand the established results in the book.

Chapter 2 is devoted to solving the multi-dimensional multi-objective variational control problem in the face of data uncertainty in the objective functional. For this purpose, first, we formulate its associated auxiliary problem with the help of the penalty function method and show the equivalence between its solution sets under convexity hypotheses. Further, we construct the robust Lagrangian for the considered constrained problem and prove that if the robust Lagrangian is convex then the constrained problem and its associated auxiliary problem attain their weak robust efficiency at the same point.

In Chap. 3, we intend to the applicability of the exact l_1 penalty function method for solving the multi-dimensional multi-objective optimization problem with the data uncertainty in the constraint functionals. In this way, we shift our attention toward the robust saddle-point criteria and prove that a saddle-point of the considered constrained problem and a weak robust efficient solution of its corresponding unconstrained problem are equivalent under the convexity assumptions.

Chapter 4 deals with the multi-dimensional multi-objective optimization problem with data uncertainty in the objective and constraint functionals. We construct the unconstrained multi-dimensional multi-objective optimization problem with the non-zero non-negative penalty parameter and non-negative penalty function. Further, we prove the equivalence between the associated robust solution sets under convexity hypotheses.

The main objective of Chap. 5 is to provide an interesting technique to solve the multi-dimensional optimization problem. Firstly, we use the modified objective function approach to simplify the uncertain multi-dimensional multi-objective optimization problem. Moreover, we show that the solutions set of the original problem and its associated modified problem are equivalent under the convexity assumption. Further, we apply the exact l_1 penalty function method to transform the modified problem into an equivalent penalized problem. Moreover, the robust saddle-point criteria are discussed to establish the relationship between the modified problem and its associated penalized problem in the face of data uncertainty.

Chapter 6 is concerned with an uncertain multi-dimensional optimization problem involving an infinite constraint functionals. We use the exact l_1 penalty function method to modulate an equivalent penalized semi-infinite multi-dimensional multi-objective optimization problem with data uncertainty. After that, we study the solution techniques to establish a relationship between the weak robust efficient solutions to the constrained problem and its corresponding penalized problem in the face of data uncertainty. In this order, firstly we use the simple convexity assumption of the involved functionals, then the convex Lagrange functional. Thereafter, we focus on the robust saddle-point criteria.

Chapter 7 deals with some new outcomes on the multi-dimensional optimization problem with data uncertainty by its dual problem. The Wolfe, Mond-Weir, and mixed type dual problems are discussed in this chapter in order to derive duality results under the convexity hypotheses.

Chapter 8 is devoted to studying the multi-dimensional multi-objective variational control problem involving second-order Partial Differential Equations (PDEs) and Inequations (PDIs). To facilitate the study of the considered problem, we modulate its auxiliary problem and develop some characterization results via robust saddle-point criteria.

Dhanbad, India Anurag Jayswal
Dhanbad, India Preeti
Bucharest, Romania Savin Treanţă

Contents

About the Authors

Anurag Jayswal is an associate professor at the Department of Mathematics and Computing, Indian Institute of Technology (Indian School of Mines) Dhanbad, India. He earned his Ph.D. degree in mathematics from Banaras Hindu University, Varanasi, India. He obtained his master in mathematics from same university and was awarded first order of merit. He received a young scientist project from the Department of Science and Technology, Government of India. He has more than 15 years of teaching experience at Birla Institute of Technology (BIT) Mesra, Ranchi, India, and IIT (ISM) Dhanbad, India. His research interest is in continuous optimization, nonsmooth optimization, generalized convexity, control theory, and variational inequalities problems. He is the author and coauthor of more than 100 research papers in the field of continuous optimization and variational inequalities and has supervised more than 10 Ph.D. students. He is on the editorial board of OPSEARCH and Advances in Variational Inequalities. He visited several countries to deliver their talks in international conferences. He is a reviewer of various international journals.

Preeti is an assistant professor at the Department of Applied Science and Humanities, Inderprastha Engineering College, Ghaziabad, India. She completed her Ph.D. degree in mathematics from the Department of Mathematics and Computing, Indian Institute of Technology (Indian School of Mines) Dhanbad, India. Her research interest is in multi-time optimization problem, generalized convexity, and control theory. She has published many scientific papers on these topics in various prominent journals.

Savin Treanţă is a lecturer at the Department of Applied Mathematics, Faculty of Applied Sciences, University Politehnica of Bucharest, Romania. His research interests include optimization theory, control theory, nonlinear and variational analysis, geometric partial differential equations, and information theory. He has published more than 80 scientific papers on these topics in various prestigious journals.

Glossary of Notations

Throughout the book, we shall denote the set of real numbers by $\mathbb{R}$ and the n-dimensional Euclidean space of real numbers by $\mathbb{R}^n$. The partial order in $\mathbb{R}^n$ leads to the following notational convention: for any two vectors $x, y \in \mathbb{R}^n$,

(i) $x = y \iff x_i = y_i, \forall i = 1, \dots, n;$
(ii) $x < y \iff x_i < y_i, \forall i = 1, \dots, n;$
(iii) $x \leqq y \iff x_i \leq y_i, \forall i = 1, \dots, n;$
(iv) $x \leq y \iff x_i \leq y_i, \forall i = 1, \dots, n$ and $x_j < y_j$ for some j.

$\mathbb{R}^n_+$	non-negative orthant of $\mathbb{R}^n$, i.e. the set $\{x \in \mathbb{R}^n : x_i \geq 0, \forall i = 1, \dots, n\}$
$\mathbb{R}^n_+ \setminus \{0\}$	non-negative orthant of $\mathbb{R}^n$ excluding the origin
$\mathbb{R}^{mn}$	set of $m \times n$ real matrices
max	maximum value of a set, function, etc.
min	minimum value of a set, function, etc.
A^T	transpose of the matrix A
$\langle ., . \rangle$	inner product on $\mathbb{R}^n$
$\lVert . \rVert$	the standard Euclidean norm
$\lVert x \rVert_\infty$	the maximum norm of $x \in \mathbb{R}^n$
$\nabla f(x)$	the gradient of a function f at x
int$C$	interior of a set C
C^1-class function	all differentiable functions whose first derivative is continuous
C^2-class function	all differentiable functions whose second derivative is continuous
C^3-class function	all differentiable functions whose third derivative is continuous

Chapter 1
Introduction

1.1 Overview of Literature

Optimization problems are as old as science itself, implicit in the behavior of humanity. These are defined by their specific objective functions, which should be maximized or minimized depending on the situation. In recent decades, the optimization problem has become an active field, as it provides a large-scale platform for modeling real-world applications that take place in everyday life. Applied science and engineering, industrial and system engineering, mathematical economics, management science, transportation and logistics, inventor planning and programming, and many other fields are included in it.

A great mathematician, named Leonhard Euler (1707–1783), gave a new dimension to the theory of optimization via calculus of variations that plays a vital role in finding the best possible solution to the problem. The trajectory of the fastest movement, minimal surfaces and the shortest distance between two points are some examples in this respect. Later, Hanson [1], following Berkovitz [2], established the equivalence between optimization problems and variational problems. In this regard, remarkable research works have been conducted to find the solution to many engineering problems with the help of the variational problem (see, for instance, Pereira [3, 4]). Thereafter, the variational problem was extended over multi-time, and it is known as the multi-time variational problem. There is a long story about using the concept of *multi-time*. We make dishonesty by mentioning only a part: the multi-temporal equations have been used for the first time in Physics by Dirac *et al.* [5], where an individual time is introduced for each of the n particles: "Besides the common time T and the field time t an individual time $t_s = t_1, t_2, \ldots, t_n$ is introduced for each particle". Later, the term *many-time* appears explicitly in Tomonaga [6]: "The equation (...) is the starting point of the *many-time* theory. In this theory, one introduces then the function $\Phi(q_1, t_1, q_2, t_2, \ldots, q_N, t_N)$ containing so many time variables $t_1, t_2, \ldots, t_N$ as the number of the particles ...". We often operate in physical problems with a *two-time*, $t = (t^1, t^2)$, where one component means the intrinsic time and the other one represents the observer time, not having any preference for

A. Jayswal et al., *Multi-dimensional Control Problems*, Industrial and Applied Mathematics, https://doi.org/10.1007/978-981-19-6561-6_1

one of the two components. For more contributions and various approaches, we refer to Petrat and Tumulka [7], Lienert and Nickel [8], Deckert and Nickel [9], Keppeler and Sieber [10] and Teufel and Tumulka [11]. The *multi-time* notion was also used in Mathematics: Friedman [12], Friedman and Littman [13], Yurchuk [14], Kendall [15], Saunders [16], Bouziani [17], Khoshnevisan *et al.* [18], Bayen *et al.* [19], Motta and Rampazzo [20], Udrişte and Ţevy [21], Prepeliţă [22], Cardin and Viterbo [23], Atanasiu and Neagu [24], Benrabah *et al.* [25], Damian [26], Ghiu [27], Vu [28], Mititelu and Treanţă [29], Treanţă [30–33], Jayswal *et al.* [34, 35] and so on. Pitea *et al.* [36] considered the multi-time multi-objective fractional variational problem involving geometrical language and curvilinear integrals. Then, the authors proved the necessary conditions and established the duality results. Mititelu and Postolache [37] examined the multi-time vector fractional and nonfractional variational problems involving multiple integrals on Riemannian manifolds and derived the duality results. Later, Pitea and Anctzak [38] studied a new class of nonconvex multi-time multi-objective variational problems and established sufficient optimality conditions for efficiency and proper efficiency under the univexity.

The notion of multi-time is extended over variational control problems, which give a broad application in diverse areas of the experimental sciences and technology like in economics as process control, in engineering as robotics and automation, in psychology as impulse control disorders, in medicine as bladder control and in biology as population ecosystems. From the theoretical and application aspects, these optimization problems have been intensively analyzed in the last few years (see, for example, [39, 40]). Treanţă and Mititelu [41] examined the multi-dimensional multi-objective fractional control problems and derived the duality results under (ρ, b)-quasi-invexity assumptions. For more contribution, we refer to [29, 42–47] and references therein.

On the other hand, during the last few years, multi-objective optimization (also referred to as multi-objective programming, vector optimization or multi-criteria optimization) has received great interest from researchers in the areas of optimization theory. This is a consequence of the fact that multi-objective optimization is applied in many mathematical disciplines, and it is a useful mathematical model in order to investigate many real-world problems with conflicting objectives, arising from human decision-making, engineering, mechanics, economics, logistics, optimization, information technology and others, where optimal decisions need to be taken in the presence of trade-offs between two or more conflicting objectives.

The present book focuses on that part of the calculus of variations and related applications in the presence of uncertainty which combines tools and methods from partial differential equations with multi-dimensional optimal control theory. More specifically, this book is devoted to uncertain nonlinear optimization problems coming from different areas, with particular reference to those introducing new techniques capable of solving a wide range of problems. With various examples and applications to complement and substantiate the mathematical developments, the present book is a valuable guide for researchers, engineers and students in the field of mathematics, optimal control science, artificial intelligence, management science and economics.

1.2 Pontryagin Maximum Principle for Single-time Optimal Control Problem with Second-order ODE Constraints

According to Treanţă [48], let us consider an optimal control problem governed by a simple integral cost functional and second-order ODE constraints:

$$\max_{(u(\cdot),x_{t_0})} \left\{ I\left(u(\cdot)\right) = \int_0^{t_0} X\left(t, x(t), \dot{x}(t), u(t)\right) dt \right\} \tag{1.2.1}$$

$$\text{subject to} \quad \ddot{x}^i(t) = X^i\left(t, x(t), \dot{x}(t), u(t)\right), \quad i = \overline{1, n}, \tag{1.2.2}$$

$$x(0) = x_0,\ x(t_0) = x_{t_0},\ \dot{x}(0) = \tilde{x}_0,\ \dot{x}(t_0) = \tilde{x}_{t_0},\ u(t) \in \mathcal{U},\ \forall t \in [0, t_0]. \tag{1.2.3}$$

Terminology and notations: $t \in [0, t_0]$ is a single parameter of evolution, or *single-time*; $[0, t_0] \subset \mathbb{R}_+$ is the *time interval*; $x(t) = (x^i(t)),\ i = \overline{1, n}$, is a C^3-class function, called *state vector*; $u(t) = (u^\alpha(t)),\ \alpha = \overline{1, k}$, is a *continuous control vector*; the *running cost* $X\left(t, x(t), \dot{x}(t), u(t)\right)$ is a C^1-class function, called *non-autonomous Lagrangian*.

Further, the summation over the repeated indices is assumed. Also, we remark that the differential system (1.2.1) can be rewritten as follows:

$$\dot{x}^i(t) := z^i(t),\ \dot{z}^i(t) = X^i\left(t, x(t), z(t), u(t)\right), \quad i = \overline{1, n}. \tag{1.2.4}$$

Using the Lagrange function (*Lagrangian*),

$$L\left(t, x(t), \dot{x}(t), z(t), \dot{z}(t), u(t), p(t), q(t)\right) = X\left(t, x(t), z(t), u(t)\right)$$
$$+\ p_i(t)\left[z^i(t) - \dot{x}^i(t)\right] + q_i(t)\left[X^i\left(t, x(t), z(t), u(t)\right) - \dot{z}^i(t)\right],$$

where $p(t) = (p_i(t)),\ q(t) = (q_i(t)),\ i = \overline{1, n}$, are called *co-state variables* or *Lagrange multipliers*; we build the *control Hamiltonian*,

$$H\left(t, x(t), z(t), u(t), p(t), q(t)\right) = X\left(t, x(t), z(t), u(t)\right) + p_i(t)z^i(t)$$
$$+\ q_i(t)X^i\left(t, x(t), z(t), u(t)\right),$$

or, equivalently, $H = L + p_i\dot{x}^i + q_i\dot{z}^i$ (*modified Legendrian duality*).

The next theorem formulates the necessary optimality conditions under the form of the *Pontryagin Maximum Principle*.

Theorem 1.2.0.1 ***(Simplified single-time Pontryagin maximum principle)*** *Let $(x, \hat{u})$ be an optimal pair in (1.2.1), subject to (1.2.2) and (1.2.3). Then there exist a C^1-class co-state variable $p = (p_i)$, respectively, a C^2-class co-state variable $q = (q_i)$, defined over $[0, t0]$ such that*

$$\dot{x}^j(t) = \frac{\partial H}{\partial p_j}\left(t, x(t), \dot{x}(t), \hat{u}(t), p(t), q(t)\right) \tag{1.2.5}$$

$$\ddot{x}^j(t) = \frac{\partial H}{\partial q_j}\left(t, x(t), \dot{x}(t), \hat{u}(t), p(t), q(t)\right), \quad \forall t \in [0, t_0], \quad j = \overline{1, n} \tag{1.2.6}$$

$$x(0) = x_0, \quad \dot{x}(0) = \tilde{x}_0, \tag{1.2.7}$$

the functions $p = (p_i)$, $q = (q_i)$ *satisfying*

$$\dot{p}_j(t) = -H_{x^j}\left(t, x(t), \dot{x}(t), \hat{u}(t), p(t), q(t)\right), \quad p_j(t_0) = 0, \tag{1.2.8}$$

$$\dot{q}_j(t) = -H_{\dot{x}^j}\left(t, x(t), \dot{x}(t), \hat{u}(t), p(t), q(t)\right), \quad q_j(t_0) = 0, \tag{1.2.9}$$

and the critical point conditions are

$$H_{u^\alpha}\left(t, x(t), \dot{x}(t), \hat{u}(t), p(t), q(t)\right) = 0, \quad \forall t \in [0, t_0], \quad \alpha = \overline{1, k} \tag{1.2.10}$$

and

$$\frac{\partial H}{\partial x^j}\left(t, x(t), \dot{x}(t), \hat{u}(t), p(t), q(t)\right)$$
$$- \frac{d}{dt}\left[\frac{\partial H}{\partial \dot{x}^j}\left(t, x(t), \dot{x}(t), \hat{u}(t), p(t), q(t)\right) - p_j(t)\right] + \frac{d^2}{dt^2}\left[-q_j(t)\right] = 0, \quad \forall t \in [0, t_0].$$

1.3 Pontryagin Maximum Principle for Multi-time Optimal Control Problem with Distribution-type Constraints

Next, in accordance with Treanţă [49], we consider a multi-time optimal control problem that involves as basic tools a multiple integral cost functional and constraints of distribution type:

$$\max_{u(\cdot)} \left\{ J\left(u(\cdot)\right) = \int_{\Omega_{t_0,t_1}} X\left(t, x(t), x_{\alpha_1}(t), \ldots, x_{\alpha_1\alpha_2\ldots\alpha_{s-1}}(t), u(t)\right)\omega \right\} \tag{1.3.1}$$

subject to $\quad dx^i_{\alpha_1\alpha_2\ldots\alpha_{s-1}}(t) = X^i_\beta\left(t, x(t), x_{\alpha_1}(t), \ldots, x_{\alpha_1\alpha_2\ldots\alpha_{s-1}}(t), u(t)\right) dt^\beta,$ (1.3.2)

$$u(t) \in \mathcal{U}, \ \forall t \in \Omega_{t_0,t_1}; \quad x(t_\xi) = x_\xi, \ x_{\alpha_1\ldots\alpha_j}(t_\xi) = \tilde{x}_{\alpha_1\ldots\alpha_j\xi},$$
$$i = \overline{1, n}, \quad \alpha_\zeta \in \{1, \ldots, m\}, \quad \zeta, j = \overline{1, s-1}, \quad \xi = 0, 1. \tag{1.3.3}$$

This kind of problem may appear when we want to describe the torsion of prismatic bars in the elastic or elastic-plastic case.

We have the following: $t = (t^\alpha) \in \Omega_{t_0,t_1}$ is a *multi-time*; $\Omega_{t_0,t_1} \subset \mathbb{R}^m_+$ is determined by the diagonally opposite points t_0 and t_1 in $\mathbb{R}^m_+$; $x(t) = (x^i(t)) : \Omega_{t_0,t_1} \to \mathbb{R}^n$, is a C^{s+1}-class function, called *state variable*; $u(t) = (u^a(t)),\ a = \overline{1, k}$, is a *continuous control variable*; the *running cost* $X\left(t, x(t), x_{\alpha_1}(t), \ldots, x_{\alpha_1\alpha_2\ldots\alpha_{s-1}}(t), u(t)\right)\omega$ is a *non-autonomous Lagrangian m-form*; $\omega = dt^1 \wedge \cdots \wedge dt^m$ is the *volume form* in $\mathbb{R}^m_+$; the equations in (1.3.2) are *distribution-type equations*; the vector fields $X_\beta = \left(X^i_\beta\right)$, $\beta = \overline{1, m}$, are functions of C^1-class which have independent variables.

Further, let us consider the *Lagrange multiplier tensors* or *co-state tensors*, $p_\gamma(t) = p^\alpha_{i\gamma}(t)\dfrac{\partial}{\partial t^\alpha} \otimes dx^i$, $\gamma = \overline{1, s}$, and the $(m-1)$-forms $\omega_\lambda = \dfrac{\partial}{\partial t^\lambda} \rfloor \omega$ (see $Y\rfloor\omega$ as the *contraction* between Y and ω). Now, we build a new Lagrangian m-form

$$
\begin{aligned}
&L\left(t, v_1(t), v_2(t), \ldots, v_s(t), dv_1(t), dv_2(t), \ldots, dv_s(t), u(t), p_1(t), \ldots, p_s(t)\right)\\
&= X\left(t, v_1(t), v_2(t), \ldots, v_s(t), u(t)\right)\omega + p^\lambda_{i1}(t)\left[v^i_{1\alpha_1}(t)dt^{\alpha_1} - dv^i_1(t)\right] \wedge \omega_\lambda\\
&\quad + \cdots + p^\lambda_{is-1}(t)\left[v^i_{s-1\alpha_{s-1}}(t)dt^{\alpha_{s-1}} - dv^i_{s-1}(t)\right] \wedge \omega_\lambda\\
&\quad + p^\lambda_{is}(t)\left[X^i_\beta\left(t, v_1(t), v_2(t), \ldots, v_s(t), u(t)\right)dt^\beta - dv^i_s(t)\right] \wedge \omega_\lambda.
\end{aligned}
$$

Also, we introduce the *control Hamiltonian m-form*

$$
\begin{aligned}
&H\left(t, v_1(t), v_2(t), \ldots, v_s(t), u(t), p_1(t), \ldots, p_s(t)\right)\\
&= X\left(t, v_1(t), v_2(t), \ldots, v_s(t), u(t)\right)\omega + p^\lambda_{i1}(t)v^i_{1\alpha_1}(t)dt^{\alpha_1} \wedge \omega_\lambda\\
&\quad + \cdots + p^\lambda_{is-1}(t)v^i_{s-1\alpha_{s-1}}(t)dt^{\alpha_{s-1}} \wedge \omega_\lambda + p^\lambda_{is}(t)X^i_\beta\left(t, v_1(t), v_2(t), \ldots, v_s(t), u(t)\right)dt^\beta \wedge \omega_\lambda\\
&= \left[X(\cdot) + p^{\alpha_1}_{i1}(t)v^i_{1\alpha_1}(t) + \cdots + p^\beta_{is}(t)X^i_\beta(\cdot)\right]\omega\\
&= H_1\left(t, v_1(t), v_2(t), \ldots, v_s(t), u(t), p_1(t), \ldots, p_s(t)\right)\omega,
\end{aligned}
$$

or, equivalently, $H = L + p^\lambda_{i1}(t)dv^i_1(t) \wedge \omega_\lambda + \cdots + p^\lambda_{is}(t)dv^i_s(t) \wedge \omega_\lambda$ (*modified Legendrian duality*) that permits us to rewrite the foregoing multi-time optimal control problem into the next equivalent form

$$
\begin{aligned}
\max_{u(\cdot)} \Bigg\{ &\int_{\tilde{\Omega}} H\left(t, x(t), x_{\alpha_1}(t), \ldots, x_{\alpha_1\ldots\alpha_{s-1}}(t), u(t), p_1(t), \ldots, p_s(t)\right)\\
&- \int_{\tilde{\Omega}} \left[p^\lambda_{i1}(t)dv^i_1(t) \wedge \omega_\lambda + \cdots + p^\lambda_{is}(t)dv^i_s(t) \wedge \omega_\lambda\right]\Bigg\}
\end{aligned}
$$

with

$$\tilde{\Omega} := \left(\Omega_{t_0,t_1}, x(\Omega_{t_0,t_1}), \ldots, x_{\alpha_1\ldots\alpha_{s-1}}(\Omega_{t_0,t_1})\right)$$

$$u(t) \in \mathscr{U}, \quad \{p_1(t), \ldots, p_s(t)\} \subseteq \mathscr{P}, \quad \forall t \in \Omega_{t_0,t_1}$$

$$x(t_\xi) = x_\xi, \quad x_{\alpha_1\ldots\alpha_j}(t_\xi) = \tilde{x}_{\alpha_1\ldots\alpha_j\xi}, \quad \xi = 0, 1$$

$$\alpha_\zeta \in \{1, \ldots, m\}, \quad \zeta, j = \overline{1, s-1},$$

and the set $\mathscr{P}$ of co-state tensors will be defined later.

Remark 1.3.0.1 The relation $\tilde{\Omega} \subset \mathbb{R}^m_+ \times \mathbb{R}^n \times \cdots \times \mathbb{R}^{[nm(m+1)\cdots(m+s-2)]/(s-1)!}$ is satisfied.

The necessary conditions of optimality for the aforementioned optimal control problem can be formulated as follows.

Theorem 1.3.0.2 ***(Simplified multi-time Pontryagin maximum principle)*** *Let us assume that the problem of maximizing the functional (1.3.1), constrained by (1.3.2) and (1.3.3), has an interior optimal solution* $\hat{u}(t) \in Int\,\mathcal{U}$ *which determines the optimal evolution* $x(t)$*. Then, there exist the* C^1*-class co-state tensors,* $p_r = \left(p^{\alpha}_{ir}\right)$, $r = \overline{1, s}$, *defined on* Ω_{t_0,t_1}, *such that*

$$dx^i_{\alpha_1\ldots\alpha_{r-1}}(t) \wedge \omega_\lambda = \frac{\partial H}{\partial p^\lambda_{ir}}\left(t, \ldots, x_{\alpha_1\ldots\alpha_{s-1}}(t), \hat{u}(t), p(t)\right), \quad \forall t \in \Omega_{t_0,t_1}, \tag{1.3.4}$$

and the function $p = (p_r)$, $r = \overline{1, s}$, *is the unique solution for the following Pfaff system:*

$$dp^\lambda_{j1}(t) \wedge \omega_\lambda = -H_{x^j}\left(t, x(t), \ldots, x_{\alpha_1\ldots\alpha_{s-1}}(t), \hat{u}(t), p(t)\right) \tag{1.3.5}$$

$$dp^\lambda_{j2}(t) \wedge \omega_\lambda = -H_{x^j_{\alpha_1}}\left(t, x(t), \ldots, x_{\alpha_1\ldots\alpha_{s-1}}(t), \hat{u}(t), p(t)\right)$$

$$\vdots$$

$$dp^\lambda_{js}(t) \wedge \omega_\lambda = -H_{x^j_{\alpha_1\ldots\alpha_{s-1}}}\left(t, x(t), \ldots, x_{\alpha_1\ldots\alpha_{s-1}}(t), \hat{u}(t), p(t)\right), \quad \forall t \in \Omega_{t_0,t_1}$$

$$\delta_{\alpha\beta}\, p^\alpha_{j1}(t)\eta^\beta(t) = 0, \ldots, \delta_{\mu\nu}\, p^\mu_{js}(t)\zeta^\nu(t) = 0 \quad (orthogonality/tangency)$$

and satisfies the critical point conditions

$$H_{u^a}\left(t, x(t), \ldots, x_{\alpha_1\ldots\alpha_{s-1}}(t), \hat{u}(t), p(t)\right) = 0, \quad \forall t \in \Omega_{t_0,t_1}. \tag{1.3.6}$$

1.4 General Construction of Multi-dimensional Control Problems

Multi-dimensional control problems find applications in various branches of mathematical, economical and engineering sciences, especially in mechanical engineering due to the fact that the curvilinear integral objectives have a physical meaning in mechanical work. Various real-world and application-oriented problems arising in diverse fields of science and engineering such as shape-optimization in fluid mechanics and medicine, material inversion in geophysics, structural optimization, optimal control of processes and data assimilation in regional weather prediction modeling require optimization problems with partial differential inequations and equations (PDI and PDE) as constraints. Therefore, PDI- and PDE-constrained multi-dimensional control problems, which present significant reasoning and computational challenges, have been given considerable interest in recent years.

Now, we introduce some mathematical notations and tools to formulate the multi-dimensional control problem:

- $\mathbb{R}^m$, $\mathbb{R}^n$, $\mathbb{R}^r$ and $\mathbb{R}^q$ are Euclidean spaces of dimension m, n, r and q, respectively.
- $\Omega = \Omega_{t_0,t_1} \subset \mathbb{R}^r$ is a hyperparallelepiped fixed by the diagonally opposite points $t_0 = (t_0^\alpha)$ and $t_1 = (t_1^\alpha)$, $\alpha = \overline{1, r}$.
- $\mathcal{X}$ is the space of piecewise smooth state functions $x : \Omega_{t_0,t_1} \to \mathbb{R}^m$.
- C is the space of piecewise continuous control functions $c : \Omega_{t_0,t_1} \to \mathbb{R}^n$.
- $dt = dt^1 \ldots dt^r$ is the volume element on $\mathbb{R}^r \supset \Omega_{t_0,t_1}$.
- For u, $v \in \mathbb{R}^q$, we use the following convention for inequalities and equalities:

$$\begin{aligned}
(i)\ & \mathrm{u} < v \Leftrightarrow \mathrm{u}_i < v_i, && \forall i = \overline{1, q}, \\
(ii)\ & \mathrm{u} = v \Leftrightarrow \mathrm{u}_i = v_i, && \forall i = \overline{1, q}, \\
(iii)\ & \mathrm{u} \leqq v \Leftrightarrow \mathrm{u}_i \leq v_i, && \forall i = \overline{1, q}, \\
(iv)\ & \mathrm{u} \leq v \Leftrightarrow \mathrm{u}_i \leq v_i, && \forall i = \overline{1, q} \text{ and } \mathrm{u}_i < v_i \text{ for some } i.
\end{aligned}$$

The general formulation of a multi-objective multi-dimensional control problem is as follows:

$$\begin{aligned}
\text{(MCP)} \quad & \min_{(x(\cdot),c(\cdot))} && \int_\Omega \Big(f_1(t, x(t), c(t)), \ldots, f_k(t, x(t), c(t))\Big) dt \\
& \text{subject to} && g_i(t, x(t), c(t)) \leqq 0, \\
& && \frac{\partial x^j}{\partial t^\alpha} = h_j^\alpha(t, x(t), c(t)), \\
& && x(t_0) = x_0, \quad x(t_1) = x_1,
\end{aligned}$$

where the objective functional $f = (f_l) : \Omega \times \mathscr{X} \times \mathscr{C} \to \mathbb{R}^k$, $l = \overline{1, k}$, the inequality constraint $g = (g_i) : \Omega \times \mathscr{X} \times \mathscr{C} \to \mathbb{R}^u$, $i = \overline{1, u}$ and the equality constraint $h = (h_j) : \Omega \times \mathscr{X} \times \mathscr{C} \to \mathbb{R}^m$, $j = \overline{1, m}$ are C^∞-class functionals. The functions h^α satisfy the closeness conditions (complete integrability conditions) $D_\beta h^\alpha = D_\alpha h^\beta$, $\alpha, \beta = \overline{1, r}$, $\alpha \neq \beta$, where $D_\alpha := \dfrac{\partial}{\partial t^\alpha}$.

A point $(\bar{x}(t), \bar{c}(t)) \in \mathscr{X} \times \mathscr{C}$ is said to be a feasible point/solution to the optimization problem (MCP) if all constraint functionals are satisfied. The collection of all feasible points is called a feasible set or feasible region. Mathematically, the feasible set is defined as

$$\begin{aligned}
\mathcal{D} = \{(x(t), c(t)) \in \mathscr{X} \times \mathscr{C} :\ & g_i(t, x(t), c(t)) \leqq 0,\ h_j^\alpha(t, x(t), c(t)) = \frac{\partial x^j}{\partial t^\alpha}, \\
& x(t_0) = x_0,\ x(t_1) = x_1\}.
\end{aligned}$$

In multi-objective programming, several conflicting and non-commensurate objective (criterion) functions have to be optimized over a feasible set determined by constraint functions. Due to the conflicting nature of the objective functions, a unique feasible solution optimizing all the objectives, in general, does not exist. The concepts of a weak Pareto solution (a weakly efficient solution) and a Pareto solution (an efficient solution) play useful roles in analyzing and solving this type of optimization problem. Mathematically, the definition of an efficient and a weak efficient solution for (MCP) are defined as follows.

Definition 1.4.0.1 A point $(\bar{x}(t), \bar{c}(t)) \in \mathcal{D}$ is said to be an efficient solution to the problem (MCP) if there does not exist $(x(t), c(t)) \in \mathcal{D}$ such that

$$\int_{\Omega} f(t, x(t), c(t))dt \leqq \int_{\Omega} f(t, \bar{x}(t), \bar{c}(t))dt,$$

with at least one strict inequality.

Definition 1.4.0.2 A point $(\bar{x}(t), \bar{c}(t)) \in \mathcal{D}$ is said to be a weak efficient solution to the problem (MCP) if there does not exist $(x(t), c(t)) \in \mathcal{D}$ such that

$$\int_{\Omega} f(t, x(t), c(t))dt < \int_{\Omega} f(t, \bar{x}(t), \bar{c}(t))dt.$$

1.5 Classification of the Multi-dimensional Cost Functionals

Mathematically, the multi-dimensional functional is categorized as follows.

1.5.1 Curvilinear Integral Cost Functional

$$\textbf{(MCP1)} \qquad \min_{(x(\cdot),c(\cdot))} \int_{\Gamma} f_{\alpha}(t, x(t), c(t))dt$$
$$= \left(\int_{\Gamma} f_{\alpha}^{1}(t, x(t), c(t))dt, \ldots, \int_{\Gamma} f_{\alpha}^{k}(t, x(t), c(t))dt \right)$$

where $f_{\alpha}^{l} : \Omega \times \mathscr{X} \times \mathscr{C} \to \mathbb{R}, \alpha = \overline{1, r}, l = \overline{1, k}$, are continuously differentiable functionals and Γ is a piecewise smooth curve, included in Ω, joining the points t_0 and t_1.

Example 1.5.1.1 For $k = \alpha = 2,\ \Omega = [0, 2] \times [0, 2] = [0, 2]^2$ and $\Omega \supset \Upsilon : t \in \overline{t_1, t_2}$ a piecewise smooth curve joining the points $t_1 = (0, 0)$ and $t_2 = (2, 2)$ in Ω, we consider

$$\mathscr{X} \times \mathscr{C} = \{(x, c) \mid x : \Omega \to \mathbb{R},\ x = \text{affine piecewise smooth function};$$

$$c : \Omega \to \left[-\frac{1}{2}, \frac{1}{2}\right],\ c = \text{piecewise continuous function};$$

$$\frac{\partial x}{\partial t^1} = \frac{\partial x}{\partial t^2} = 2 - c(t),\ 16 - x^2(t) \le 0,\ x(0,0) = 4,\ x(2,2) = 10\}.$$

For $(x, c) \in \mathscr{X} \times \mathscr{C}$, we consider the curvilinear integral cost functional

$$\begin{aligned} F(x, c) &= \int_\Upsilon f_\alpha(t, x(t), c(t))\, dt^\alpha = \int_\Upsilon \left([c(t) - 5]^2, c^2(t)\right) dt^\alpha \\ &= \left(\int_\Upsilon [c(t) - 5]^2\, dt^\alpha, \int_\Upsilon c^2(t) dt^\alpha\right) \\ &= \left(\int_\Upsilon [c(t) - 5]^2\, dt^1 + [c(t) - 5]^2\, dt^2, \int_\Upsilon c^2(t) dt^1 + c^2(t) dt^2\right), \end{aligned}$$

where we used $f^1 = \left(f_1^1, f_2^1\right) = \left([c(t) - 5]^2, [c(t) - 5]^2\right)$ and $f^2 = \left(f_1^2, f_2^2\right) = \left(c^2(t), c^2(t)\right)$.

1.5.2 Multiple Integral Cost Functional

$$\textbf{(MCP2)} \qquad \min_{(x(\cdot), c(\cdot))} \int_\Omega f(t, x(t), c(t)) dt = \left(\int_\Omega f_1(t, x(t), c(t)) dt, \ldots, \int_\Omega f_k(t, x(t), c(t)) dt\right)$$

where $f : \Omega \times \mathscr{X} \times \mathscr{C} \to \mathbb{R}^k$ is a vector-valued continuously differentiable functional.

Example 1.5.2.1 For $k = 2$ and $\Omega = [0, 3] \times [0, 3] = [0, 3]^2$, we formulate the following optimization problem:

$$\begin{aligned} &\min_{(x,c)} \left(\int_\Omega [c(t) - 4]^2 dt^1 dt^2, \int_\Omega c^2(t) dt^1 dt^2\right) \\ &\text{subject to} \quad 81 - x^2(t) \le 0, \quad t = \left(t^1, t^2\right) \in \Omega \\ &\qquad \frac{\partial x}{\partial t^1}(t) = \frac{\partial x}{\partial t^2}(t) = 3 - c(t), \quad t = \left(t^1, t^2\right) \in \Omega \\ &\qquad x(0) = x(0, 0) = 6, \quad x(3) = x(3, 3) = 8. \end{aligned} \tag{1.5.1}$$

The aforementioned bidimensional bi-objective control problem means to find the control function $c : \Omega \to \mathbb{R}$ [that determines the state function $x : \Omega \to \mathbb{R}$] to bring the (PDEs) dynamical system 1.5.1 from the initial point $x(0) = x(0, 0) = 6$, at two-time $(t^1, t^2) = (0, 0)$, to the terminal point $x(3) = x(3, 3) = 8$, at two-time $(t^1, t^2) = (3, 3)$, such as to minimize the objective vector functional.

In our illustrative example, we have

$$f : \Omega \times \mathscr{X} \times \mathscr{C} \to \mathbb{R}^2,$$

$$f(t, x(t), c(t)) = (f_1(t, x(t), c(t)), f_2(t, x(t), c(t))) = \left((c(t) - 4)^2, c^2(t)\right),$$

continuously differentiable functions.

1.6 Necessary and Sufficient Efficiency Conditions

The necessary and sufficient optimality conditions satisfied at optimal solutions form the foundation for both the development and analysis of efficient numerical methods for the practical solution to a given optimization problem. Differential calculus performs a crucial role in determining the optimal solutions to optimization problems. In this regard, the classical methods of optimization are also known as calculus methods.

Fritz John necessary efficiency conditions for multi-objective multi-dimensional control problems are defined as follows:

Theorem 1.6.0.1 (Fritz John necessary efficiency conditions) *Let* $(\bar{x}(t), \bar{c}(t)) \in \mathcal{D}$ *be an efficient solution to* (MCP). *Then, there exist the Lagrange multipliers* (*piecewise smooth functions*) $\theta_l \in \mathbb{R}, l = \overline{1, k},\ \mu_i(t) \in \mathbb{R}_+, i = \overline{1, u}$, *and* $\gamma_\alpha^j(t) \in \mathbb{R}, \alpha = \overline{1, r}, j = \overline{1, m}$, *satisfying the following conditions* (*with summation over the repeated indices*)

$$\theta_l \frac{\partial f_l}{\partial x^\varsigma}(t, \bar{x}(t), \bar{c}(t)) + \mu_i(t) \frac{\partial g_i}{\partial x^\varsigma}(t, \bar{x}(t), \bar{c}(t)) + \gamma_\alpha^j(t) \frac{\partial h_j^\alpha}{\partial x^\varsigma}(t, \bar{x}(t), \bar{c}(t)) + \frac{\partial \gamma_\alpha^j}{\partial t^\alpha}(t) = 0, \quad \varsigma = \overline{1, m},$$

$$\theta_l \frac{\partial f_l}{\partial c^\tau}(t, \bar{x}(t), \bar{c}(t)) + \mu_i(t) \frac{\partial g_i}{\partial c^\tau}(t, \bar{x}(t), \bar{c}(t)) + \gamma_\alpha^j(t) \frac{\partial h_j^\alpha}{\partial c^\tau}(t, \bar{x}(t), \bar{c}(t)) = 0, \quad \tau = \overline{1, n},$$

$$\mu_i(t)g_i(t, \bar{x}(t), \bar{c}(t)) = 0, \quad \theta \geq 0, \quad \mu(t) \geq 0,$$

for all $t \in \Omega$, *except at discontinuities.*

In Fritz John's necessary conditions, it is not necessary to impose restrictions on the objective and constraint functions apart from differentiability. If the Lagrange multiplier is associated with the objective function $\theta = 0$, then we cannot get any information about the gradient of the objective function via Fritz John necessary conditions. Hence, some restrictions are imposed on the constraint functions to ensure that $\theta > 0$, and these restrictions are known as constraint qualifications. There are many constraint qualifications that help turn Fritz John's necessary conditions into Kuhn-Tucker's necessary conditions.

Theorem 1.6.0.2 (Kuhn-Tucker necessary efficiency conditions) *Let* $(\bar{x}(t), \bar{c}(t)) \in \mathcal{D}$ *be an efficient solution to the problem* (MCP). *If the constraint conditions (for the existence of the multipliers) hold, then there exist the Lagrange multipliers (piecewise smooth functions)* $\theta_l \in \mathbb{R}, l = \overline{1, k}, \ \mu_i(t) \in \mathbb{R}, i = \overline{1, u}, \ \gamma_\alpha^j(t) \in \mathbb{R}, \alpha = \overline{1, r}, j = \overline{1, m}$ *satisfying the following conditions* (*with summation over the repeated indices*)

$$\theta_l \frac{\partial f_l}{\partial x^\varsigma}(t, \bar{x}(t), \bar{c}(t)) + \mu_i(t) \frac{\partial g_i}{\partial x^\varsigma}(t, \bar{x}(t), \bar{c}(t)) + \gamma_\alpha^j(t) \frac{\partial h_j^\alpha}{\partial x^\varsigma}(t, \bar{x}(t), \bar{c}(t)) + \frac{\partial \gamma_\alpha^j}{\partial t^\alpha}(t) = 0, \quad \varsigma = \overline{1, m}, \tag{1.6.1}$$

$$\theta_l \frac{\partial f_l}{\partial c^\tau}(t, \bar{x}(t), \bar{c}(t)) + \mu_i(t) \frac{\partial g_i}{\partial c^\tau}(t, \bar{x}(t), \bar{c}(t)) + \gamma_\alpha^j(t) \frac{\partial h_j^\alpha}{\partial c^\tau}(t, \bar{x}(t), \bar{c}(t)) = 0, \quad \tau = \overline{1, n}, \tag{1.6.2}$$

$$\mu_i(t)g_i(t, \bar{x}(t), \bar{c}(t)) = 0, \quad \theta_l > 0, \quad \mu(t) \geq 0, \tag{1.6.3}$$

for all $t \in \Omega$, *except at discontinuities.*

A point that satisfies the above Kuhn-Tucker conditions is not necessarily the optimal point of a given optimization problem. Therefore, Kuhn and Tucker proved the sufficiency of these conditions to guarantee the optimality of a feasible point if the considered objective and constraint functions are convex.

Theorem 1.6.0.3 (Kuhn-Tucker sufficient efficiency conditions) *Let* $(\bar{x}(t), \bar{c}(t)) \in \mathcal{D}$ *and the necessary efficiency conditions given in* (1.6.1)–(1.6.3) *be also satisfied at* $(\bar{x}(t), \bar{c}(t))$. *Then,* $(\bar{x}(t), \bar{c}(t))$ *is an efficient solution to the problem* (MCP) *if the involved functionals are convex at* $(\bar{x}(t), \bar{c}(t))$.

1.7 Robust Approach

In general, optimization problems are usually modeled as deterministic optimization problems, whereas, in real-world applications, optimization problems are framed with uncertain data. The data are often uncertain in real-world process models, for example, due to estimation errors, prediction errors or lack of information (for example, optimization problems arising in industry or commerce might involve various costs, financial returns and future demands that might be unknown at the time of the decision). Therefore, many decisions that we make in real life cannot be modeled easily in deterministic terms because of the inaccuracy of the data involved. Recently, there is an increasing interest in developing methods which are used for solving optimization problems with uncertainty. This is a consequence of the fact that many real-world problems from various areas of human activity should be modeled as uncertain optimization problems.

There have been many issues with uncertain data over the past decades. Generally, there are no circumstances in which anyone actually gathers specific information that interprets the exact figure of the object. No one can neglect that an optimal solution of the problem can be affected by a small uncertainty in the data. Therefore, it is important to support the decision maker's strategy for handling such an unpredictable optimization problem. The robust approach played a significant role in analyzing the optimization problem with data uncertainty. The practical aspects of this approach have been widely studied in logistics, finance, water management, energy management, machine learning, etc.

Mathematically, the uncertainty is involved in the multi-dimensional control problem in the following ways.

Case I: The multi-dimensional control problem (MCP) with data uncertainty in the objective functional is modulated as follows:

$$
\begin{aligned}
\text{(UCP1)} \quad & \min_{(x(\cdot),c(\cdot))} \left(\int_\Omega f_1(t, x(t), c(t), a_1)dt, \dots, \int_\Omega f_k(t, x(t), c(t), a_k)dt \right) \\
& \text{subject to} \quad g_i(t, x(t), c(t)) \leqq 0, \\
& \qquad \frac{\partial x^j}{\partial t^\alpha} = h_j^\alpha(t, x(t), c(t)), \\
& \qquad x(t_0) = x_0, \quad x(t_1) = x_1,
\end{aligned}
$$

where $a = (a_k)$ is the uncertain parameter for some convex compact subset of $\mathcal{A} = (\mathcal{A}_k) \subset \mathbb{R}^k$. $f = (f_l) : \Omega \times \mathcal{X} \times \mathcal{C} \times \mathcal{A} \to \mathbb{R}^k, l = \overline{1, k}, g = (g_i) : \Omega \times \mathcal{X} \times \mathcal{C} \to \mathbb{R}^u, i = \overline{1, u}, h = (h_j^\alpha) : \Omega \times \mathcal{X} \times \mathcal{C} \to \mathbb{R}^{mr}, \alpha = \overline{1, r}, j = \overline{1, m}$, are C^∞-class functionals.

The purpose of the robust approach is to ensure the existence of the solution in every possible uncertain situation. We can say that in this method, we obtain a robust solution that immunized against all the possible uncertain scenarios of the problem. Therefore, the robust counterpart for (UCP1) is given as

$$\text{(RCP1)} \quad \min_{(x(\cdot),c(\cdot))} \Big(\int_\Omega \max_{a_1 \in \mathcal{A}_1} f_1(t, x(t), c(t), a_1)dt, \ldots, \int_\Omega \max_{a_k \in \mathcal{A}_k} f_k(t, x(t), c(t), a_k)dt \Big)$$

$$\text{subject to} \quad g_i(t, x(t), c(t)) \leqq 0,$$

$$\frac{\partial x^j}{\partial t^\alpha} = h_j^\alpha(t, x(t), c(t)),$$

$$x(t_0) = x_0, \quad x(t_1) = x_1,$$

where f, g and h are the same as defined in (UCP1).
Let $\mathcal{D}1$ be the set of all feasible solutions to the problem (UCP1), which is defined as

$$\mathcal{D}1 = \{(x(t), c(t)) \in (\mathcal{X}, \mathcal{C}) : g_i(t, x(t), c(t)) \leqq 0, i = \overline{1, u}, \frac{\partial x^j}{\partial t^\alpha} = h_j^\alpha(t, x(t), c(t)),$$
$$\alpha = \overline{1, r}, j = \overline{1, m}, x(t_0) = x_0, \ x(t_1) = x_1\}.$$

The definitions of a robust efficient solution and a weak robust efficient solution for (UCP1) are formulated as follows.

Definition 1.7.0.1 A robust feasible point $(\bar{x}(t), \bar{c}(t))$ is said to be a robust efficient solution to (UCP1) if and only if it is an efficient solution to (RCP1), i.e.

$$\int_\Omega \max_{a \in \mathcal{A}} f(t, \bar{x}(t), \bar{c}(t), a)dt \leqq \int_\Omega \max_{a \in \mathcal{A}} f(t, x(t), c(t), a)dt$$

holds for all $(x(t), c(t)) \in \mathcal{D}1$, with at least one strict inequality.

Definition 1.7.0.2 A robust feasible point $(\bar{x}(t), \bar{c}(t))$ is said to be a weak robust efficient solution to (UCP1) if and only if it is a weak efficient solution to (RCP1), i.e.

$$\int_\Omega \max_{a \in \mathcal{A}} f(t, \bar{x}(t), \bar{c}(t), a)dt < \int_\Omega \max_{a \in \mathcal{A}} f(t, x(t), c(t), a)dt$$

holds for all $(x(t), c(t)) \in \mathcal{D}1$.

Case II: The multi-dimensional control problem (MCP) with data uncertainty in the constraint functionals is modulated as follows:

$$\text{(UCP2)} \quad \min_{(x(\cdot),c(\cdot))} \Big(\int_\Omega f_1(t, x(t), c(t))dt, \ldots, \int_\Omega f_k(t, x(t), c(t))dt \Big)$$

$$\text{subject to} \quad g_i(t, x(t), c(t), b) \leqq 0,$$

$$\frac{\partial x^j}{\partial t^\alpha} = h_j^\alpha(t, x(t), c(t), d),$$

$$x(t_0) = x_0, \quad x(t_1) = x_1,$$

where $b = (b_i)$, $d = (d_j^\alpha)$ are the uncertain parameter for some convex compact subset of $\mathscr{B} = (\mathscr{B}_i) \subset \mathbb{R}^u$ and $\mathfrak{D} = (\mathfrak{D}_j^\alpha) \subset \mathbb{R}^{mr}$. $f = (f_l) : \Omega \times \mathscr{X} \times \mathscr{C} \to \mathbb{R}^k, l = \overline{1,k}$, $g = (g_i) : \Omega \times \mathscr{X} \times \mathscr{C} \times \mathscr{B} \to \mathbb{R}^u$, $i = \overline{1,u}$, $h = (h_j^\alpha) : \Omega \times \mathscr{X} \times \mathscr{C} \times \mathfrak{D} \to \mathbb{R}^{mr}$, $\alpha = \overline{1,r}$, $j = \overline{1,m}$ are C^∞-class functionals.
Therefore, the robust counterpart for (UCP2) is given as

$$
\begin{aligned}
\text{(RCP2)} \quad & \min_{(x(\cdot),c(\cdot))} \left(\int_\Omega f_1(t, x(t), c(t))dt, \ldots, \int_\Omega f_k(t, x(t), c(t))dt \right) \\
& \text{subject to} \quad g_i(t, x(t), c(t), b) \leqq 0, \ \forall b \in \mathscr{B}, \\
& \qquad \frac{\partial x^j}{\partial t^\alpha} = h_j^\alpha(t, x(t), c(t), d), \ \forall d \in \mathfrak{D}, \\
& \qquad x(t_0) = x_0, \ x(t_1) = x_1,
\end{aligned}
$$

where f, g and h are the same as defined in (UCP2).

Let $\mathcal{D}2$ be the set of all feasible solutions of the (UCP2), which is defined as

$$
\begin{aligned}
\mathcal{D}2 = \{&(x(t), c(t)) \in (\mathscr{X}, \mathscr{C}) : g_i(t, x(t), c(t), b) \leqq 0, i = \overline{1,u}, \forall b \in \mathscr{B}, h_j^\alpha(t, x(t), c(t), d) \\
&= \frac{\partial x^j}{\partial t^\alpha}, \alpha = \overline{1,r}, j = \overline{1,m}, \ \forall d \in \mathfrak{D}, x(t_0) = x_0, x(t_1) = x_1\}.
\end{aligned}
$$

The definitions of a robust efficient solution and a weak robust efficient solution for (UCP2) are formulated as follows.

Definition 1.7.0.3 A robust feasible point $(\bar{x}(t), \bar{c}(t))$ is said to be a robust efficient solution to (UCP2) if and only if it is an efficient solution to (RCP2), i.e.

$$
\int_\Omega f(t, \bar{x}(t), \bar{c}(t))dt \leq \int_\Omega f(t, x(t), c(t))dt
$$

holds for all $(x(t), c(t)) \in \mathcal{D}2$.

Definition 1.7.0.4 A robust feasible point $(\bar{x}(t), \bar{c}(t))$ is said to be a weak robust efficient solution to (UCP2) if and only if it is a weak efficient solution to (RCP2), i.e.

$$
\int_\Omega f(t, \bar{x}(t), \bar{c}(t))dt < \int_\Omega f(t, x(t), c(t))dt
$$

holds for all $(x(t), c(t)) \in \mathcal{D}2$.

Case III: The multi-dimensional control problem with data uncertainty in the objective as well as constraint functionals is modulated as follows:

$$\begin{aligned}
\text{(UCP3)} \quad & \min_{(x(\cdot),c(\cdot))} \left(\int_\Omega f_1(t,x(t),c(t),a_1)dt, \ldots, \int_\Omega f_k(t,x(t),c(t),a_k)dt \right) \\
& \text{subject to} \quad g_i(t,x(t),c(t),b) \leqq 0, \\
& \qquad \frac{\partial x^j}{\partial t^\alpha} = h_j^\alpha(t,x(t),c(t),d), \\
& \qquad x(t_0) = x_0, \; x(t_1) = x_1,
\end{aligned}$$

where $a = (a_k)$, $b = (b_i)$ and $d = (d_j^\alpha)$ are the uncertain parameter for some convex compact subset of $\mathcal{A}_k \subset \mathbb{R}^k$, $\mathcal{B} = (\mathcal{B}_i) \subset \mathbb{R}^u$ and $\mathfrak{D} = (\mathfrak{D}_j^\alpha) \subset \mathbb{R}^{mr}$, respectively. $f = (f_l) : \Omega \times \mathcal{X} \times \mathcal{C} \times \mathcal{A} \to \mathbb{R}^k, l = \overline{1,k}$, $g = (g_i) : \Omega \times \mathcal{X} \times \mathcal{C} \times \mathcal{B} \to \mathbb{R}^u$, $i = \overline{1,u}$, $h = (h_j^\alpha) : \Omega \times \mathcal{X} \times \mathcal{C} \times \mathfrak{D} \to \mathbb{R}^{mr}, \alpha = \overline{1,r}$, $j = \overline{1,m}$, are C^∞-class functionals.
Therefore, the robust counterpart for (UCP3) is given as

$$\begin{aligned}
\text{(RCP3)} \quad & \min_{(x(\cdot),c(\cdot))} \left(\int_\Omega \max_{a_1 \in \mathcal{A}_1} f_1(t,x(t),c(t),a_1)dt, \ldots, \int_\Omega \max_{a_k \in \mathcal{A}_k} f_k(t,x(t),c(t),a_k)dt \right) \\
& \text{subject to} \quad g_i(t,x(t),c(t),b) \leqq 0, \; \forall b \in \mathcal{B}, \\
& \qquad \frac{\partial x^j}{\partial t^\alpha} = h_j^\alpha(t,x(t),c(t),d), \; \forall d \in \mathfrak{D}, \\
& \qquad x(t_0) = x_0, \; x(t_1) = x_1,
\end{aligned}$$

where f, g and h are the same as defined in (UCP3).
Let $\mathcal{D}3$ be the set of all feasible solutions of the (UCP3), which is defined as

$$\begin{aligned}
\mathcal{D}3 = \{ & (x(t),c(t)) \in (\mathcal{X},\mathcal{C}) : g_i(t,x(t),c(t),b) \leqq 0, i = \overline{1,u}, \forall b \in \mathcal{B}, h_j^\alpha(t,x(t),c(t),d) \\
& = \frac{\partial x^j}{\partial t^\alpha}, \alpha = \overline{1,r}, j = \overline{1,m}, \; \forall d \in \mathfrak{D}, x(t_0) = x_0, x(t_1) = x_1 \}.
\end{aligned}$$

The definitions of a robust efficient solution and a weak robust efficient solution for (UCP3) are given as follows.

Definition 1.7.0.5 A robust feasible point $(\bar{x}(t), \bar{c}(t))$ is said to be a robust efficient solution to (UCP3) if and only if it is an efficient solution to (RCP3), i.e.

$$\int_\Omega \max_{a \in \mathcal{A}} f(t,\bar{x}(t),\bar{c}(t),a)dt \leqq \int_\Omega \max_{a \in \mathcal{A}} f(t,x(t),c(t),a)dt$$

holds for all $(x(t), c(t)) \in \mathcal{D}3$, with at least one strict inequality.

Definition 1.7.0.6 A robust feasible point $(\bar{x}(t), \bar{c}(t))$ is said to be a weak robust efficient solution to (UCP3) if and only if it is a weak efficient solution to (RCP3), i.e.

$$\int_{\Omega} \max_{a \in \mathcal{A}} f(t, \bar{x}(t), \bar{c}(t), a)dt < \int_{\Omega} \max_{a \in \mathcal{A}} f(t, x(t), c(t), a)dt$$

holds for all $(x(t), c(t)) \in \mathcal{D}3$.

1.8 Penalty Function Method

An attractive approach to solving a nonlinear mathematical programming problem is to determine an exact penalty function, defined in terms of the objective function and both inequality and equality constraints and possibly derivatives thereof, which is minimized by an optimal solution to the original extremum problem. This holds out the possibility that the solution can be found by a single application of an unconstrained minimization technique to the exact penalized function. This is a consequence of the fact that these methods transform the original constrained optimization problem into a single unconstrained optimization problem by adding a term to the objective function that reflects the measure of violation of constraints. Exact penalty functions have quite different features. One important property that distinguishes these functions is the exactness of the penalization.

Attaining and maintaining feasibility with respect to nonlinear constraints are potentially infinite (if not impossible) tasks in a constrained optimization problem. The penalty function method is a technique that is used to convert the constrained problem into an equivalent unconstrained problem or into a problem having simple bound constraints. It means this method generates an unconstrained problem by adding the constraint functions into the objective function in the form of a penalty function that prescribes a high cost to the infeasible point for given restricted problems. A positive penalty occurs if constraints are violated, otherwise no penalty. During the transforming process, this method preserves the optimality of the original constrained problem which is an important point of this method.

Further, the penalty method is categorized in the following way: one is the exact penalty method and the other is the inexact penalty method. The solution to a constrained problem can be obtained by solving just a single unconstrained optimization problem associated with the considered one in exact penalty methods, whereas, in Inexact penalty methods, the constrained problem is transformed into a sequence of unconstrained problems whose solutions ultimately converge to the solution to the constrained optimization problem.

Mathematically, the unconstrained problem corresponding to the multi-dimensional control problem (MCP) can be modulated as

$$(\text{MCP})_\rho \quad \text{Minimize}\, \mathcal{P}(x(t), c(t), \rho) = \int_{\Omega} \Big\{ f(t, x(t), c(t)) + \rho p(t, x(t), c(t))e \Big\} dt,$$

where $e = (1, \dots, 1) \in \mathbb{R}^k$, ρ is a non-zero positive penalty parameter and p is a penalty function which is defined as

$$p(t, x(t), c(t)) = \sum_{i=1}^{u} \phi(g_i(t, x(t), c(t))) + \sum_{\alpha=1}^{r} \sum_{j=1}^{m} \varphi\Big(h_j^\alpha(t, x(t), c(t)) - \frac{\partial x^j}{\partial t^\alpha}\Big),$$

where ϕ and φ are continuous functions. Mathematically, these functions are defined as

$$\phi(g_i(t, x(t), c(t))) = \begin{cases} 0, & \text{if } g_i(t, x(t), c(t)) \leqq 0, \\ g_i(t, x(t), c(t), & \text{if } g_i(t, x(t), c(t)) > 0, \end{cases}$$

and

$$\varphi\Big(h_j^\alpha(t, x(t), c(t)) - \frac{\partial x^j}{\partial t^\alpha}\Big) = \begin{cases} 0, & \text{if } h_j^\alpha(t, x(t), c(t)) - \frac{\partial x^j}{\partial t^\alpha} = 0, \\ h_j^\alpha(t, x(t), c(t)) - \frac{\partial x^j}{\partial t^\alpha}, & \text{if } h_j^\alpha(t, x(t), c(t)) - \frac{\partial x^j}{\partial t^\alpha} \neq 0. \end{cases}$$

Typically, ϕ and φ are described as

$$\phi(g_i(t, x(t), c(t))) = \max\{0, g_i(t, x(t), c(t))\},$$

$$\varphi\Big(h_j^\alpha(t, x(t), c(t)) - \frac{\partial x^j}{\partial t^\alpha}\Big) = \Big|h_j^\alpha(t, x(t), c(t)) - \frac{\partial x^j}{\partial t^\alpha}\Big|.$$

Thus, the unconstrained optimization problem can be written as

$$(\text{MCP})_\rho \quad \min \mathcal{P}(x(t), c(t), \rho) = \int_\Omega \Big\{f(t, x(t), c(t)) + \rho\Big[\sum_{i=1}^{u} \max\{0, g_i(t, x(t), c(t))\} + \Big|h_j^\alpha(t, x(t), c(t)) - \frac{\partial x^j}{\partial t^\alpha}\Big|\Big]e\Big\}dt.$$

Courant [50] gave an ideology for solving constrained problems by converting them into unconstrained problems. In this way, the penalty function approach is the conventional approach for converting a constrained problem to a sequence of unconstrained problems. In the literature, many penalty functions have been introduced, in which the exponential penalty function is one of them introduced by Motzkin [51]. Another penalty function is the exact penalty function which was investigated by Zangwill [52].

1.9 Saddle-point Criterion

In general, a saddle-point is a point on the surface of the graph of a function where the slopes (derivatives) in orthogonal directions are all zero, but these points are not a local extremum of the given function. A saddle-point is also known as a minimax point or a critical point.

The theory of a saddle-point has an important place in optimization theory to obtain the optimality of the problem. It is well known that the saddle-point of Lagrange functional for the constrained optimization problem is equivalent to its solution under suitable assumptions. One of the features of saddle-point criteria is that it helps to establish the relationship between the solution set to the constrained problem and its associated unconstrained problem.
Now, we recall the definition of Lagrange functional and saddle-point for (MCP).

Definition 1.9.0.7 The Lagrange functional $L : \mathscr{X} \times \mathscr{C} \times \mathbb{R}^u_+ \times \mathbb{R}^{rm} \to \mathbb{R}^k$ for the considered problem (MCP) is defined by

$$L(x(t), c(t), \mu, \gamma) = \int_\Omega \Big\{ f(t, x(t), c(t)) + \mu_i^T g_i(t, x(t), c(t))e + (\gamma_j^\alpha)^T \Big(h_j^\alpha(t, x(t), c(t)) - \frac{\partial x^j}{\partial t^\alpha} \Big) e \Big\} dt.$$

Definition 1.9.0.8 A point $(\bar{x}(t), \bar{c}(t), \bar{\mu}, \bar{\gamma}) \in \mathcal{D} \times \mathbb{R}^u_+ \times \mathbb{R}^{rm}$ is said to be a saddle-point for the considered control problem (MCP), iff

(i) $L(\bar{x}(t), \bar{c}(t), \mu, \gamma) \leqq L(\bar{x}(t), \bar{c}(t), \bar{\mu}, \bar{\gamma}), \ \forall \mu \in \mathbb{R}^u_+, \ \ \forall \gamma \in \mathbb{R}^{rm}$,
(ii) $L(x(t), c(t), \bar{\mu}, \bar{\gamma}) \geqq L(\bar{x}(t), \bar{c}(t), \bar{\mu}, \bar{\gamma}), \ \ \forall (x(t), c(t)) \in \mathscr{X} \times \mathscr{C}$.

1.10 Duality Theory

The notion of duality is an important part to study optimization problems. In 1961, Wolfe formulated a dual problem by using the Kuhn-Tucker conditions for a nonlinear problem in the spirit of duality in linear problems. The purpose of defining the dual problem is to give a lower bound on the optimal value of the primal problem. After that in 1981, Mond and Weir modified the Wolfe dual and introduced a new dual which is known as the Mond-Weir dual. The important fact of the Mond-Weir dual is that the objective function of the Mond-Weir dual problem is the same as the primary problem and that the outcomes of duality are obtained by further relaxation of the conditions for invexity.

In short, duality is basically a unifying principle that develops the relationship between two constrained optimization problems, one of them known as the primal problem, which is a problem of minimization (maximization), and the other one referred to as the dual problem, which is a problem of maximization (minimization). In the duality theory, two problems are associated in such a way that the existence of a solution to one problem ensures the solution of the other one under certain assumptions. Further, a number of properties relating to the two problems hold, often under suitable convexity assumptions. The most well-known are the weak and strong duality results.

• **Weak duality**: The weak duality provides a lower bound for the objective value of the primal problem for any feasible solution. That means the optimal value of the primal problem is greater than or equal to the optimal value of the dual problem. If the inequality holds as a strict inequality, then it is said that there exists a duality gap.

• **Strong duality**: The strong duality states that there is no duality gap between the primal and dual problems if some convexity/generalized convexity assumptions and constraint qualifications are satisfied.

1.10.1 Wolfe Type Dual Problem

In 1947, Neumann investigated the duality theorems for linear optimization. Thereafter, Wolfe extended the duality theory for linear programming to convex nonlinear programming problems and introduced the Wolfe dual problem in optimization theory.

For a constrained minimization problem, the Wolfe dual problem associated with the primal problem (MCP) is formulated in the following way:

$$
\begin{aligned}
\text{(DCP1)} \quad \max_{(y(\cdot),z(\cdot))} \int_{\Omega_{t_0,t_1}} & \Big\{ f_l(t, y(t), z(t)) + \mu_i^T g_i(t, y(t), z(t))e \\
& + (\gamma_\alpha^j)^T \Big(h_j^\alpha(t, y(t), z(t)) - \frac{\partial y^j}{\partial t^\alpha} \Big) e \Big\} dt \\
\text{subject to} \quad & \theta_l^T \frac{\partial f_l}{\partial x^\varsigma}(t, y(t), z(t)) + \mu_i^T \frac{\partial g_i}{\partial x^\varsigma}(t, y(t), z(t)) \\
& + (\gamma_\alpha^j)^T \frac{\partial h_j^\alpha}{\partial x^\varsigma}(t, y(t), z(t)) + \frac{\partial \gamma_\alpha}{\partial t^\alpha} = 0, \ \varsigma = \overline{1, m}, \\
& \theta_l^T \frac{\partial f_l}{\partial c^\tau}(t, y(t), z(t), p) + \mu_i^T \frac{\partial g_i}{\partial c^\tau}(t, y(t), z(t)) \\
& + (\gamma_\alpha^j)^T \frac{\partial h_j^\alpha}{\partial c^\tau}(t, y(t), z(t)) = 0, \ \tau = \overline{1, n}, \\
& \mu_i g_i(t, y(t), z(t)) \geqq 0, \\
& \frac{\partial y^j}{\partial t^\alpha} = h_j^\alpha(t, y(t), z(t)), \\
& y(t_0) = x_0, \ y(t_1) = x_1, \\
& \theta \in \mathbb{R}_+^k, \mu \in \mathbb{R}_+^u, \gamma_\alpha^j \in \mathbb{R}^{rm}.
\end{aligned}
$$

1.10.2 Mond-Weir Type Dual Problem

The Mond-Weir dual problem associated with the primal problem (MCP) is formulated in the following way:

$$
\begin{aligned}
\text{(DCP2)} \quad & \max_{(y(\cdot),z(\cdot))} \int_{\Omega_{t_0,t_1}} f(t, y(t), z(t))dt \\
\text{subject to} \quad & \theta_l^T \frac{\partial f_l}{\partial x^\varsigma}(t, y(t), z(t)) + \mu_i^T \frac{\partial g_i}{\partial x^\varsigma}(t, y(t), z(t)) \\
& + (\gamma_\alpha^j)^T \frac{\partial h_j^\alpha}{\partial x^\varsigma}(t, y(t), z(t)) + \frac{\partial \gamma_\alpha}{\partial t^\alpha} = 0, \ \varsigma = \overline{1, m}, \\
& \theta_l^T \frac{\partial f_l}{\partial c^\tau}(t, y(t), z(t), p) + \mu_i^T \frac{\partial g_i}{\partial c^\tau}(t, y(t), z(t)) \\
& + (\gamma_\alpha^j)^T \frac{\partial h_i^\alpha}{\partial c^\tau}(t, y(t), z(t)) = 0, \ \tau = \overline{1, n}, \\
& \mu_i g_i(t, y(t), z(t)) \geqq 0, \\
& \frac{\partial y^j}{\partial t^\alpha} = h_j^\alpha(t, y(t), z(t)), \\
& y(t_0) = x_0, \ y(t_1) = x_1, \\
& \theta \in \mathbb{R}_+^k, \mu \in \mathbb{R}_+^u, \gamma_\alpha^j \in \mathbb{R}^{rm}.
\end{aligned}
$$

1.10.3 Mixed Type Dual Problem

The mixed dual problem associated with the primal problem (MCP) is formulated in the following way:

$$
\begin{aligned}
\text{(DCP3)} \quad & \max_{(y(\cdot),z(\cdot))} \int_{\Omega_{t_0,t_1}} \Big\{ f_l(t, y(t), z(t)) + \mu_i^T g_i(t, y(t), z(t))e \\
& \qquad + (\gamma_\alpha^j)^T \Big(h_j^\alpha(t, y(t), z(t)) - \frac{\partial y^j}{\partial t^\alpha} \Big) e \Big\} dt \\
\text{subject to} \quad & \theta_l^T \frac{\partial f_l}{\partial x^\varsigma}(t, y(t), z(t)) + \mu_i^T \frac{\partial g_i}{\partial x^\varsigma}(t, y(t), z(t)) \\
& + (\gamma_\alpha^j)^T \frac{\partial h_j^\alpha}{\partial x^\varsigma}(t, y(t), z(t)) + \frac{\partial \gamma_\alpha}{\partial t^\alpha} = 0, \ \varsigma = \overline{1, m}, \\
& \theta_l^T \frac{\partial f_l}{\partial c^\tau}(t, y(t), z(t), p) + \mu_i^T \frac{\partial g_i}{\partial c^\tau}(t, y(t), z(t)) \\
& + (\gamma_\alpha^j)^T \frac{\partial h_j^\alpha}{\partial c^\tau}(t, y(t), z(t)) = 0, \ \tau = \overline{1, n}, \\
& \mu_i g_i(t, y(t), z(t)) \geqq 0, \\
& \frac{\partial y_t}{\partial t^\alpha} = h_j^\alpha(t, y(t), z(t)), \\
& y(t_0) = x_0, \ y(t_1) = x_1, \\
& \theta \in \mathbb{R}_+^k, \mu \in \mathbb{R}_+^u, \gamma_\alpha^j \in \mathbb{R}^{rm}.
\end{aligned}
$$

For a number of reasons, the analysis of duality theory is important in optimization theory:

• Duality is the backbone of optimization theory. To completely understand the optimization theory, one should have a strong grasp of the principle of underlying duality.

• In various scenarios, it gives a broader appreciation of optimal conditions and thus provides valuable insights for designing efficient computational methods and algorithms.

• It gives a meaningful explanation of many practical problems in the real world, which in turn deepens the perception of the optimum solution underlying them.

• The dual problem also has a strong mathematical, geometric or computational structure which is helpful to find the computational solutions to both the primal and the dual problems.

References

1. M.A. Hanson, Bounds for functionally convex optimal control problems. J. Math. Anal. Appl. **8**, 84–89 (1964)
2. L.D. Berkovitz, Variational methods in problems of control and programming. J. Math. Anal. Appl. **3**, 145–169 (1961)
3. F.L. Pereira, A maximum principle for impulsive control problems with state constraints. Comput. Appl. Math. **19**, 1–19 (2000)
4. F.L. Pereira, Control design for autonomous vehicles: a dynamic optimization perspective. Eur. J. Control **7**, 178–202 (2001)
5. P.A.M. Dirac, V.A. Fock, B. Podolski, On quantum electrodynamics. Physikalische Zeitschrift der Sowjetunion **2**(6), 468–479 (1932)
6. S. Tomonaga, On a relativistically invariant formulation of the quantum theory of wave fields. Progress of Theoret. Phys. **1**, 27–42 (1946)
7. S. Petrat, R. Tumulka, Multi-time wave functions for quantum field theory. Ann. Phys. **345**, 17–54 (2014)
8. M. Lienert, L. Nickel, A simple explicitly solvable interacting relativistic N-particle model. J. Phys. A: Math. Theor. **48**, 325301 (2015)
9. D.A. Deckert, L. Nickel, Consistency of multi-time Dirac equations with general interaction potentials. J. Math. Phys. **57**, 072301 (2016)
10. S. Keppeler, M. Sieber, Particle creation and annihilation at interior boundaries: One-dimensional models, arXiv:1511.03071
11. S. Teufel, R. Tumulka, New type of Hamiltonians without ultraviolet divergence for quantum field theories, arXiv:1505.04847v1
12. A. Friedman, The Cauchy problem in several time variables. J. Math. Mech. (Indiana Univ. Math. J.) **11**, 859–889 (1962),
13. A. Friedman, W. Littman, Partially characteristic boundary problems for hyperbolic equations. J. Math. Mech. (Indiana Univ. Math. J.) **12**, 213–224 (1963)
14. N.I. Yurchuk, A partially characteristic mixed boundary value problem with Goursat initial conditions for linear equations with two-dimensional time. Diff. Uravn. **5**, 898–910 (1969)
15. W.S. Kendall, Contours of Brownian processes with several-dimensional times. Probab. Theory Relat. Fields **52**, 267–276 (1980)
16. D.J. Saunders, *The Geometry of Jet Bundles*, London Math. Soc. Lecture Notes Series 142. (Cambridge University Press, Cambridge, 1989)

17. A. Bouziani, Strong solution for a mixed problem with nonlocal condition for certain pluriparabolic equations. Hiroshima Math. J. **27**, 373–390 (1997)
18. D. Khoshnevisan, Y. Xiao, Y. Zhong, Local times of additive Levy processe. Stoch. Process. Appl. **104**, 193–216 (2003)
19. A.M. Bayen, R.L. Raffard, C.J. Tomlin, Adjoint-based constrained control of Eulerian transportation networks: Application to air trac control, in *Proceedings of the American Control Conference* (Boston, June, 2004)
20. M. Motta, F. Rampazzo, Nonsmooth multi-time Hamilton-Jacobi systems. Indiana Univ. Math. J. **55**, 1573–1614 (2006)
21. C. Udrişte, I. Ţevy, Multi-time Euler-Lagrange-Hamilton theory. WSEAS Trans. Math. **6**, 701–709 (2007)
22. V. Prepeliţă, Stability of a class of multidimensional continuous-discrete linear systems. Math. Reports **9**(59), 387–98 (2007)
23. F. Cardin, C. Viterbo, Commuting Hamiltonians and Hamilton-Jacobi multi-time equations. Duke Math. J. **144**, 235–284 (2008)
24. Gh. Atanasiu, M. Neagu, Canonical nonlinear connections in the multi-time Hamilton geometry. Balkan J. Geom. Appl. **14**, 1–12 (2009)
25. A. Benrabah, F. Rebbani, N. Boussetila, A study of the multitime evolution equation with time-nonlocal conditions. Balkan J. Geom. Appl. **16**, 13–24 (2011)
26. V. Damian, *Multitime Stochastic Optimal Control*, PhD Thesis, University "Politehnica" of Bucharest, (2011)
27. C. Ghiu, *Controllability of Multitime Linear PDEs Systems*, PhD Thesis, University "Politehnica" of Bucharest, 2013
28. Q.P. Vu, Stability and asymptotic behavior of systems with multi-time. Vietnam J. Math. **43**, 417–437 (2015)
29. Şt. Mititelu and S. Treanţă, Efficiency conditions in vector control problems governed by multiple integrals, J. Appl. Math. Comput. **57**, 647–665 (2018)
30. S. Treanţă, Multiobjective fractional variational problem on higher-order jet bundles. Commun. Math. Stat. **4**, 323–340 (2016)
31. S. Treanţă, On Modified Interval-Valued Variational Control Problems with First-Order PDE Constraints. Symmetry-Basel **12**, 472 (2020)
32. S. Treanţă, On a modified optimal control problem with first-order PDE constraints and the associated saddle-point optimality criterion. Eur. J. Control **51**, 1–9 (2020)
33. S. Treanţă, Saddle-point optimality criteria in modified variational control problems with PDE constraints. Optimal Control. Appl. Meth. **41**, 1160–1175 (2020)
34. A. Jayswal, T. Antczak, S. Jha, On equivalence between a variational problem and its modified variational problem with the η?objective function under invexity. Int. Trans. Op. Res. **26**, 2053–2070 (2019)
35. A. Jayswal, T. Antczak, S. Jha, Modified objective function approach for multitime variational problems. Turkish J. Math. **42**, 1111–1129 (2018)
36. A. Pitea, C. Udrişte, Şt. Mititelu, PDI&PDE-constrained optimization problems with curvilinear functional quotients as objective vectors, Balkan J. Geom. Appl. **14**, 75–88 (2009)
37. Şt. Mititelu, M. Postolache, Efficiency and duality for multitime vector fractional variational problems on manifolds, Balkan J. Geom. Appl. **16**, 90–101 (2011)
38. A. Pitea, T. Antczak, Proper efficiency and duality for a new class of nonconvex multitime multiobjective variational problems. J. Inequal. Appl. **2014**, 333 (2014)
39. K.J. Astrom, R.M. Murray, *Feedback Systems: An Introduction for Scientists and Engineers*. (Princeton University Press, 2008)
40. C. Udrişte, Simplified multitime maximum principle. Balkan J. Geom. Appl. **14**, 102–119 (2009)
41. S. Treanţă, Şt. Mititelu, Duality with (ρ, b)-quasiinvexity for multidimensional vector fractional control problems, J. Inform. Optim. Sci. **40**, 1429–1445 (2019)
42. E. Adida, G. Perakis, A robust optimization approach to dynamic pricing and inventory control with no backorders. Math. Program. **107**, 97–129 (2006)

43. M. Arana-Jiménez, R. Osuna-Gómez, A. Rufián-Lizana, G. Ruiz-Garzón, KT-invex control problem. Appl. Math. Comput. **197**, 489–496 (2008)
44. S. Treanţă, Constrained variational problems governed by second-order Lagrangians. Appl. Anal. **99**, 1467–1484 (2020)
45. S. Treanţă, Efficiency in uncertain variational control problems. Neural Comput. Appl. **33**, 5719–5732 (2021)
46. A. Jayswal, Preeti, An exact l_1 penalty function method for multi-dimensional first-order PDE constrained control optimisation problem. Eur. J. Control **52**, 34–41 (2020)
47. A. Jayswal, Preeti, Saddle point criteria for multi-dimensional control optimisation problem involving first-order PDE constraints. Internat. J. Control **94**, 1567–1576 (2021)
48. S. Treanţă, C. Udrişte, Optimal control problems with higher order ODEs constraints. Balkan J. Geo. Appl. **18**, 71–86 (2013)
49. S. Treanţă, Optimal control problems on higher order jet bundles, BSG Proceedings 21, The International Conference "Differential Geometry - Dynamical Systems", DGDS-2013, October 10–13, 2013. Bucharest-Romania **21**, 181–192 (2014)
50. A. Correia, J. Matias, P. Mestre, C. Serôdio, Classification of some penalty methods. Integr. Methods Sci. Eng. **2**, 131–140 (2010)
51. T.S. Motzkin, New technique for linear inequalities and optimization, in: *Project SCOOP Symposium on Linear Inequalities and Programming, Planning Research Division.* (U.S. Air Force, Washington D.C., 1952)
52. W.I. Zangwill, Nonlinear programming via penalty functions. Manag. Sci. **13**, 344–358 (1967)

Chapter 2
Multi-dimensional Variational Control Problem with Data Uncertainty in Objective Functional

2.1 Introduction

The multi-dimensional optimization model provides the mathematical framework which deals with applied science and engineering, especially in mechanical engineering, due to the fact that the curvilinear integral objective functionals have a physical meaning of mechanical work. The literature on various types of multi-dimensional optimization problems is very rich, which plays a versatile role in mathematics, economics and engineering sciences. Remarkable research works have been achieved toward the multi-dimensional optimization problems by several authors under suitable assumptions. Mititelu [1] obtained the optimality conditions for the multi-dimensional optimization problem and then derived the duality results for it. Treanţă and Arana-Jiménez [2] also studied the multi-dimensional optimization problem under KT-pseudoinvexity assumptions. Further, Treanţă [3] extended his study under the generalized V-KT-pseudoinvexity on the multi-dimensional control problem. Jayswal and Preeti [4] studied the multi-dimensional control problem involving first-order PDE constraints with the help of saddle-point criteria. Then, Jayswal *et al.* [5] extended the multi-dimensional control problem involving first-order PDE constrained under uncertain data and showed that the constraints problem and its associated penalized problem attain their optimality at the same point under convexity assumptions.

2.2 Problem Description

Throughout the chapter, we are considering the following notations and working hypotheses:

• $\mathbb{R}^p$, $\mathbb{R}^q$, $\mathbb{R}^r$ and $\mathbb{R}^n$ are Euclidean spaces of dimensions p, q, r and n, respectively.

• $\Gamma_{t_0,t_1} \subset \mathbb{R}^p$ is a hyperparallelepiped fixed by the diagonally opposite points $t_0 = (t_0^\alpha)$ and $t_1 = (t_1^\alpha)$, $\alpha = \overline{1, p}$ and the point $t = ((t^\alpha), \alpha = \overline{1, p}) \in \Gamma_{t_0,t_1} \subset \mathbb{R}^p$.

A. Jayswal et al., *Multi-dimensional Control Problems*, Industrial and Applied Mathematics, https://doi.org/10.1007/978-981-19-6561-6_2

• $\mathscr{X}$ is the space of state functions (piecewise smooth) $x = (x^{\tau}) : \Gamma_{t_0,t_1} \subset \mathbb{R}^p \to \mathbb{R}^q$ and $\frac{\partial x}{\partial t^{\alpha}} = x_{\alpha}$ denotes the first-order partial derivative of x with respect to t^{α}.
• $\mathscr{C}$ is the space of control functions (piecewise continuous) $c = (c^j) : \Gamma_{t_0,t_1} \subset \mathbb{R}^p \to \mathbb{R}^r$.
• $dt = dt^1 \dots dt^p$ is the volume element on $\mathbb{R}^p \supset \Gamma_{t_0,t_1}$.
• T denotes the transpose of a vector.
• We use the following convention for inequalities and equalities: for $x, y \in \mathbb{R}^n$, we have

$$(i)\ x < y \Leftrightarrow x_i < y_i, \forall i = \overline{1,n},$$
$$(ii)\ x = y \Leftrightarrow x_i = y_i, \forall i = \overline{1,n},$$
$$(iii)\ x \leqq y \Leftrightarrow x_i \leq y_i, \forall i = \overline{1,n} \text{ and}$$
$$(iv)\ x \leq y \Leftrightarrow x_i \leq y_i, \forall i = \overline{1,n} \text{ and } x_i < y_i \text{ for some } i.$$

The multi-dimensional multi-objective optimization problem with the data uncertainty in the objective functional is defined as follows:

$$\begin{aligned}
\text{(MMCOPOU)} \quad & \min_{(x(\cdot),c(\cdot))} \left\{ \int_{\Gamma_{t_0,t_1}} f(t, x(t), c(t), w) dt \right. \\
& \left. = \left(\int_{\Gamma_{t_0,t_1}} f_1(t, x(t), c(t), w_1) dt, \dots, \int_{\Gamma_{t_0,t_1}} f_s(t, x(t), c(t), w_s) dt \right) \right\} \\
\text{subject to} \quad & G(t, x(t), c(t)) \leqq 0, \\
& x_{\alpha} = H(t, x(t), c(t)), \\
& x(t_0) = x_0, \ x(t_1) = x_1,
\end{aligned}$$

where $f = (f_1, \dots, f_s)$; $(f_k) : \Gamma_{t_0,t_1} \times \mathscr{X} \times \mathscr{C} \times \mathcal{W}_k \to \mathbb{R}^s, k = \overline{1,s}$, $G = (G_1, \dots, G_m)$; $(G_l) : \Gamma_{t_0,t_1} \times \mathscr{X} \times \mathscr{C} \to \mathbb{R}^m$, $l = \overline{1,m}$, $H = (H_{\alpha}^{\tau}) : \Gamma_{t_0,t_1} \times \mathscr{X} \times \mathscr{C} \to \mathbb{R}^{pq}$, $\alpha = \overline{1,p}$, $\tau = \overline{1,q}$ are C^{∞}-class functionals. $w = (w_k)$ is the uncertain parameter for some convex compact subsets $\mathcal{W} = (\mathcal{W}_k) \subset \mathbb{R}^k$. The functions H_{α} satisfy the closeness conditions (complete integrability conditions) $D_{\beta} H_{\alpha} = D_{\alpha} H_{\beta}, \alpha, \beta = \overline{1,p}, \alpha \neq \beta$, where the total derivative is denoted by D_{β}.

The associated robust counterpart of the multi-dimensional multi-objective optimization problem (MMCOPOU) is defined as

$$\begin{aligned}
\text{(RMMCOPOU)} \quad & \min_{(x(\cdot),c(\cdot))} \left\{ \int_{\Gamma_{t_0,t_1}} \max_{w \in \mathcal{W}} f(t, x(t), c(t), w) dt \right. \\
& \left. = \left(\int_{\Gamma_{t_0,t_1}} \max_{w_1 \in \mathcal{W}_1} f_1(t, x(t), c(t), w_1) dt, \dots, \int_{\Gamma_{t_0,t_1}} \max_{w_s \in \mathcal{W}_s} f_s(t, x(t), c(t), w_s) dt \right) \right\} \\
\text{subject to} \quad & G(t, x(t), c(t)) \leqq 0, \\
& x_{\alpha} = H(t, x(t), c(t)), \\
& x(t_0) = x_0, \ x(t_1) = x_1,
\end{aligned}$$

where f, G and H are defined as above in the multi-dimensional multi-objective optimization problem (MMCOPOU).

We denote $\mathcal{D} = \{(x, c) \in \mathcal{X} \times \mathcal{C} : G(t, x(t), c(t)) \leqq 0, x_\alpha = H(t, x(t), c(t)), x(t_0) = x_0, x(t_1) = x_1\}$ as the set of all feasible solutions in (RMMCOPOU), and we say that it is the robust feasible solution set to the problem (MMCOPOU).

From now onwards, to simplify our presentation, we introduce some notations as follows: $x = x(t), \hat{x} = \hat{x}(t), \bar{x} = \bar{x}(t), c = c(t), \hat{c} = \hat{c}(t), \bar{c} = \bar{c}(t), \pi = (t, x(t), c(t)), \bar{\pi} = (t, \bar{x}(t), \bar{c}(t)), \hat{\pi} = (t, \hat{x}(t), \hat{c}(t)), \rho = \rho(t), \bar{\rho} = \bar{\rho}(t)$ and $\Gamma = \Gamma_{t_0,t_1}$.

The partial derivatives of Lagrange multiplier γ and the partial derivatives associated with f are defined as

$$\gamma_t = \begin{bmatrix} \frac{\partial \gamma_1^\tau}{\partial t^1} & \cdots & \frac{\partial \gamma_1^\tau}{\partial t^p} \\ \vdots & \ddots & \vdots \\ \frac{\partial \gamma_p^\tau}{\partial t^1} & \cdots & \frac{\partial \gamma_p^\tau}{\partial t^p} \end{bmatrix}, \quad f_x = \begin{bmatrix} \frac{\partial f_1}{\partial x^1} & \cdots & \frac{\partial f_1}{\partial x^q} \\ \vdots & \ddots & \vdots \\ \frac{\partial f_s}{\partial x^1} & \cdots & \frac{\partial f_s}{\partial x^q} \end{bmatrix}, \quad f_c = \begin{bmatrix} \frac{\partial f_1}{\partial c^1} & \cdots & \frac{\partial f_1}{\partial c^r} \\ \vdots & \ddots & \vdots \\ \frac{\partial f_s}{\partial c^1} & \cdots & \frac{\partial f_s}{\partial c^r} \end{bmatrix}.$$

Similarly, we have G_x and G_c using matrices with m rows and H_x, H_c, x_α using matrices with p rows.

In general, for a vector optimization problem, it is difficult to find a point which optimizes more than one objective simultaneously. Therefore, the notion of an efficient solution is used to determine optimality in vector optimization problems.

Definition 2.2.0.1 A point $(\bar{x}, \bar{c}) \in \mathcal{D}$ is said to be weak robust efficient solution to the multi-dimensional multi-objective optimization problem (MMCOPOU), if there does not exist another point $(x, c) \in \mathcal{D}$ such that

$$\int_\Gamma \max_{w \in \mathcal{W}} f(\pi, w)dt < \int_\Gamma \max_{w \in \mathcal{W}} f(\bar{\pi}, w)dt.$$

Definition 2.2.0.2 A functional $F(x, c, \bar{w}) = \int_\Gamma f(t, x(t), x_\sigma(t), c(t), \bar{w})dt$ is said to be convex at $(\bar{x}, \bar{c})$, if the following inequality

$$\begin{aligned} F(x, c, \bar{w}) - F(\bar{x}, \bar{c}, \bar{w}) \geqq & \int_\Gamma (x(t) - \bar{x}(t)) f_x(t, \bar{x}(t), \bar{x}_\sigma(t), \bar{c}(t), \bar{w})dt \\ & + \int_\Gamma (x_\sigma(t) - \bar{x}_\sigma(t)) f_{x_\sigma}(t, \bar{x}(t), \bar{x}_\sigma(t), \bar{c}(t), \bar{w})dt \\ & + \int_\Gamma (c(t) - \bar{c}(t)) f_c(t, \bar{x}(t), \bar{x}_\sigma(t), \bar{c}(t), \bar{w})dt \end{aligned}$$

holds for all $(x(t), c(t))$ on $\mathcal{X} \times \mathcal{C}$.

Following the same lines formulated in Mititelu [1], we establish the following result. Thus, the proof is omitted. The proof can be read (it is free) at the following link: https://www.journal.fairpartners.ro/volume-22009-no-1_3.html

Theorem 2.2.0.1 *Let $(\bar{x}, \bar{c}) \in \mathcal{D}$ be a weak robust efficient solution to the problem* (MMOPOCU). *Further, assume that* $\max_{w \in \mathcal{W}} f(\pi, w) = f(\pi, \bar{w})$. *If the constraint conditions (for the existence of the multipliers) hold, then there exist the scalar vector*

$\bar{\mu} \in \mathbb{R}^s$, the piecewise smooth Lagrange multipliers $\bar{\nu} = (\bar{\nu}_l(t)) \in \mathbb{R}_+^\Lambda, l \in \mathbb{R}^\Lambda, \bar{\gamma} = (\bar{\gamma}_\alpha^\tau(t)) \in \mathbb{R}^{pq}, \tau = \overline{1, q}, \alpha = \overline{1, p}$ and the uncertain parameters $\bar{u} \in \mathcal{U}, \bar{v} \in \mathcal{V}$ such that $(\bar{x}, \bar{c})$ satisfies the following conditions:

$$\bar{\mu}^T f_x(\bar{\pi}, \bar{w}) + \bar{\nu}^T G_x(\bar{\pi}) + \bar{\gamma}^T H_x(\bar{\pi}) + \bar{\gamma}_t = 0, \tag{2.2.1}$$

$$\bar{\mu}^T f_c(\bar{\pi}, \bar{w}) + \bar{\nu}^T G_c(\bar{\pi}) + \bar{\gamma}^T H_c(\bar{\pi}) = 0, \tag{2.2.2}$$

$$\bar{\nu}^T G(\bar{\pi}) = 0, \tag{2.2.3}$$

$$\bar{\mu} \geq 0, \bar{\nu} \geqq 0 \tag{2.2.4}$$

hold for all $t \in \Gamma$, except at discontinuities.

Remark 2.2.0.1 The conditions (2.2.1)–(2.2.4) are known as robust necessary efficiency conditions for the multi-dimensional multi-objective optimization problem (MMCOPOU).

2.3 Robust Sufficient Efficiency Conditions

In this section, we will derive the robust sufficient efficiency conditions for the multi-dimensional multi-objective optimization problem (MMCOPOU) which states that any robust feasible solution to the problem (MMCOPOU) will be a weak robust efficient solution to the considered problem (MMCOPOU) if the involved functionals are convex.

Theorem 2.3.0.2 *Let $(\bar{x}, \bar{c})$ be a robust feasible solution to the multi-dimensional multi-objective optimization problem* (MMCOPOU) *and the robust necessary efficiency conditions (2.2.1)–(2.2.4) be satisfied at $(\bar{x}, \bar{c})$ and $\max_{w \in \mathcal{W}} f(\pi, w) = f(\pi, \bar{w})$. Further, assume that $\int_\Gamma \bar{\mu}^T f(\pi, \bar{w}) dt$, $\int_\Gamma \bar{\nu}^T G(\pi, \bar{u}) dt$, and $\int_\Gamma \bar{\gamma}^T (H(\pi, \bar{v}) - x_\alpha) dt$ are convex at $(\bar{x}, \bar{c})$. Then, $(\bar{x}, \bar{c})$ is a weak robust efficient solution to the multi-dimensional multi-objective optimization problem* (MMCOPOU).

Proof Contrary to the result, assume that $(\bar{x}, \bar{c})$ is not a weak robust efficient solution to the multi-dimensional multi-objective optimization problem (MMCOPOU). Then, there exists $(\hat{x}, \hat{c}) \in \mathcal{D}$ such that

$$\int_\Gamma \max_{w \in \mathcal{W}} f(\hat{\pi}, w) dt < \int_\Gamma \max_{w \in \mathcal{W}} f(\bar{\pi}, w) dt.$$

By assumption, $\max_{w \in \mathcal{W}} f(\pi, w) = f(\pi, \bar{w})$, therefore, we get

$$\int_\Gamma f(\hat{\pi}, \bar{w}) dt < \int_\Gamma f(\bar{\pi}, \bar{w}) dt. \tag{2.3.1}$$

Since $(\bar{x}, \bar{c})$ satisfies the conditions (2.2.1)–(2.2.4), therefore, multiplying Eq. (2.2.1) and (2.2.2) by $(\hat{x} - \bar{x})$ and $(\hat{c} - \bar{c})$, respectively, and integrating them, we get

$$\int_{\Gamma}(\hat{x}-\bar{x})\Big\{\bar{\mu}^T f_x(\bar{\pi},\bar{w})+\bar{\nu}^T G_x(\bar{\pi})+\bar{\gamma}^T H_x(\bar{\pi})+\bar{\gamma}_t\Big\}dt$$
$$+\int_{\Gamma}(\hat{c}-\bar{c})\Big\{\bar{\mu}^T f_c(\bar{\pi},\bar{w})+\bar{\nu}^T G_c(\bar{\pi})+\bar{\gamma}^T H_c(\bar{\pi})\Big\}dt=0. \quad (2.3.2)$$

On the other hand, from the assumption that $\int_{\Gamma}\bar{\mu}^T f(\pi,\bar{w})dt$ is convex at $(\bar{x},\bar{c})$, we have

$$\int_{\Gamma}\Big\{\bar{\mu}^T f(\hat{\pi},\bar{w})-\bar{\mu}^T f(\bar{\pi},\bar{w})\Big\}dt \geqq \int_{\Gamma}(\hat{x}-\bar{x})\bar{\mu}^T f_x(\bar{\pi},\bar{w})dt+\int_{\Gamma}(\hat{c}-\bar{c})\bar{\mu}^T f_c(\bar{\pi},\bar{w})dt,$$

which along with the inequality (2.3.1) gives

$$\int_{\Gamma}(\hat{x}-\bar{x})\bar{\mu}^T f_x(\bar{\pi},\bar{w})dt+\int_{\Gamma}(\hat{c}-\bar{c})\bar{\mu}^T f_c(\bar{\pi},\bar{w})dt<0. \quad (2.3.3)$$

Again, by assumption that $\int_{\Gamma}\bar{\nu}^T G(\pi)dt$ is convex at $(\bar{x},\bar{c})$, we obtain

$$\int_{\Gamma}\Big\{\bar{\nu}^T G(\hat{\pi})-\bar{\nu}^T G(\bar{\pi})\Big\}dt \geqq \int_{\Gamma}(\hat{x}-\bar{x})\bar{\nu}^T G_x(\bar{\pi})dt+\int_{\Gamma}(\hat{c}-\bar{c})\bar{\nu}^T G_c(\bar{\pi})dt.$$

The above inequality together with the robust feasibility of $(\hat{x},\hat{c})$ in the multi-dimensional multi-objective optimization problem (MMCOPOU) and the condition (2.2.3) yields

$$\int_{\Gamma}(\hat{x}-\bar{x})\bar{\nu}^T G_x(\bar{\pi})dt+\int_{\Gamma}(\hat{c}-\bar{c})\bar{\nu}^T G_c(\bar{\pi})dt \leqq 0. \quad (2.3.4)$$

Further, from the assumption that $\int_{\Gamma}\bar{\gamma}^T(H(\pi)-x_{\alpha})dt$ is convex at $(\bar{x},\bar{c})$ and the robust feasibility of $(\hat{x},\hat{c})$ in the multi-dimensional multi-objective optimization problem (MMCOPOU), we obtain

$$\int_{\Gamma}(\hat{x}-\bar{x})\Big\{\bar{\gamma}^T H_x(\bar{\pi})+\bar{\gamma}_t\Big\}dt+\int_{\Gamma}(\hat{c}-\bar{c})\bar{\gamma}^T H_c(\bar{\pi})dt \leqq 0. \quad (2.3.5)$$

On adding inequalities (2.3.3), (2.3.4) and (2.3.5), we have

$$\int_{\Gamma}(\hat{x}-\bar{x})\Big\{\bar{\mu}^T f_x(\bar{\pi},\bar{w})+\bar{\nu}^T G_x(\bar{\pi})+\bar{\gamma}^T H_x(\bar{\pi})+\bar{\gamma}_t\Big\}dt$$
$$+\int_{\Gamma}(\hat{c}-\bar{c})\Big\{\bar{\mu}^T f_c(\bar{\pi},\bar{w})+\bar{\nu}^T G_c(\bar{\pi})+\bar{\gamma}^T H_c(\bar{\pi})\Big\}dt<0,$$

which contradicts the relation (2.3.2). Hence, this completes the proof.

The following example verifies the robust sufficient efficiency conditions for a weak robust efficient solution to the multi-dimensional multi-objective optimization problem (MMCOPOU).

Example 2.3.0.1 Let $p = 2, q = r = 1, k = 2, \mathcal{W}_1 = [1, 2], \mathcal{W}_2 = [0, 1]$ and Γ_{t_0,t_1} is fixed by the diagonally opposite points $t_0 = (t_0^1, t_0^2)$ and $t_1 = (t_1^1, t_1^2)$ in $\mathbb{R}^2$.

We consider the following multi-dimensional multi-objective optimization problem in the face of data uncertainty:

$$\text{(MMCOPOU1)} \quad \min_{(x(\cdot),c(\cdot))} \int_{\Gamma_{0,\frac{1}{2}}} (w_1 + c^2,\ w_2 x e^{2x+\frac{1}{2}})\, dt^1 dt^2$$

$$\text{subject to} \quad x^2 + x - 2 \leq 0, \tag{2.3.6}$$

$$x_{t^1} = x_{t^2} = 1 - 2c, \tag{2.3.7}$$

$$x(0, 0) = 0,\ x(\frac{1}{2}, \frac{1}{2}) = \frac{1}{3}, \tag{2.3.8}$$

where $t = (t^1, t^2) \in \Gamma_{0,\frac{1}{2}}$.
Therefore, the associated robust counterpart for the multi-dimensional multi-objective optimization problem (MMCOPOU1) is defined as

$$\text{(RMMCOPOU1)} \quad \min_{(x(\cdot),c(\cdot))} \int_{\Gamma_{0,\frac{1}{2}}} \Big(\max_{w_1 \in \mathcal{W}_1} (w_1 + c^2),\ \max_{w_2 \in \mathcal{W}_2} (w_2 x e^{2x+\frac{1}{2}}) \Big) dt^1 dt^2,$$

$$\text{subject to} \quad x^2 + x - 2 \leq 0,$$

$$x_{t^1} = x_{t^2} = 1 - 2c,$$

$$x(0, 0) = 0,\ x(\frac{1}{2}, \frac{1}{2}) = \frac{1}{3},$$

where $t = (t^1, t^2) \in \Gamma_{0,\frac{1}{2}}$. Clearly $\mathcal{D} = \{(x, c) \in \mathscr{X} \times \mathscr{C} : -2 \leq x \leq 1 \wedge x_{t^1} = x_{t^2} = 1 - 2c, x(0, 0) = 0, x(\frac{1}{2}, \frac{1}{2}) = \frac{1}{3}\}$ is a robust feasible solution set to (MMCOPOU1).
Now, we are interested to find a weak robust efficient solution to the problem (MMCOPOU1). It means that we have to find the control function $\bar{c} : \Gamma_{0,\frac{1}{2}} \to \mathbb{R}$ (that determines the state function $\bar{x} : \Gamma_{0,\frac{1}{2}} \to \mathbb{R}$) satisfying dynamical system (2.3.7) with respect to boundary conditions (2.3.8). Additionally, we assume that the state function is affine.

Let $(\bar{x}, \bar{c})$ be a robust feasible solution to the problem (MMCOPOU1). The necessary efficiency conditions (2.2.1)–(2.2.3) at $(\bar{x}, \bar{c})$ with the Lagrange multipliers $\bar{\mu}_1, \bar{\mu}_2, \bar{\nu}, \bar{\gamma}_1^1, \bar{\gamma}_2^1$ and $\max_{w \in \mathcal{W}} f(\pi, w) = f(\pi, \bar{w})$ are as follows:

$$\bar{\mu}_2 \bar{w}_2 (e^{2\bar{x}+\frac{1}{2}} + \frac{x}{2} e^{2\bar{x}+\frac{1}{2}}) + \bar{\nu}(2\bar{x} + 1) + (\bar{\gamma}_1^1)_{t^1} + (\bar{\gamma}_2^1)_{t^2} = 0, \tag{2.3.9}$$

$$\bar{\mu}_1 2\bar{c} - 2(\bar{\gamma}_1^1 + \bar{\gamma}_2^1) = 0, \tag{2.3.10}$$

$$\bar{\nu}(\bar{x}^2 + \bar{x} - 2) = 0, \tag{2.3.11}$$

where relation (2.3.11) is satisfied if either $\bar{\nu} = 0$ or $(\bar{x}^2 + \bar{x} - 2) = 0$, for all $t = (t^1, t^2) \in \text{int}\Gamma_{0,\frac{1}{2}}$. Let $\bar{\nu} \neq 0$, then $\bar{x}^2 + \bar{x} - 2 = 0$ implies that $\bar{x} = 1, -2$. Clearly, the boundary conditions (2.3.8) are failing at $\bar{x} = 1, -2$. Consequently, $\bar{\nu} = 0$.
Case 1. Assume that the Lagrange multipliers $\bar{\gamma}_\alpha^\tau$, $\tau = \overline{1, q}, \alpha = \overline{1, p}$ are constant functions, say $\bar{\gamma}_1^1 = d_1$ and $\bar{\gamma}_2^1 = d_2$ where $d_1, d_2 \in \mathbb{R}_+$. Therefore, Eq. (2.3.10) gives

$$\bar{c} = \frac{d_1 + d_2}{\bar{\mu}_1}, \ \forall t \in \Gamma_{0,\frac{1}{2}},$$

which shows that $\bar{c}$ is a constant function on $\Gamma_{0,\frac{1}{2}}$. Taking into account equality constraints (2.3.7), it follows that

$$\bar{x} = \Big(1 - 2\frac{d_1 + d_2}{\bar{\mu}_1}\Big)t^1 + \Big(1 - 2\frac{d_1 + d_2}{\bar{\mu}_1}\Big)t^2 + d_3, \ d_3 \in \mathbb{R}_+.$$

By using the boundary condition (2.3.8) along with Lagrange multiplier $\bar{\mu}_1 = \frac{1}{2}, \bar{\mu}_2 = \frac{1}{2}$ and the uncertain parameters $\bar{w}_1 = 2, \bar{w}_2 = 1$, we get $d_3 = 0$ and $d_1 + d_2 = \frac{1}{6}$, and therefore

$$\bar{x} = \frac{1}{3}(t^1 + t^2), \ \bar{c} = \frac{1}{3}, \ \forall t \in \Gamma_{0,\frac{1}{2}}.$$

One can easily verify that the necessary efficiency conditions (2.3.9), (2.3.10) and (2.3.11) are satisfied at $(\bar{x}, \bar{c}) = (\frac{1}{3}(t^1 + t^2), \frac{1}{3})$ at $t^1 = t^2 = 0$, i.e. $(0, \frac{1}{3})$ with Lagrange multipliers $\bar{\mu}_1 = \frac{1}{2}, \bar{\mu}_2 = \frac{1}{2}, \bar{\nu} = 0, \bar{\gamma}_1^1 + \bar{\gamma}_2^1 = \frac{1}{6}$ and uncertain parameters $\bar{w}_1 = 2, \bar{w}_2 = 1$. Further, it can also be easily verified that the involved functionals are convex at $(\bar{x}, \bar{c}) = (0, \frac{1}{3})$. Hence, all the conditions of Theorem 2.3.0.2 are satisfied, which ensure that $(0, \frac{1}{3})$ is also a weak robust efficient solution to the problem (MMCOPOU1).
Case 2. Assume that the Lagrange multipliers $\bar{\gamma}_\alpha^\tau$ are functions of t, say $\bar{\gamma}_1^1 = t^2, \bar{\gamma}_2^1 = t^1$. So, from Eq. (2.3.10), we have

$$\bar{c} = \frac{t^2 + t^1}{2\bar{\mu}_1}, \ \forall t \in \Gamma_{0,\frac{1}{2}}. \tag{2.3.12}$$

On solving Eq. (2.3.7) with the help of Eq. (2.3.12) together with Lagrange multiplier $\bar{\mu}_1 = \frac{1}{2}$, we have

$$\bar{x} = (1 - 4t^2)t^1 + (1 - 4t^1)t^2 - 2(t^1)^2 - 2(t^2)^2 + d_3, \ d_3 \in \mathbb{R}_+,$$

after simplifying which along with boundary conditions (2.3.8), we get

$$\bar{x} = (1 - 4t^2)t^1 + (1 - 4t^1)t^2 - 2(t^1)^2 - 2(t^2)^2.$$

This shows that the state function is not an affine function. Consequently, we get a contradiction with the hypothesis.

2.4 The Exact l_1 Penalty Function Method

In this section, we use the exact l_1 penalty function method to construct a penalized multi-dimensional multi-objective optimization problem associated with the multi-dimensional multi-objective optimization problem (MMCOPOU). This is defined as

$$(\text{MMCOPOU})_\rho \quad \min_{(x(\cdot),c(\cdot))} \Big\{\mathscr{P}(t, x(t), c(t), w, \rho(t)) = \int_\Gamma \Big\{f(t, x(t), c(t), w)$$
$$+ \rho(t)\Big[\sum_{l=1}^{m} G_l^+(t, x(t), c(t)) + \sum_{\tau=1}^{q}\sum_{\alpha=1}^{p}\Big|H_\alpha^\tau(t, x(t), c(t)) - \frac{\partial x^\tau}{\partial t^\alpha}\Big|\Big]e\Big\}dt\Big\},$$

where $e = (1, \ldots, 1) \in \mathbb{R}^s$, $w \in \mathcal{W}$ is the uncertain parameter, $\rho(t)$ is a non-zero non-negative penalty parameter and the penalty function $G_l^+(t, x(t), c(t))$ is defined as

$$G_l^+(t, x(t), c(t)) = \begin{cases} 0, & \text{if } G_l(t, x(t), c(t)) \leqq 0, \\ G_l(t, x(t), c(t)), & \text{if } G_l(t, x(t), c(t)) > 0. \end{cases} \tag{2.4.1}$$

Thus, the associated robust counterpart to the problem $(\text{MMCOPOU})_\rho$ is given by

$$(\text{RMMCOPOU})_\rho \quad \min_{(x(\cdot),c(\cdot))} \Big\{\max_{w\in\mathcal{W}} \mathscr{P}(t, x(t), c(t), w, \rho(t)) = \int_\Gamma \Big\{\max_{w\in\mathcal{W}} f(t, x(t), c(t), w)$$
$$+ \rho(t)\Big[\sum_{l=1}^{m} G_l^+(t, x(t), c(t)) + \sum_{\tau=1}^{q}\sum_{\alpha=1}^{p}\Big|H_\alpha^\tau(t, x(t), c(t)) - \frac{\partial x^\tau}{\partial t^\alpha}\Big|\Big]e\Big\}dt\Big\}.$$

Definition 2.4.0.3 A point $(\bar{x}, \bar{c}) \in \mathscr{X} \times \mathscr{C}$ is said to be a weak robust efficient solution to the problem $(\text{MMCOPOU})_\rho$, if there does not exist another point $(x, c) \in \mathscr{X} \times \mathscr{C}$ such that

$$\max_{w\in\mathcal{W}} \mathscr{P}(\pi, w, \rho) < \max_{w\in\mathcal{W}} \mathscr{P}(\bar{\pi}, w, \rho).$$

Theorem 2.4.0.3 *Let $(\bar{x}, \bar{c})$ be a robust feasible solution to the multi-dimensional multi-objective optimization problem* (MMCOPOU) *and the conditions (2.2.1)–(2.2.4) be satisfied at $(\bar{x}, \bar{c})$ with respect to the considered Lagrange multipliers and* $\max_{w\in\mathcal{W}} f(\pi, w) = f(\pi, \bar{w})$. *Further, assume that $\int_\Gamma \bar{\mu}^T f(\pi, \bar{w})dt$, $\int_\Gamma \bar{\nu}^T G(\pi)dt$ and $\int_\Gamma \bar{\gamma}^T (H(\pi) - x_\alpha)dt$ are convex at $(\bar{x}, \bar{c})$ on $\mathscr{X} \times \mathscr{C}$. If $\rho \geq \max\{\bar{\nu}, |\bar{\gamma}|\}$, then,*

$(\bar{x}, \bar{c})$ is a weak robust efficient solution to the penalized multi-dimensional multi-objective optimization problem (MMCOPOU)ρ.

Proof Contrary to the results, we assume that $(\bar{x}, \bar{c})$ is not a weak robust efficient solution to the problem (MMCOPOU)$_\rho$. Then, there exists another point $(\hat{x}, \hat{c}) \in \mathscr{X} \times \mathscr{C}$ such that

$$\max_{w \in \mathcal{W}} \mathscr{P}(\hat{\pi}, w, \rho) < \max_{w \in \mathcal{W}} \mathscr{P}(\bar{\pi}, w, \rho).$$

By the definition of the penalized multi-dimensional multi-objective optimization problem (MMCOPOU)$_\rho$, we get

$$\int_\Gamma \Big\{ \max_{w \in \mathcal{W}} f(\hat{\pi}, w) + \rho \Big[\sum_{l=1}^{m} G_l^+(\hat{\pi}) + \sum_{\tau=1}^{q} \sum_{\alpha=1}^{p} \Big| H_\alpha^\tau(\hat{\pi}) - \frac{\partial x^\tau}{\partial t^\alpha} \Big| \Big] e \Big\} dt$$
$$< \int_\Gamma \Big\{ \max_{w \in \mathcal{W}} f(\bar{\pi}, w) + \rho \Big[\sum_{l=1}^{m} G_l^+(\bar{\pi}) + \sum_{\tau=1}^{q} \sum_{\alpha=1}^{p} \Big| H_\alpha^\tau(\bar{\pi}) - \frac{\partial x^\tau}{\partial t^\alpha} \Big| \Big] e \Big\} dt.$$

By assumption that $\max_{p \in \mathcal{W}} f(\pi, w) = f(\pi, \bar{w})$, we get

$$\int_\Gamma \Big\{ f(\hat{\pi}, \bar{w}) + \rho \Big[\sum_{l=1}^{m} G_l^+(\hat{\pi}) + \sum_{\tau=1}^{q} \sum_{\alpha=1}^{p} \Big| H_\alpha^\tau(\hat{\pi}) - \frac{\partial x^\tau}{\partial t^\alpha} \Big| \Big] e \Big\} dt$$
$$< \int_\Gamma \Big\{ f(\bar{\pi}, \bar{w}) + \rho \Big[\sum_{l=1}^{m} G_l^+(\bar{\pi}) + \sum_{\tau=1}^{q} \sum_{\alpha=1}^{p} \Big| H_\alpha^\tau(\bar{\pi}) - \frac{\partial x^\tau}{\partial t^\alpha} \Big| \Big] e \Big\} dt.$$

With the above inequality together with the robust feasibility of $(\bar{x}, \bar{c})$ to the problem (MMCOPOU) and (2.4.1), we have

$$\int_\Gamma \Big\{ f(\hat{\pi}, \bar{w}) + \rho \Big[\sum_{l=1}^{m} G_l^+(\hat{\pi}) + \sum_{\tau=1}^{q} \sum_{\alpha=1}^{p} \Big| H_\alpha^\tau(\hat{\pi}) - \frac{\partial x^\tau}{\partial t^\alpha} \Big| \Big] e \Big\} dt < \int_\Gamma f(\bar{\pi}, \bar{w}) dt.$$

On multiplying the above inequality by scalar vector $\bar{\mu}$ on both sides, we get

$$\int_\Gamma \Big\{ \bar{\mu}^T f(\hat{\pi}, \bar{w}) + \bar{\mu}^T \rho \Big[\sum_{l=1}^{m} G_l^+(\hat{\pi}) + \sum_{\tau=1}^{q} \sum_{\alpha=1}^{p} \Big| H_\alpha^\tau(\hat{\pi}) - \frac{\partial x^\tau}{\partial t^\alpha} \Big| \Big] e \Big\} dt$$
$$< \int_\Gamma \bar{\mu}^T f(\bar{\pi}, \bar{w}) dt.$$

As $\bar{\mu}^T e = 1$, it follows that

$$\int_{\Gamma}\Big\{\bar{\mu}^T f(\hat{\pi},\bar{w})+\rho\Big[\sum_{l=1}^{m}G_l^{+}(\hat{\pi})+\sum_{\tau=1}^{q}\sum_{\alpha=1}^{p}\Big|H_{\alpha}^{\tau}(\hat{\pi})-\frac{\partial x^{\tau}}{\partial t^{\alpha}}\Big|\Big]\Big\}dt$$
$$<\int_{\Gamma}\bar{\mu}^T f(\bar{\pi},\bar{w})dt.$$

Since $\rho \geq \max\{\bar{v}, |\bar{\gamma}|\}$, thus the above inequality yields

$$\int_{\Gamma}\Big\{\bar{\mu}^T f(\hat{\pi},\bar{w})+\sum_{l=1}^{m}\bar{v}_l G_l^{+}(\hat{\pi})+\sum_{\tau=1}^{q}\sum_{\alpha=1}^{p}\bar{\gamma}_{\alpha}^{\tau}\Big|H_{\alpha}^{\tau}(\hat{\pi})-\frac{\partial x^{\tau}}{\partial t^{\alpha}}\Big|\Big\}dt$$
$$<\int_{\Gamma}\bar{\mu}^T f(\bar{\pi},\bar{w})dt. \tag{2.4.2}$$

From the condition (2.2.3) along with the robust feasibility of $(\bar{x},\bar{c})$ to the problem (MMCOPOU), the inequality (2.4.2) can be written as

$$\int_{\Gamma}\Big\{\bar{\mu}^T f(\hat{\pi},\bar{w})+\sum_{l=1}^{m}\bar{v}_l G_l(\hat{\pi})+\sum_{\tau=1}^{q}\sum_{\alpha=1}^{p}\bar{\gamma}_{\alpha}^{\tau}\Big(H_{\alpha}^{\tau}(\hat{\pi})-\frac{\partial x^{\tau}}{\partial t^{\alpha}}\Big)\Big\}dt$$
$$<\int_{\Gamma}\Big\{\bar{\mu}^T f(\bar{\pi},\bar{w})+\sum_{l=1}^{m}\bar{v}_l G_l(\bar{\pi})+\sum_{\tau=1}^{q}\sum_{\alpha=1}^{p}\bar{\gamma}_{\alpha}^{\tau}\Big(H_{\alpha}^{\tau}(\bar{\pi})-\frac{\partial x^{\tau}}{\partial t^{\alpha}}\Big)\Big\}dt.$$

Equivalently,

$$\int_{\Gamma}\Big\{\bar{\mu}^T f(\hat{\pi},\bar{w})+\bar{v}^T G(\hat{\pi})+\bar{\gamma}^T(H(\hat{\pi})-x_{\alpha})\Big\}dt$$
$$<\int_{\Gamma}\Big\{\bar{\mu}^T f(\bar{\pi},\bar{w})+\bar{v}^T G(\bar{\pi})+\bar{\gamma}^T(H(\bar{\pi})-x_{\alpha})\Big\}dt. \tag{2.4.3}$$

On the other hand, by assumption that $\int_{\Gamma}\bar{\mu}^T f(\pi,\bar{w})dt$, $\int_{\Gamma}\bar{v}^T G(\pi)dt$ and $\int_{\Gamma}\bar{\gamma}^T (H(\pi)-x_{\alpha})dt$ are convex at $(\bar{x},\bar{c})$ on $\mathscr{X}\times\mathscr{C}$, we have

$$\int_{\Gamma}\Big\{\bar{\mu}^T f(\hat{\pi},\bar{w})-\bar{\mu}^T f(\bar{\pi},\bar{w})\Big\}dt \geqq \int_{\Gamma}(\hat{x}-\bar{x})\bar{\mu}^T f_x(\bar{\pi},\bar{w})dt+\int_{\Gamma}(\hat{c}-\bar{c})\bar{\mu}^T f_c(\bar{\pi},\bar{w})dt, \tag{2.4.4}$$

$$\int_{\Gamma}\Big\{\bar{v}^T G(\hat{\pi})-\bar{v}^T G(\bar{\pi})\Big\}dt \geqq \int_{\Gamma}(\hat{x}-\bar{x})\bar{v}^T G_x(\bar{\pi})dt+\int_{\Gamma}(\hat{c}-\bar{c})\bar{v}^T G_c(\bar{\pi})dt, \tag{2.4.5}$$

and

$$\int_{\Gamma}\Big\{\bar{\gamma}^T(H(\hat{\pi})-x_\alpha)-\bar{\gamma}^T(H(\bar{\pi})-x_\alpha)\Big\}dt$$
$$\geqq \int_{\Gamma}(\hat{x}-\bar{x})\Big\{\bar{\gamma}^T H_x(\bar{\pi})+\bar{\gamma}_t\Big\}dt+\int_{\Gamma}(\hat{c}-\bar{c})\bar{\gamma}^T H_c(\bar{\pi})dt. \tag{2.4.6}$$

On combining the inequalities (2.4.4)–(2.4.6) together with the conditions (2.2.1)–(2.2.2), we have

$$\int_{\Gamma}\Big\{\bar{\mu}^T f(\hat{\pi},\bar{w})+\bar{\nu}^T G(\hat{\pi})+\bar{\gamma}^T(H(\hat{\pi})-x_\alpha)\Big\}dt$$
$$\geqq \int_{\Gamma}\Big\{\bar{\mu}^T f(\bar{\pi},\bar{w})+\bar{\nu}^T G(\bar{\pi})+\bar{\gamma}^T(H(\bar{\pi})-x_\alpha)\Big\}dt,$$

which contradicts the inequality (2.4.3). Hence, this completes the proof.

Let us consider Example 2.3.0.1 to show the relationship between the solution set of a multi-dimensional multi-objective optimization problem and its associated penalized problem.

Example 2.4.0.1 Now, the penalized multi-dimensional multi-objective optimization problem in the face of data uncertainty $(\text{MMCOPOU1})_\rho$ associated with the multi-dimensional multi-objective optimization problem in the face of data uncertainty (MMCOPOU1) is defined as

$$(\text{MMCOPOU1})_\rho \quad \min_{(x(\cdot),c(\cdot))}\Bigg(\int_{\Gamma_{0,\frac{1}{2}}}\Big\{(w_1+c^2)+\rho\Big[\max\{0,x^2+x-2\}+\Big|\frac{\partial x}{\partial t^1}+2c-1\Big|$$
$$+\Big|\frac{\partial x}{\partial t^2}+2c-1\Big|\Big]\Big\}dt^1dt^2,\int_{\Gamma_{0,\frac{1}{2}}}\Big\{(w_2xe^{2x+\frac{1}{2}})+\rho\Big[\max\{0,x^2+x-2\}$$
$$+\Big|\frac{\partial x}{\partial t^1}+2c-1\Big|+\Big|\frac{\partial x}{\partial t^2}+2c-1\Big|\Big]\Big\}dt^1dt^2\Bigg).$$

Therefore, the robust counterpart for $(\text{MMCOPOU1})_\rho$ is formulated as

$$(\text{MMCOPOU1})_\rho \quad \min_{(x(\cdot),c(\cdot))}\Bigg(\int_{\Gamma_{0,\frac{1}{2}}}\Big\{\max_{w_1\in\mathcal{W}_1}(w_1+c^2)+\rho\Big[\max\{0,x^2+x-2\}$$
$$+\Big|\frac{\partial x}{\partial t^1}+2c-1\Big|+\Big|\frac{\partial x}{\partial t^2}+2c-1\Big|\Big]\Big\}dt^1dt^2,\int_{\Gamma_{0,\frac{1}{2}}}\Big\{\max_{w_2\in\mathcal{W}_2}(w_2xe^{2x+\frac{1}{2}})$$
$$+\rho\Big[\max\{0,x^2+x-2\}+\Big|\frac{\partial x}{\partial t^1}+2c-1\Big|+\Big|\frac{\partial x}{\partial t^2}+2c-1\Big|\Big]\Big\}dt^1dt^2\Bigg).$$

In virtue of Definition 2.4.0.3, it can be observed that the inequality $\mathscr{P}(\pi,w,\rho)-\mathscr{P}(\bar{\pi},w,\rho)\geq 0$ at $(\bar{x},\bar{c})=(0,\frac{1}{3})$. This shows that $(0,\frac{1}{3})$ is a weak robust efficient solution to $(\text{MMCOPOU1})_\rho$ with $\rho\geq 1$. Hence, Theorem 2.4.0.3 is verified.

Proposition 2.4.0.1 Let $(\bar{x}, \bar{c}) \in \mathscr{X} \times \mathscr{C}$ be a weak robust efficient solution to the problem $(\text{MMCOPOU})_{\bar{\rho}}$ and $\max_{w \in \mathcal{W}} f(\pi, w) = f(\pi, \bar{w})$. Then, there does not exist another point $(x, c) \in \mathcal{D}$ such that

$$\int_{\Gamma} f(\pi, \bar{w}) dt < \int_{\Gamma} f(\bar{\pi}, \bar{w}) dt.$$

Proof Since $(\bar{x}, \bar{c})$ is a weak robust efficient solution to the problem $(\text{MMCOPOU})_{\bar{\rho}}$, then there does not exist another point $(x, c) \in \mathscr{X} \times \mathscr{C}$ such that

$$\max_{w \in \mathcal{W}} \mathscr{P}(\pi, w, \bar{\rho}) < \max_{w \in \mathcal{W}} \mathscr{P}(\bar{\pi}, w, \bar{\rho}).$$

By using the definition of the penalized semi-infinite multi-dimensional multi-objective optimization problem $(\text{MMCOPOU})_{\rho}$, we get

$$\int_{\Gamma} \Big\{ \max_{w \in \mathcal{W}} f(\pi, w) + \bar{\rho}\Big[\sum_{l=1}^{m} G_l^{+}(\pi) + \sum_{\tau=1}^{q} \sum_{\alpha=1}^{p} \Big| H_{\alpha}^{\tau}(\pi) - x_{\alpha} \Big| \Big] e \Big\} dt$$
$$< \int_{\Gamma} \Big\{ \max_{w \in \mathcal{W}} f(\bar{\pi}, w) + \bar{\rho}\Big[\sum_{l=1}^{m} G_l^{+}(\bar{\pi}) + \sum_{\tau=1}^{q} \sum_{\alpha=1}^{p} \Big| H_{\alpha}^{\tau}(\bar{\pi}) - x_{\alpha} \Big| \Big] e \Big\} dt.$$

Since $\max_{w \in \mathcal{W}} f(\pi, w) = f(\pi, \bar{w})$, we have

$$\int_{\Gamma} \Big\{ f(\pi, \bar{w}) + \bar{\rho}\Big[\sum_{l=1}^{m} G_l^{+}(\pi) + \sum_{\tau=1}^{q} \sum_{\alpha=1}^{p} \Big| H_{\alpha}^{\tau}(\pi) - x_{\alpha} \Big| \Big] e \Big\} dt$$
$$< \int_{\Gamma} \Big\{ f(\bar{\pi}, \bar{w}) + \bar{\rho}\Big[\sum_{l=1}^{m} G_l^{+}(\bar{\pi}) + \sum_{\tau=1}^{q} \sum_{\alpha=1}^{p} \Big| H_{\alpha}^{\tau}(\bar{\pi}) - x_{\alpha} \Big| \Big] e \Big\} dt.$$

Since $\mathcal{D} \subseteq \mathscr{X} \times \mathscr{C}$, therefore, it follows that there does not exist another point $(x, c) \in \mathcal{D}$ such that

$$\int_{\Gamma} \Big\{ f(\pi, \bar{w}) + \bar{\rho}\Big[\sum_{l=1}^{m} G_l^{+}(\pi) + \sum_{\tau=1}^{q} \sum_{\alpha=1}^{p} \Big| H_{\alpha}^{\tau}(\pi) - x_{\alpha} \Big| \Big] e \Big\} dt$$
$$< \int_{\Gamma} \Big\{ f(\bar{\pi}, \bar{w}) + \bar{\rho}\Big[\sum_{l=1}^{m} G_l^{+}(\bar{\pi}) + \sum_{\tau=1}^{q} \sum_{\alpha=1}^{p} \Big| H_{\alpha}^{\tau}(\bar{\pi}) - x_{\alpha} \Big| \Big] e \Big\} dt.$$

Equivalently, one can say that the following relation holds for all $(x, c) \in \mathcal{D}$

$$\int_{\Gamma} \Big\{ f_k(\pi, \bar{w}) + \bar{\rho} \Big[\sum_{l=1}^{m} G_l^{+}(\pi) + \sum_{\tau=1}^{q} \sum_{\alpha=1}^{p} \Big| H_{\alpha}^{\tau}(\pi) - \frac{\partial x^{\tau}}{\partial t^{\alpha}} \Big| \Big] \Big\} dt$$

$$\geqq \int_{\Gamma} \Big\{ f_k(\bar{\pi}, \bar{w}) + \bar{\rho} \Big[\sum_{l=1}^{m} G_l^{+}(\bar{\pi}) + \sum_{\tau=1}^{q} \sum_{\alpha=1}^{p} \Big| H_{\alpha}^{\tau}(\bar{\pi}) - \frac{\partial x^{\tau}}{\partial t^{\alpha}} \Big| \Big] \Big\} dt.$$

By virtue of the relation (2.4.1), we get

$$\int_{\Gamma} f_k(\pi, \bar{w}) dt \geqq \int_{\Gamma} f_k(\bar{\pi}, \bar{w}) dt.$$

Therefore, we conclude that there does not exist another point $(x, c) \in \mathcal{D}$ such that

$$\int_{\Gamma} f(\pi, \bar{w}) dt < \int_{\Gamma} f(\bar{\pi}, \bar{w}) dt.$$

This completes the proof.

Now, we prove the converse result of Theorem 2.4.0.3 without imposing any convexity assumption on the involved functionals.

Theorem 2.4.0.4 *Let* $(\bar{x}, \bar{c})$ *be a weak robust efficient solution to the multi-dimensional multi-objective optimization problem* (MMCOPOU)$_{\bar{\rho}}$. *Assume that* $\max_{w \in \mathcal{W}} \mathscr{P}(\pi, w, \rho) \geqq \max_{w \in \mathcal{W}} \mathscr{P}(\bar{\pi}, w, \rho)$ *is satisfied for any* $\rho \geq \bar{\rho}$ *and for all* $(x, c) \in \mathcal{D}$ *and* $\max_{w \in \mathcal{W}} f(\pi, w) = f(\pi, \bar{w})$. *If* $\rho \geq \max\{\bar{\nu}, |\bar{\gamma}|\}$, *then,* $(\bar{x}, \bar{c})$ *is a weak robust efficient solution to the multi-dimensional multi-objective optimization problem* (MMCOPOU).

Proof Since $(\bar{x}, \bar{c})$ is a weak robust efficient solution to the problem (MMCOPOU)$_{\bar{\rho}}$, by Proposition 2.4.0.1, there does not exist another point $(x, c) \in \mathcal{D}$ such that

$$\int_{\Gamma} f(\pi, \bar{w}) dt < \int_{\Gamma} f(\bar{\pi}, \bar{w}) dt,$$

equivalently,

$$\int_{\Gamma} f_k(\pi, \bar{w}) dt < \int_{\Gamma} f_k(\bar{\pi}, \bar{w}) dt, \ \forall k = \overline{1, s}, \tag{2.4.7}$$

which specify that the objective functional $\int_{\Gamma} f_k(\pi, \bar{w}) dt$ is bounded below in the compact subset $\mathcal{D}$.

Now, to show that $(\bar{x}, \bar{c})$ is a weak robust efficient solution to the problem (MMCOPOU), firstly, we shall prove that $(\bar{x}, \bar{c}) \in \mathcal{D}$. We proceed by contradiction and assume that $(\bar{x}, \bar{c}) \notin \mathcal{D}$. Therefore, by virtue of (2.4.1), we have

$$\int_{\Gamma} \Big\{ \sum_{l=1}^{m} G_l^{+}(\bar{\pi}) + \sum_{\tau=1}^{q} \sum_{\alpha=1}^{p} \Big| H_{\alpha}^{\tau}(\bar{\pi}) - \frac{\partial x^{\tau}}{\partial t^{\alpha}} \Big| \Big\} e \, dt > 0. \tag{2.4.8}$$

Let (x, c) be any robust feasible solution to the problem (MMCOPOU). By assumption, the following inequality

$$\max_{w\in\mathcal{W}} \mathscr{P}(\pi, w, \rho)dt \geqq \max_{w\in\mathcal{W}} \mathscr{P}(\bar{\pi}, w, \rho)dt$$

holds for any $(x, c) \in \mathcal{D}$ and for all $\rho \geq \bar{\rho}$.
By the definition of the penalized multi-dimensional multi-objective optimization problem (MMCOPOU), we get

$$\int_{\Gamma}\Big\{\max_{w\in\mathcal{W}} f(\pi, w) + \rho\Big[\sum_{l=1}^{m} G_l^{+}(\pi) + \sum_{\tau=1}^{q}\sum_{\alpha=1}^{p}\Big|H_{\alpha}^{\tau}(\pi) - \frac{\partial x^{\tau}}{\partial t^{\alpha}}\Big|\Big]e\Big\}dt$$
$$\geqq \int_{\Gamma}\Big\{\max_{w\in\mathcal{W}} f(\bar{\pi}, w) + \rho\Big[\sum_{l=1}^{m} G_l^{+}(\bar{\pi}) + \sum_{\tau=1}^{q}\sum_{\alpha=1}^{p}\Big|H_{\alpha}^{\tau}(\bar{\pi}) - \frac{\partial x^{\tau}}{\partial t^{\alpha}}\Big|\Big]e\Big\}dt.$$

Since $\max_{w\in\mathcal{W}} f(\pi, w) = f(\pi, \bar{w})$, we get

$$\int_{\Gamma}\Big\{f(\pi, \bar{w}) + \rho\Big[\sum_{l=1}^{m} G_l^{+}(\pi) + \sum_{\tau=1}^{q}\sum_{\alpha=1}^{p}\Big|H_{\alpha}^{\tau}(\pi) - \frac{\partial x^{\tau}}{\partial t^{\alpha}}\Big|\Big]e\Big\}dt$$
$$\geqq \int_{\Gamma}\Big\{f(\bar{\pi}, \bar{w}) + \rho\Big[\sum_{l=1}^{m} G_l^{+}(\bar{\pi}) + \sum_{\tau=1}^{q}\sum_{\alpha=1}^{p}\Big|H_{\alpha}^{\tau}(\bar{\pi}) - \frac{\partial x^{\tau}}{\partial t^{\alpha}}\Big|\Big]e\Big\}dt. \tag{2.4.9}$$

On the other hand, if we set

$$\rho > \max\left\{\frac{\int_{\Gamma}\{f(\pi, \bar{w}) - f(\bar{\pi}, \bar{w})\}dt}{\int_{\Gamma}\Big\{\sum_{l=1}^{m} G_l^{+}(\bar{\pi}) + \sum_{\tau=1}^{q}\sum_{\alpha=1}^{p}\Big|H_{\alpha}^{\tau}(\bar{\pi}) - \frac{\partial x^{\tau}}{\partial t^{\alpha}}\Big|\Big\}e\,dt}, \bar{\rho};\ (x, c) \in \mathcal{D}\right\},$$

the inequalities (2.4.7) and (2.4.8) imply that ρ is a positive real number. Then, we have

$$\int_{\Gamma}\Big\{f(\bar{\pi}, \bar{w}) + \rho\Big[\sum_{l=1}^{m} G_l^{+}(\bar{\pi}) + \sum_{\tau=1}^{q}\sum_{\alpha=1}^{p}\Big|H_{\alpha}^{\tau}(\bar{\pi}) - \frac{\partial x^{\tau}}{\partial t^{\alpha}}\Big|\Big]e\Big\}dt > \int_{\Gamma} f(\pi, \bar{w})dt.$$

Further, by the robust feasibility of (x, c) to the problem (MMCOPOU) and in virtue of (2.4.1), the above inequality can be written as

$$\int_{\Gamma} \Big\{f(\bar{\pi}, \bar{w}) + \rho\Big[\sum_{l=1}^{m} G_l^{+}(\bar{\pi}) + \sum_{\tau=1}^{q}\sum_{\alpha=1}^{p} \Big|H_{\alpha}^{\tau}(\bar{\pi}) - \frac{\partial x^{\tau}}{\partial t^{\alpha}}\Big|\Big]e\Big\}dt$$
$$> \int_{\Gamma} \Big\{f(\pi, \bar{w}) + \rho\Big[\sum_{l=1}^{m} G_l^{+}(\pi) + \sum_{\tau=1}^{q}\sum_{\alpha=1}^{p} \Big|H_{\alpha}^{\tau}(\pi) - \frac{\partial x^{\tau}}{\partial t^{\alpha}}\Big|\Big]e\Big\}dt,$$

which contradicts the inequality (2.4.9). Therefore, $(\bar{x}, \bar{c}) \in \mathcal{D}$, i.e. $(\bar{x}, \bar{c})$ is a robust feasible solution to the problem (MMCOPOU). Hence, the inequality (2.4.7) directly concludes that $(\bar{x}, \bar{c})$ is a weak robust efficient solution to the problem (MMCOPOU).

2.5 Uncertain Lagrange Functional

In this section, we introduce the uncertain Lagrange functional for (MMCOPOU) and prove the relationship between a weak robust efficient solution to the problem (MMCOPOU) and (MMCOPOU)$_{\rho}$ under convexity assumption of the uncertain Lagrange functional.

Definition 2.5.0.1 The uncertain Lagrange functional $L(x, c, \mu, \nu, \gamma, w) : \mathcal{X} \times \mathcal{C} \times \mathbb{R}^{s} \times \mathbb{R}_{+}^{m} \times \mathbb{R}^{pq} \times \mathcal{W} \to \mathbb{R}$ to the problem (MMCOPOU) is defined as

$$L(x, c, \mu, \nu, \gamma, w) = \int_{\Gamma} \Big\{\mu^{T} f(\pi, w) + \nu^{T} G(\pi) + \gamma^{T}(H(\pi) - x_{\alpha})\Big\}\, dt.$$

Now, we show that (MMCOPOU) and (MMCOPOU)$_{\rho}$ consist of the same weak robust efficient solution set under the convexity assumption of the uncertain Lagrange functional.

Theorem 2.5.0.1 *Let $(\bar{x}, \bar{c})$ be a robust feasible solution to the multi-dimensional multi-objective optimization problem* (MMCOPOU) *at which the robust necessary efficiency conditions (2.2.1)–(2.2.4) hold and* $\max_{w \in \mathcal{W}} f(\pi, w) = f(\pi, \bar{w})$. *Further, assume that if the Lagrange functional $L(x, c, \mu, \nu, \gamma, w)$ is convex at $(\bar{x}, \bar{c})$ on $\mathcal{X} \times \mathcal{C}$ and* $\rho \geq \max\{\bar{\nu}, |\bar{\gamma}|\}$, *then, $(\bar{x}, \bar{c})$ is also a weak robust efficient solution to the multi-dimensional multi-objective optimization problem* (MMCOPOU)$_{\rho}$.

Proof We proceed by contradiction and assume that $(\bar{x}, \bar{c})$ is not a weak robust efficient solution to the problem (MMCOPOU)$_{\rho}$. Then, there exists another point $(\hat{x}, \hat{c}) \in \mathcal{X} \times \mathcal{C}$ such that

$$\max_{w \in \mathcal{W}} \mathcal{P}(\hat{\pi}, w, \rho) < \max_{w \in \mathcal{W}} \mathcal{P}(\bar{\pi}, w, \rho).$$

Then, on the line of Theorem 2.4.0.3, we get

$$\int_{\Gamma}\Big\{\sum_{k=1}^{s}\bar{\mu}_k f_k(\hat{\pi},\bar{w}_k)+\sum_{l=1}^{m}\bar{\nu}_l G_l(\hat{\pi})+\sum_{\tau=1}^{q}\sum_{\alpha=1}^{p}\bar{\gamma}_\alpha^\tau\Big(H_\alpha^\tau(\hat{\pi})-\frac{\partial x^\tau}{\partial t^\alpha}\Big)\Big\}dt$$
$$<\int_{\Gamma}\Big\{\sum_{k=1}^{s}\bar{\mu}_k f_k(\bar{\pi},\bar{w}_k)+\sum_{l=1}^{m}\bar{\nu}_l G_l(\bar{\pi})+\sum_{\tau=1}^{q}\sum_{\alpha=1}^{p}\bar{\gamma}_\alpha^\tau\Big(H_\alpha^\tau(\bar{\pi})-\frac{\partial x^\tau}{\partial t^\alpha}\Big)\Big\}dt,$$

equivalently,

$$\int_{\Gamma}\Big\{\bar{\mu}^T f(\hat{\pi},\bar{w})+\bar{\nu}^T G(\hat{\pi})+\bar{\gamma}^T(H(\hat{\pi})-x_\alpha)\Big\}dt$$
$$<\int_{\Gamma}\Big\{\bar{\mu}^T f(\bar{\pi},\bar{w})+\bar{\nu}^T G(\bar{\pi})+\bar{\gamma}^T(H(\bar{\pi})-x_\alpha)\Big\}dt. \tag{2.5.1}$$

On the other hand, by assumption that the uncertain Lagrange functional is convex at $(\bar{x},\bar{c})$ on $\mathscr{X}\times\mathscr{C}$, we have

$$L(\hat{x},\hat{c},\bar{\mu},\bar{\nu},\bar{\gamma},\bar{w})-L(\bar{x},\bar{c},\bar{\mu},\bar{\nu},\bar{\gamma},\bar{w})\geqq(\hat{x}-\bar{x})L_x(\bar{x},\bar{c},\bar{\mu},\bar{\nu},\bar{\gamma},\bar{w})$$
$$+(\hat{c}-\bar{c})L_c(\bar{x},\bar{c},\bar{\mu},\bar{\nu},\bar{\gamma},\bar{w}).$$

By the definition of the Lagrange functional, we get

$$\int_{\Gamma}\Big\{\bar{\mu}^T f(\hat{\pi},\bar{w})+\bar{\nu}^T G(\hat{\pi})+\bar{\gamma}^T(H(\hat{\pi})-x_\alpha)\Big\}dt-\int_{\Gamma}\Big\{\bar{\mu}^T f(\bar{\pi},\bar{w})+\bar{\nu}^T G(\bar{\pi})$$
$$+\bar{\gamma}^T(H(\bar{\pi})-x_\alpha)\Big\}dt\geqq\int_{\Gamma}(\hat{x}-\bar{x})\Big\{\bar{\mu}^T f_x(\bar{\pi},\bar{w})+\bar{\nu}^T G_x(\bar{\pi})+\bar{\gamma}^T H_x(\bar{\pi})+\bar{\gamma}_t\Big\}dt$$
$$+\int_{\Gamma}(\hat{c}-\bar{c})\Big\{\bar{\mu}^T f_c(\bar{\pi},\bar{w})+\bar{\nu}^T G_c(\bar{\pi})+\bar{\gamma}^T H_c(\bar{\pi})\Big\}dt.$$

Using the robust necessary efficiency conditions (2.2.1) and (2.2.2), the above inequality can be written as

$$\int_{\Gamma}\Big\{\bar{\mu}^T f(\hat{\pi},\bar{w})+\bar{\nu}^T G(\hat{\pi})+\bar{\gamma}^T(H(\hat{\pi})-x_\alpha)\Big\}dt$$
$$\geqq\int_{\Gamma}\Big\{\bar{\mu}^T f(\bar{\pi},\bar{w})+\bar{\nu}^T G(\bar{\pi})+\bar{\gamma}^T(H(\bar{\pi})-x_\alpha)\Big\}dt,$$

which contradicts the inequality (2.5.1). Hence, this completes the proof.

Now, we give the following example to validate Theorem 2.5.0.1.

Example 2.5.0.1 Let $p=2, q=r=1, k=2, \mathcal{W}_1=[0,1], \mathcal{W}_2=[0,2]$ and Γ_{t_0,t_1} is fixed by the diagonally opposite points $t_0=(t_0^1,t_0^2)$ and $t_1=(t_1^1,t_1^2)$ in $\mathbb{R}^2$.

We consider the following multi-dimensional multi-objective optimization problem as

$$\text{(MMCOPOU2)} \quad \min_{(x(\cdot),c(\cdot))} \int_{\Gamma_{0,1}} (e^x - x + w_1 c^2,\ x\tan^{-1} x + w_2)\, dt^1 dt^2$$

$$\text{subject to} \quad x^2 + 2x - 3 \leqq 0, \tag{2.5.2}$$

$$x_{t^1} = x_{t^2} = 2 - c, \tag{2.5.3}$$

$$x(0,0) = 0,\ x(1,1) = 3, \tag{2.5.4}$$

where $t = (t^1, t^2) \in \Gamma_{0,1}$.

Therefore, the robust counterpart for the problem (MMCOPOU2) is defined as

$$\text{(RMMCOPOU2)} \quad \min_{(x(\cdot),c(\cdot))} \int_{\Gamma_{0,1}} \Big(\max_{w_1 \in \mathcal{W}_1} (e^x - x + w_1 c^2),\ \max_{w_2 \in \mathcal{W}_2} (x\tan^{-1} x + w_2) \Big)\, dt^1 dt^2$$

$$\text{subject to} \quad x^2 + 2x - 3 \leqq 0,$$

$$x_{t^1} = x_{t^2} = 2 - c,$$

$$x(0,0) = 0,\ x(1,1) = 3,$$

where $t = (t^1, t^2) \in \Gamma_{0,1}$. Clearly $\mathcal{D} = \{(x, c) \in \mathcal{X} \times \mathcal{C} : -3 \leq x \leq 1 \wedge x_{t^1} = x_{t^2} = 2 - c, x(0,0) = 0, x(1,1) = 3\}$ is a robust feasible solution set to (MMCOPOU2).

Now, we find a weak robust efficient solution to the problem (MMCOPOU2), i.e. find the control function $\bar{c} : \Gamma_{0,1} \to \mathbb{R}$ (that determines the state function $\bar{x} : \Gamma_{0,1} \to \mathbb{R}$) satisfying the dynamical system (2.5.3) with respect to boundary conditions (2.5.4). Additionally, we assume that the state function is affine.

Let $(\bar{x}, \bar{c})$ be a robust feasible solution to the problem (MMCOPOU2). The robust necessary efficiency conditions (2.2.1)–(2.2.3) at $(\bar{x}, \bar{c})$ with Lagrange multipliers $\bar{\mu}_1, \bar{\mu}_2, \bar{\nu}, \bar{\gamma}_1^1, \bar{\gamma}_2^1$ and the uncertain assumption $\max_{w \in \mathcal{W}} f(\pi, w) = f(\pi, \bar{w})$ are

$$\bar{\mu}_1(e^{\bar{x}} - 1) + \bar{\mu}_2\Big(\tan^{-1}\bar{x} + \frac{\bar{x}}{1 + \bar{x}^2}\Big) + 2\bar{\nu}(\bar{x} + 1) + (\bar{\gamma}_1^1)_{t^1} + (\bar{\gamma}_2^1)_{t^2} = 0, \tag{2.5.5}$$

$$2\bar{\mu}_1 \bar{c}\bar{w}_1 - \bar{\gamma}_1^1 - \bar{\gamma}_2^1 = 0, \tag{2.5.6}$$

$$\bar{\nu}(\bar{x}^2 + 2\bar{x} - 3) = 0, \tag{2.5.7}$$

where relation (2.5.7) is satisfied if either $\bar{\nu} = 0$ or $(\bar{x}^2 + 2\bar{x} - 3) = 0$, for all $t = (t^1, t^2) \in \mathrm{int}\Gamma_{0,1}$ (interior of hyperparallelepiped Γ). Let $\bar{\nu} \neq 0$, then $\bar{x}^2 + 2\bar{x} - 3 = 0$ implies that $\bar{x} = -3, 1$. Clearly, the boundary conditions (2.5.4) are failing at $\bar{x} = -3, 1$. Consequently, $\bar{\nu} = 0$.

Case 1. Assume that the Lagrange multipliers $\bar{\gamma}_\alpha^\tau$, $\tau = \overline{1, q}, \alpha = \overline{1, p}$ are constant functions, say $\bar{\gamma}_1^1 = h_1$ and $\bar{\gamma}_2^1 = h_2$ where $h_1, h_2 \in \mathbb{R}_+$. Therefore, Eq. (2.5.6) gives

$$\bar{c} = \frac{h_1 + h_2}{2\bar{\mu}_1 w_1},\ \forall t \in \Gamma_{0,1},$$

which shows that $\bar{c}$ is a constant function on $\Gamma_{0,1}$. Taking into account equality constraints (2.5.3), it follow that

$$\bar{x} = \left(2 - \frac{h_1 + h_2}{2\bar{\mu}_1 w_1}\right)t^1 + \left(2 - \frac{h_1 + h_2}{2\bar{\mu}_1 w_1}\right)t^2 + h_3, \ h_3 \in \mathbb{R}_+.$$

By using the boundary condition (2.5.4) along with Lagrange multiplier $\bar{\mu}_1 = 1$, and the uncertain parameters $\bar{w}_1 = 1$, we get $h_3 = 0$ and $h_1 + h_2 = 1$, and therefore

$$\bar{x} = \frac{3}{2}(t^1 + t^2), \ \bar{c} = \frac{1}{2}, \ \forall t \in \Gamma_{0,1}.$$

Now, we shall show that $(\bar{x}, \bar{c}) = (\frac{3}{2}(t^1 + t^2), \frac{1}{2})$ at $t^1 = t^2 = 0$, i.e. $(\bar{x}, \bar{c}) = (0, \frac{1}{2})$ is a weak robust efficient solution to the problem (MMCOPOU2). In virtue of Definition 2.2.0.1, for all $(x, c) \in \mathcal{D}$, we have

$$\begin{aligned}
&\int_{\Gamma_{0,1}} \max_{w \in \mathcal{W}} f(\pi, w)dt - \int_{\Gamma_{0,1}} \max_{w \in \mathcal{W}} f(\bar{\pi}, w)dt \\
&= \int_{\Gamma_{0,1}} f(\pi, \bar{w})dt - \int_{\Gamma_{0,1}} f(\bar{\pi}, \bar{w})dt \\
&= \int_{\Gamma_{0,1}} \left(e^x - x + \bar{w}_1 c^2, \ x \tan^{-1} x + \bar{w}_2\right)dt^1 dt^2 \\
&\quad - \int_{\Gamma_{0,1}} \left(e^{\bar{x}} - \bar{x} + \bar{w}_1 \bar{c}^2, \ \bar{x} \tan^{-1} \bar{x} + \bar{w}_2\right)dt^1 dt^2 \\
&= \int_{\Gamma_{0,1}} \left(e^x - x + \bar{w}_1 c^2, \ x \tan^{-1} x + \bar{w}_2\right)dt^1 dt^2 - \int_{\Gamma_{0,1}} \left(\frac{3}{4}, 2\right)dt^1 dt^2 \\
&= \int_{\Gamma_{0,1}} \left\{\left(e^x - x + \bar{w}_1 c^2 - \frac{3}{4}\right), \left(x \tan^{-1} x + \bar{w}_2 - 2\right)\right\}dt^1 dt^2. \quad (2.5.8)
\end{aligned}$$

On solving Eq. (2.5.3), we have $c = (2 - x_{t^1})$. Therefore, the above Eq. (2.5.8) can be written as

$$\begin{aligned}
\int_{\Gamma_{0,1}} \max_{w \in \mathcal{W}} f(\pi, w)dt - \int_{\Gamma_{0,1}} \max_{w \in \mathcal{W}} f(\bar{\pi}, w)dt = \int_{\Gamma_{0,1}} \Big\{\left(e^x - x + \bar{w}_1 (2 - x_{t^1})^2 - \frac{3}{4}\right), \\
\times \left(x \tan^{-1} x + \bar{w}_2 - 2\right)\Big\}dt^1 dt^2.
\end{aligned} \quad (2.5.9)$$

Now, it remains to find the state function x which satisfies the boundary conditions (2.5.4).

The equality constraints (2.5.3) follow that

$$x_{t^1} = x_{t^2}, \ \forall t \in \Gamma_{0,1}.$$

Applying the necessary Euler-Lagrange PDE (see, for instance, [6]) for Eq. (2.5.9), it follows that

$$x = \phi(t^1 + t^2) + \psi, \ \phi, \psi \in \mathbb{R}, \ \forall t \in \Gamma_{0,1}.$$

The above solution together with the boundary conditions (2.5.4) imply that $\phi = \frac{3}{2}$ and $\psi = 0$. Therefore, $x = \frac{3}{2}(t^1 + t^2)$. Hence, Eq. (2.5.9) yields

$$\begin{aligned}
&\int_{\Gamma_{0,1}} \max_{w \in \mathcal{W}} f(\pi, w)dt - \int_{\Gamma_{0,1}} \max_{w \in \mathcal{W}} f(\bar{\pi}, w)dt \\
&= \int_{\Gamma_{0,1}} \Big\{ \Big(e^{(\frac{3}{2}(t^1+t^2))} - \frac{3}{2}(t^1 + t^2) + \bar{w}_1(2 - \frac{3}{2})^2 - \frac{5}{4} \Big), \\
&\quad \Big(\frac{3}{2}(t^1 + t^2)\tan^{-1}\Big(\frac{3}{2}(t^1 + t^2)\Big) + \bar{w}_2 - 2 \Big) \Big\} dt^1 dt^2 \\
&= \Big(2.8876,\ 1.5216\Big) \geq 0.
\end{aligned}$$

Similarly, on solving Eq. (2.5.3), we have $c = (2 - x_{t^2})$ and the inequality (2.5.8) concludes that $\int_{\Gamma_{0,1}} \max_{w \in \mathcal{W}} f(\pi, w)dt - \int_{\Gamma_{0,1}} \max_{w \in \mathcal{W}} f(\bar{\pi}, w)dt \geq 0$. In both the cases, it shows that $(0, \frac{1}{2})$ is a weak robust efficient solution to the problem (MMCOPOU2) at which the necessary efficiency conditions (2.5.5)–(2.5.7) are satisfied with Lagrange multipliers $\bar{\mu}_1 = 1, \bar{\mu}_2 = 0, \bar{\nu} = 0$ and $\bar{\gamma}_1^1 + \bar{\gamma}_2^1 = 1$ and uncertain parameter $\bar{w}_1 = 1, \bar{w}_2 = 2$. Further, it can be seen that the Lagrange functional is also convex at $(\bar{x}, \bar{c}) = (0, \frac{1}{2})$ on $\mathscr{X} \times \mathscr{C}$.

Now, we formulate the penalized multi-dimensional multi-objective optimization problem in the face of data uncertainty $(\text{MMCOPOU2})_\rho$ associated with the multi-dimensional multi-objective optimization problem in the face of data uncertainty (MMCOPOU2) as follows:

$$\begin{aligned}
(\text{MMCOPOU2})_\rho \quad &\min_{(x(\cdot),c(\cdot))} \Big(\int_{\Gamma_{0,1}} \Big\{ (e^x - x + w_1 c^2) + \rho\Big[\max\{0, x^2 + 2x - 3\} \\
&+ |x_{t^1} - 2 + c| + |x_{t^2} - 2 + c| \Big]\Big\} dt^1 dt^2, \int_{\Gamma_{0,1}} \Big\{ (x \tan^{-1} x + w_2) \\
&+ \rho\Big[\max\{0, x^2 + 2x - 3\} + |x_{t^1} - 2 + c| + |x_{t^2} - 2 + c| \Big]\Big\} dt^1 dt^2 \Big).
\end{aligned}$$

Therefore, the robust counterpart for $(\text{MMCOPOU2})_\rho$ is defined as

$$\begin{aligned}
(\text{RMMCOPOU2})_\rho \quad &\min_{(x(\cdot),c(\cdot))} \Big(\int_{\Gamma_{0,1}} \Big\{ \max_{w_1 \in \mathcal{W}_1} (e^x - x + w_1 c^2) + \rho\Big[\max\{0, x^2 + 2x - 3\} \\
&+ |x_{t^1} - 2 + c| + |x_{t^2} - 2 + c| \Big]\Big\} dt^1 dt^2, \int_{\Gamma_{0,1}} \Big\{ \max_{w_2 \in \mathcal{W}_2} (x \tan^{-1} x + w_2) \\
&+ \rho\Big[\max\{0, x^2 + 2x - 3\} + |x_{t^1} - 2 + c| + |x_{t^2} - 2 + c| \Big]\Big\} dt^1 dt^2 \Big).
\end{aligned}$$

In virtue of Definition (2.4.0.3), it can be observed that the inequality $\mathscr{P}(\pi, w, \rho) - \mathscr{P}(\bar{\pi}, w, \rho) \geq 0$ at $(\bar{x}, \bar{c}) = (0, \frac{1}{2})$ for all $(x, c) \in \mathscr{X} \times \mathscr{C}$. This shows that $(0, \frac{1}{2})$ is a weak robust efficient solution to the penalized multi-dimensional optimization problem $(\text{MMCOPOU2})_\rho$ with $\rho \geq 1$. Hence, Theorem 2.5.0.1 is verified.

Case 2. Assume that the Lagrange multipliers $\bar{\gamma}_\alpha^\tau$ are functions of t, say $\bar{\gamma}_1^1 = t^2, \bar{\gamma}_2^1 = t^1$. Thus, by Eq. (2.5.6), we have

$$\bar{c} = \frac{t^2 + t^1}{2\bar{\mu}_1 \bar{w}_1}, \quad \forall t \in \Gamma_{0,1}. \tag{2.5.10}$$

On solving Eq. (2.5.3) with the help of Eq. (2.5.10), we have

$$\bar{x} = \left\{2t^1 + 2t^2 - \frac{1}{4}(t^1)^2 - \frac{1}{4}(t^2)^2 - t^1 t^2\right\} + h_4, \; h_4 \in \mathbb{R}_+.$$

On simplify the above equation with boundary conditions (2.5.4), we get

$$\bar{x} = \left\{2t^1 + 2t^2 - \frac{1}{4}(t^1)^2 - \frac{1}{4}(t^2)^2 - t^1 t^2\right\},$$

which shows that the state function is not an affine function. Consequently, we get a contradiction with the hypothesis.

References

1. Şt. Mititelu, Optimality and duality for invex multi-dimensional control problems with mixed constraints. J. Adv. Math. Stud. **2**, 25–34 (2009)
2. S. Treanţă, M. Arana-Jiménez, KT-pseudoinvex multidimensional control problem. Optim. Control Appl. Meth. **39**, 1291–1300 (2018)
3. S. Treanţă, Efficiency in generalized V-KT-pseudoinvex control problems. Internat. J. Control **93**, 611–618 (2020)
4. A.J. Preeti, Saddle point criteria for multi-dimensional control optimisation problem involving first-order PDE constraints. Internat. J. Control **94**, 1567–1576 (2021)
5. A.J. Preeti, M. Arana-Jiménez, An exact l_1 penalty function method for a multi-dimensional optimization problem with data uncertainty. Optim. Control Appl. Meth. **41**, 1705–1717 (2020)
6. S. Treanţă, Higher-order Hamilton dynamics and Hamilton-Jacobi divergence PDE. Comput. Math. Appl. **75**, 547–560 (2018)

Chapter 3
Multi-dimensional Variational Control Problem with Data Uncertainty in Constraint Functionals

3.1 Introduction

In the last few years, the applicability of the penalty function method, initiated by Zangwill [1] for the constrained optimization problem, has grown significantly. The penalty function approach transforms the constrained optimization problem into an unconstrained optimization problem and preserves the optimality of the original one. In this way, the solution sets of the unconstrained optimization problems ideally converge to the solution sets of the constrained optimization problems. The idea behind the penalty function approach (the convergence of the solution sets for constrained optimization problem and its associated unconstrained optimization problem) encourages the researchers to establish the equivalence between the solution set of constrained and unconstrained problems under suitable assumptions for different kinds of optimization problems. Antczak [2] used an exact l_1 penalty function method in convex nondifferentiable multi-objective optimization problem and established the equivalence between the solution set of the original problem and its associated penalized problem. Also, Alvarez [3], Antczak [4] and Liu and Feng [5] explored the exponential penalty function method for multi-objective optimization problem and established the relationships between the constrained and unconstrained optimization problems. On the other hand, Li *et al.* [6] used the penalty function method to solve the continuous inequality constrained optimal control problem. Thereafter, Jayswal and Preeti [7] extended the applicability of the penalty function method for the multi-dimensional optimization problem. Moreover, Jayswal *et al.* [8] explored the same for uncertain optimization problem under convexity assumptions.

A. Jayswal et al., *Multi-dimensional Control Problems*, Industrial and Applied Mathematics, https://doi.org/10.1007/978-981-19-6561-6_3

3.2 Problem Description

We are considering the following notations and working hypotheses:

• $\mathbb{R}^p, \mathbb{R}^q, \mathbb{R}^r$ and $\mathbb{R}^n$ are Euclidean spaces of dimensions p, q, r and n, respectively.

• $\Gamma_{t_0,t_1} \subset \mathbb{R}^p$ is a hyperparallelepiped fixed by the diagonally opposite points $t_0 = (t_0^\alpha)$ and $t_1 = (t_1^\alpha)$, $\alpha = \overline{1,p}$ and the point $t = ((t^\alpha), \alpha = \overline{1,p}) \in \Gamma_{t_0,t_1} \subset \mathbb{R}^p$.

• $\mathscr{X}$ is the space of state functions (piecewise smooth) $x = (x^\tau) : \Gamma_{t_0,t_1} \subset \mathbb{R}^p \to \mathbb{R}^q$ and $\frac{\partial x}{\partial t^\alpha} = x_\alpha$ denotes the partial derivative of x with respect to t^α.

• $\mathscr{C}$ is the space of control functions (piecewise continuous) $c = (c^j) : \Gamma_{t_0,t_1} \subset \mathbb{R}^p \to \mathbb{R}^r$.

• $dt = dt^1 \dots dt^p$ is the volume element on $\mathbb{R}^p \supset \Gamma_{t_0,t_1}$.

• T denotes the transpose of a vector.

• We use the following convention for inequalities and equalities: for $x, y \in \mathbb{R}^n$, we have

$$
\begin{aligned}
&(i)\ x < y \Leftrightarrow x_i < y_i, \forall i = \overline{1,n},\\
&(ii)\ x = y \Leftrightarrow x_i = y_i,\ \forall i = \overline{1,n},\\
&(iii)\ x \leqq y \Leftrightarrow x_i \leq y_i,\ \forall i = \overline{1,n},\\
&(iv)\ x \leq y \Leftrightarrow x_i \leq y_i,\ \forall i = \overline{1,n} \text{ and } x_i < y_i \text{ for some } i.
\end{aligned}
$$

The multi-dimensional multi-objective optimization problem with the data uncertainty in constraints is defined as follows:

$$
\begin{aligned}
\text{(MMCOPCU)} \quad & \min_{(x(\cdot),c(\cdot))} \left\{ \int_{\Gamma_{t_0,t_1}} f(t, x(t), c(t))dt \right.\\
&= \left.\left(\int_{\Gamma_{t_0,t_1}} f_1(t, x(t), c(t))dt, \dots, \int_{\Gamma_{t_0,t_1}} f_s(t, x(t), c(t))dt \right) \right\}\\
\text{subject to} \quad & G(t, x(t), c(t), u) \leqq 0,\\
& x_\alpha = H(t, x(t), c(t), v),\\
& x(t_0) = x_0,\ x(t_1) = x_1,
\end{aligned}
$$

where $f = (f_1, \dots, f_s)$; $(f_k) : \Gamma_{t_0,t_1} \times \mathscr{X} \times \mathscr{C} \to \mathbb{R}^s, k = \overline{1,s}, G = (G_1, \dots, G_m)$; $(G_l) : \Gamma_{t_0,t_1} \times \mathscr{X} \times \mathscr{C} \times \mathcal{U}_l \to \mathbb{R}^m,\ l = \overline{1,m},\ H = (H_\alpha^\tau) : \Gamma_{t_0,t_1} \times \mathscr{X} \times \mathscr{C} \times \mathcal{V}_\alpha^\tau \to \mathbb{R}^{pq}, \alpha = \overline{1,p}, \tau = \overline{1,q}$ are C^∞-class functionals. $u = (u_l)$ and $v = (v_\alpha^\tau)$ are the uncertain parameters for some convex compact subsets $\mathcal{U} = (\mathcal{U}_l) \subset \mathbb{R}^m$ and $\mathcal{V} = (\mathcal{V}_\alpha^\tau) \subset \mathbb{R}^{pq}$, respectively. The functions H_α satisfy the closeness conditions (complete integrability conditions) $D_\beta H_\alpha = D_\alpha H_\beta, \alpha, \beta = \overline{1,p}, \alpha \neq \beta$, where the total derivative is denoted by D_β.

The associated robust counterpart of the multi-dimensional multi-objective optimization problem (MMCOPCU) is defined as follows:

$$\text{(RMMCOPCU)} \quad \min_{(x(\cdot),c(\cdot))} \Big\{ \int_{\Gamma_{t_0,t_1}} f(t,x(t),c(t))dt = \Big(\int_{\Gamma_{t_0,t_1}} f_1(t,x(t),c(t))dt, \ldots, \int_{\Gamma_{t_0,t_1}} f_s(t,x(t),c(t))dt \Big) \Big\}$$

$$\text{subject to} \quad G(t,x(t),c(t),u) \leqq 0, \ \forall u \in \mathcal{U},$$
$$x_\alpha = H(t,x(t),c(t),v), \ \forall v \in \mathcal{V},$$
$$x(t_0) = x_0, \ x(t_1) = x_1,$$

where f, G and H are defined as above in the multi-dimensional multi-objective optimization problem (MMCOPCU).

We denote $\mathcal{D} = \{(x,c) \in \mathcal{X} \times \mathcal{C} : \ G(t,x(t),c(t),u) \leqq 0, x_\alpha = H(t,x(t), c(t),v), x(t_0) = x_0, x(t_1) = x_1\}$ the set of all feasible solutions in (RMMCOPCU) and we say that it is the robust feasible solution set to the problem (MMCOPCU).

From now on, to simplify our presentation, we introduce some notations as follows: $x = x(t), \hat{x} = \hat{x}(t), \bar{x} = \bar{x}(t), c = c(t), \hat{c} = \hat{c}(t), \bar{c} = \bar{c}(t), \pi = (t, x(t), c(t)), \bar{\pi} = (t, \bar{x}(t), \bar{c}(t)), \hat{\pi} = (t, \hat{x}(t), \hat{c}(t)), \rho = \rho(t), \bar{\rho} = \bar{\rho}(t)$ and $\Gamma = \Gamma_{t_0,t_1}$.

The partial derivatives of Lagrange multiplier γ and the partial derivatives associated with f are defined as

$$\gamma_t = \begin{bmatrix} \frac{\partial \gamma_1^\tau}{\partial t^1} & \cdots & \frac{\partial \gamma_1^\tau}{\partial t^p} \\ \vdots & \ddots & \vdots \\ \frac{\partial \gamma_p^\tau}{\partial t^1} & \cdots & \frac{\partial \gamma_p^\tau}{\partial t^p} \end{bmatrix}, \quad f_x = \begin{bmatrix} \frac{\partial f_1}{\partial x^1} & \cdots & \frac{\partial f_1}{\partial x^q} \\ \vdots & \ddots & \vdots \\ \frac{\partial f_s}{\partial x^1} & \cdots & \frac{\partial f_s}{\partial x^q} \end{bmatrix}, \quad f_c = \begin{bmatrix} \frac{\partial f_1}{\partial c^1} & \cdots & \frac{\partial f_1}{\partial c^r} \\ \vdots & \ddots & \vdots \\ \frac{\partial f_s}{\partial c^1} & \cdots & \frac{\partial f_s}{\partial c^r} \end{bmatrix}.$$

Similarly, we interpret G_x, G_c using matrices with m rows and H_x, H_c, x_α using matrices with p rows.

In general, for a vector optimization problem, it is not so easy to find a point which optimizes more than one objective simultaneously. Therefore, the notion of an efficient solution is used to determine an optimality in vector optimization problems.

Definition 3.2.0.1 A point $(\bar{x}, \bar{c}) \in \mathcal{D}$ is said to be weak robust efficient solution to the multi-dimensional multi-objective optimization problem (MMCOPCU), if there does not exist another point $(x, c) \in \mathcal{D}$ such that

$$\int_\Gamma f(\pi)dt < \int_\Gamma f(\bar{\pi})dt.$$

Definition 3.2.0.2 A functional $F(x,c) = \int_\Gamma f(t,x(t),x_\sigma(t),c(t))dt$ is said to be convex at $(\bar{x}, \bar{c})$, if the following inequality:

$$F(x,c) - F(\bar{x},\bar{c}) \geqq \int_\Gamma (x(t) - \bar{x}(t)) f_x(t, \bar{x}(t), \bar{x}_\sigma(t), \bar{c}(t))dt$$
$$+ \int_\Gamma (x_\sigma(t) - \bar{x}_\sigma(t)) f_{x_\sigma}(t, \bar{x}(t), \bar{x}_\sigma(t), \bar{c}(t))dt$$

$$+\int_{\Gamma}(c(t)-\bar{c}(t))f_c(t,\bar{x}(t),\bar{x}_{\sigma}(t),\bar{c}(t))dt$$

holds for all $(x(t), c(t)) \in \mathscr{X} \times \mathscr{C}$.

Following the same lines formulated in Mititelu [9], we establish the following result. Thus, the proof is omitted. The proof can be read (it is free) at the following link: https://www.journal.fairpartners.ro/volume-22009-no-1_3.html

Theorem 3.2.0.1 *Let $(\bar{x}, \bar{c}) \in \mathcal{D}$ be a weak robust efficient solution to the problem* (MMOPCU). *If the constraint conditions (for the existence of the multipliers) hold, then there exist the scalar vector $\bar{\mu} \in \mathbb{R}^s$, the piecewise smooth Lagrange multipliers $\bar{\nu} = (\bar{\nu}_l(t)) \in \mathbb{R}^{\Lambda}_{+}, l \in \mathbb{R}^{\Lambda}, \bar{\gamma} = (\bar{\gamma}^{\tau}_{\alpha}(t)) \in \mathbb{R}^{pq}, \tau = \overline{1,q}, \alpha = \overline{1,p}$, and the uncertain parameters $\bar{u} \in \mathcal{U}, \bar{v} \in \mathcal{V}$ such that $(\bar{x}, \bar{c})$ satisfies the following conditions:*

$$\bar{\mu}^T f_x(\bar{\pi}) + \bar{\nu}^T G_x(\bar{\pi}, \bar{u}) + \bar{\gamma}^T H_x(\bar{\pi}, \bar{v}) + \bar{\gamma}_t = 0, \tag{3.2.1}$$

$$\bar{\mu}^T f_c(\bar{\pi}) + \bar{\nu}^T G_c(\bar{\pi}, \bar{v}) + \bar{\gamma}^T H_c(\bar{\pi}, \bar{v}) = 0, \tag{3.2.2}$$

$$\bar{\nu}^T G(\bar{\pi}, \bar{u}) = 0, \tag{3.2.3}$$

$$\bar{\mu} \geq 0, \bar{\nu} \geqq 0 \tag{3.2.4}$$

hold for all $t \in \Gamma$, except at discontinuities.

Remark 3.2.0.1 The conditions (3.2.1)–(3.2.4) are known as robust necessary efficiency conditions for the multi-dimensional multi-objective optimization problem (MMCOPCU).

3.3 Robust Sufficient Efficiency Conditions

In this section, we will derive the robust sufficient efficiency conditions for the multi-dimensional multi-objective optimization problem (MMCOPCU) which states that any robust feasible solution to the problem (MMCOPCU) will be a weak robust efficient solution to the considered problem (MMCOPCU) if the involved functionals are convex.

Theorem 3.3.0.1 *Let $(\bar{x}, \bar{c})$ be a robust feasible solution to the multi-dimensional multi-objective optimization problem* (MMCOPCU) *and the robust necessary efficiency conditions (3.2.1)–(3.2.4) be satisfied at $(\bar{x}, \bar{c})$. Further, assume that $\int_{\Gamma} \bar{\mu}^T f(\pi, \bar{w})dt$, $\int_{\Gamma} \bar{\nu}^T G(\pi, \bar{u})dt$, and $\int_{\Gamma} \bar{\gamma}^T (H(\pi, \bar{v}) - x_{\alpha})dt$ are convex at $(\bar{x}, \bar{c})$. Then, $(\bar{x}, \bar{c})$ is a weak robust efficient solution to the multi-dimensional multi-objective optimization problem* (MMCOPCU).

Proof Contrary to the result and assume that $(\bar{x}, \bar{c})$ is not a weak robust efficient solution to the multi-dimensional multi-objective optimization problem (MMCOPCU). Then, there exists $(\hat{x}, \hat{c}) \in \mathcal{D}$ such that

$$\int_{\Gamma} f(\hat{\pi})dt < \int_{\Gamma} f(\bar{\pi})dt. \tag{3.3.1}$$

Since $(\bar{x}, \bar{c})$ satisfies the conditions (3.2.1)–(3.2.4), therefore, multiplying equations (3.2.1) and (3.2.2) by $(\hat{x} - \bar{x})$ and $(\hat{c} - \bar{c})$, respectively, and integrating them, we get

$$\int_{\Gamma} (\hat{x} - \bar{x})\Big\{\bar{\mu}^T f_x(\bar{\pi}) + \bar{\upsilon}^T G_x(\bar{\pi}, \bar{u}) + \bar{\gamma}^T H_x(\bar{\pi}, \bar{\upsilon}) + \bar{\gamma}_t\Big\}dt$$
$$+ \int_{\Gamma} (\hat{c} - \bar{c})\Big\{\bar{\mu}^T f_c(\bar{\pi}) + \bar{\upsilon}^T G_c(\bar{\pi}, \bar{u}) + \bar{\gamma}^T H_c(\bar{\pi}, \bar{\upsilon})\Big\}dt = 0. \tag{3.3.2}$$

On the other hand, from assumption $\int_{\Gamma} \bar{\mu}^T f(\pi)dt$ is convex at $(\bar{x}, \bar{c})$, we have

$$\int_{\Gamma} \Big\{\bar{\mu}^T f(\hat{\pi}) - \bar{\mu}^T f(\bar{\pi})\Big\}dt \geqq \int_{\Gamma} (\hat{x} - \bar{x})\bar{\mu}^T f_x(\bar{\pi})dt + \int_{\Gamma} (\hat{c} - \bar{c})\bar{\mu}^T f_c(\bar{\pi})dt,$$

which along with the inequality (3.3.1) gives

$$\int_{\Gamma} (\hat{x} - \bar{x})\bar{\mu}^T f_x(\bar{\pi})dt + \int_{\Gamma} (\hat{c} - \bar{c})\bar{\mu}^T f_c(\bar{\pi})dt < 0. \tag{3.3.3}$$

Again, by assumption $\int_{\Gamma} \bar{\upsilon}^T G(\pi, \bar{u})dt$ is convex at $(\bar{x}, \bar{c})$, we obtain

$$\int_{\Gamma} \Big\{\bar{\upsilon}^T G(\hat{\pi}, \bar{u}) - \bar{\upsilon}^T G(\bar{\pi}, \bar{u})\Big\}dt \geqq \int_{\Gamma} (\hat{x} - \bar{x})\bar{\upsilon}^T G_x(\bar{\pi}, \bar{u})dt + \int_{\Gamma} (\hat{c} - \bar{c})\bar{\upsilon}^T G_c(\bar{\pi}, \bar{u})dt.$$

The above inequality together with the robust feasibility of $(\hat{x}, \hat{c})$ in the multi-dimensional multi-objective optimization problem (MMCOPCU) and condition (3.2.3) yields

$$\int_{\Gamma} (\hat{x} - \bar{x})\bar{\upsilon}^T G_x(\bar{\pi}, \bar{u})dt + \int_{\Gamma} (\hat{c} - \bar{c})\bar{\upsilon}^T G_c(\bar{\pi}, \bar{u})dt \leqq 0. \tag{3.3.4}$$

Further, from the assumption $\int_{\Gamma} \bar{\gamma}^T (H(\pi, \bar{\upsilon}) - x_{\alpha})dt$ is convex at $(\bar{x}, \bar{c})$ and the robust feasibility of $(\hat{x}, \hat{c})$ in the multi-dimensional multi-objective optimization problem (MMCOPCU), we obtain

$$\int_{\Gamma} (\hat{x} - \bar{x})\Big\{\bar{\gamma}^T H_x(\bar{\pi}, \bar{\upsilon}) + \bar{\gamma}_t\Big\}dt + \int_{\Gamma} (\hat{c} - \bar{c})\bar{\gamma}^T H_c(\bar{\pi}, \bar{\upsilon})dt \leqq 0. \tag{3.3.5}$$

On adding inequalities (3.3.3), (3.3.4) and (3.3.5), we have

$$\int_{\Gamma}(\hat{x}-\bar{x})\Big\{\bar{\mu}^T f_x(\bar{\pi})+\bar{v}^T G_x(\bar{\pi},\bar{u})+\bar{\gamma}^T H_x(\bar{\pi},\bar{v})+\bar{\gamma}_t\Big\}dt$$
$$+\int_{\Gamma}(\hat{c}-\bar{c})\Big\{\bar{\mu}^T f_c(\bar{\pi})+\bar{v}^T G_c(\bar{\pi},\bar{u})+\bar{\gamma}^T H_c(\bar{\pi},\bar{v})\Big\}dt<0,$$

which contradicts the relation (3.3.2). Hence, this completes the proof. □

The following example verifies the robust sufficient efficiency conditions for a weak robust efficient solution to the multi-dimensional multi-objective optimization problem (MMCOPCU).

Example 3.3.0.1 Let $p=2, q=r=1, k=2, \mathcal{U}=[0,\frac{1}{2}], \mathcal{V}_1^1=\mathcal{V}_2^1=[0,1]$ and Γ_{t_0,t_1} is fixed by the diagonally opposite points $t_0=(t_0^1,t_0^2)$ and $t_1=(t_1^1,t_1^2)$ in $\mathbb{R}^2$.

We consider the following multi-dimensional multi-objective optimization problem in the face of data uncertainty:

$$\text{(MMCOPCU1)}\quad \min_{(x(\cdot),c(\cdot))}\int_{\Gamma_{0,1}}(c^2+x\tan^{-1}(x),x+c)\,dt^1dt^2$$

$$\text{subject to}\quad -\frac{1}{3}ux\le 0, \tag{3.3.6}$$
$$x_{t^1}=1-cv_1^1, \tag{3.3.7}$$
$$x_{t^2}=1-cv_2^1, \tag{3.3.8}$$
$$x(0,0)=0,\ x(1,1)=\frac{1}{2}, \tag{3.3.9}$$

where $t=(t^1,t^2)\in\Gamma_{0,1}$.

Therefore, the associated robust counterpart for the multi-dimensional multi-objective optimization problem (MMCOPCU1) is defined as follows:

$$\text{(RMMCOPCU1)}\quad \min_{(x(\cdot),c(\cdot))}\int_{\Gamma_{0,1}}(c^2+x\tan^{-1}(x),x+c)\,dt^1dt^2$$
$$\text{subject to}\quad -\frac{1}{3}ux\le 0,\ \forall u\in\mathcal{U},$$
$$x_{t^1}=1-cv_1^1,\ \forall v_1^1\in\mathcal{V}_1^1,$$
$$x_{t^2}=1-cv_2^1,\ \forall v_2^1\in\mathcal{V}_2^1,$$
$$x(0,0)=0,\ x(1,1)=\frac{1}{2},$$

where $t=(t^1,t^2)\in\Gamma_{0,1}$. Clearly $\mathcal{D}=\{(x,c)\in\mathcal{X}\times\mathcal{C}: x\ge 0\wedge x_{t^1}=1-cv_1^1\wedge x_{t^2}=1-cv_2^1, x(0,0)=0, x(1,1)=\frac{1}{2}\}$ is a robust feasible solution set to (MMCOPCU1).

Now, we are interested to find a weak robust efficient solution to the problem (MMCOPCU1). It means that we have to find the control function $\bar{c}:\Gamma_{0,1}\to\mathbb{R}$ (that determines the state function $\bar{x}:\Gamma_{0,1}\to\mathbb{R}$) satisfying dynamical system (3.3.7) and

(3.3.8) with respect to boundary conditions (3.3.9). Additionally, we assume that the state function is affine.

Let $(\bar{x}, \bar{c})$ be a robust feasible solution to the problem (MMCOPCU1). The necessary optimality conditions (3.2.1)–(3.2.3) at $(\bar{x}, \bar{c})$ with Lagrange multipliers $\bar{\mu}_1, \bar{\mu}_2, \bar{\nu}, \bar{\gamma}_1^1, \bar{\gamma}_2^1$ and the uncertain parameters $\bar{u}, \bar{v}_1^1, \bar{v}_2^1$ are

$$\bar{\mu}_1\Big(\tan^{-1}(\bar{x}) + \frac{\bar{x}}{\bar{x}+1}\Big) + \bar{\mu}_2 - \frac{1}{3}\bar{\nu}\bar{u} + (\bar{\gamma}_1^1)_{t^1} + (\bar{\gamma}_2^1)_{t^2} = 0, \tag{3.3.10}$$

$$2\bar{\mu}_1\bar{c} + \bar{\mu}_2 - \bar{\gamma}_1^1\bar{v}_1^1 - \bar{\gamma}_2^1\bar{v}_2^1 = 0, \tag{3.3.11}$$

$$-\bar{\nu}(\frac{1}{3}\bar{u}\bar{x}) = 0, \tag{3.3.12}$$

where relation (3.3.12) is satisfied if either $\bar{\nu} = 0$ or $\frac{1}{3}\bar{u}\bar{x} = 0$, for all $t = (t^1, t^2) \in \operatorname{int}\Gamma_{0,1}$. We assume that $\bar{\nu} = 0$.

Case 1. Assume that the Lagrange multipliers $\bar{\gamma}_\alpha^\tau$, $\tau = \overline{1,q}, \alpha = \overline{1,p}$ are constant functions, say $\bar{\gamma}_1^1 = d_1$ and $\bar{\gamma}_2^1 = d_2$, where $d_1, d_2 \in \mathbb{R}_+$. Therefore, Eq. (3.3.11) gives

$$\bar{c} = \frac{d_1\bar{v}_1^1 + d_2\bar{v}_2^1 - \bar{\mu}_2}{2\bar{\mu}_1}, \quad \forall t \in \Gamma_{0,1},$$

which shows that $\bar{c}$ is a constant function on $\Gamma_{0,1}$. Taking into account the equality constraints (3.3.7) and (3.3.8), it follows that

$$\bar{x} = \Big(1 - \frac{d_1\bar{v}_1^1 + d_2\bar{v}_2^1 - \bar{\mu}_2}{2\bar{\mu}_1}\bar{v}_1^1\Big)t^1 + \Big(1 - \frac{d_1\bar{v}_1^1 + d_2\bar{v}_2^1 - \bar{\mu}_2}{2\bar{\mu}_1}\bar{v}_2^1\Big)t^2 + d_3, \ d_3 \in \mathbb{R}_+.$$

By using the boundary condition (3.3.9) along with Lagrange multiplier $\bar{\mu}_1 = 1, \bar{\mu}_2 = 0$ and the uncertain parameter $\bar{v}_1^1 = 1, \bar{v}_2^1 = 1$, we get $d_3 = 0$ and $d_1 + d_2 = \frac{3}{2}$, and therefore

$$\bar{x} = \frac{1}{4}(t^1 + t^2), \ \bar{c} = \frac{3}{4}, \quad \forall t \in \Gamma_{0,1}.$$

One can easily verify that the necessary efficiency conditions (3.3.10), (3.3.11) and (3.3.12) are satisfied at $(\bar{x}, \bar{c}) = (\frac{1}{4}(t^1 + t^2), \frac{3}{4})$ at $t^1 = t^2 = 0$, i.e. $(0, \frac{3}{4})$ with the Lagrange multipliers $\bar{\mu}_1 = 1, \bar{\mu}_2 = 0, \bar{\nu} = 0, \bar{\gamma}_1^1 = 1, \bar{\gamma}_2^1 = \frac{1}{2}$ and the uncertain parameters $\bar{u} = \frac{1}{2}, \bar{v}_1^1 = \bar{v}_2^1 = 1$. Further, it can also be easily verified that the involved functionals are convex at $(\bar{x}, \bar{c}) = (0, \frac{3}{4})$. Hence, all the conditions of Theorem 3.3.0.1 are satisfied, which ensure that $(0, \frac{3}{4})$ is also a weak robust efficient solution to (MMCOPCU1).

Case 2. Assume that the Lagrange multipliers $\bar{\gamma}_\alpha^\tau$ are functions of t, say $\bar{\gamma}_1^1 = t^2, \bar{\gamma}_2^1 = t^1$. So, from Eq. (3.3.11), we have

$$\bar{c} = \frac{t^2\bar{v}_1^1 + t^1\bar{v}_2^1 - \bar{\mu}_2}{2\bar{\mu}_1}, \ \forall t \in \Gamma_{0,1}. \tag{3.3.13}$$

On solving Eq. (3.3.7) and (3.3.8) with the help of Eq. (3.3.13) together with Lagrange multiplier $\bar{\mu}_1 = 1$ and $\bar{\mu}_2 = 0$, we have

$$\bar{x} = t^1 + t^2 - \frac{1}{2}v_1^1\Big(\frac{(t^1)^2}{2} + t^1t^2\Big) - \frac{1}{2}v_2^1\Big(\frac{(t^2)^2}{2} + t^1t^2\Big) + d_3, \ d_3 \in \mathbb{R}_+,$$

and, by considering the boundary conditions (3.3.9), we get

$$\bar{x} = t^1 + t^2 - \frac{1}{2}v_1^1\Big(\frac{(t^1)^2}{2} + t^1t^2\Big) - \frac{1}{2}v_2^1\Big(\frac{(t^2)^2}{2} + t^1t^2\Big).$$

This shows that the state function is not an affine function. Consequently, we get a contradiction with the hypothesis.

3.4 Robust Saddle-point Criteria and Exact l_1 Penalty Function Method

In this section, we use the exact l_1 penalty function method to construct an unconstrained multi-dimensional multi-objective optimization problem associated with the multi-dimensional multi-objective optimization problem (MMCOPCU). This is defined as

$$(\text{MMCOPCU})_\rho \quad \min_{(x(\cdot),c(\cdot))} \Big\{\mathscr{P}(t, x(t), c(t), u, v, \rho(t)) = \int_\Gamma \Big\{f(t, x(t), c(t))$$
$$+ \rho(t)\Big[\sum_{l=1}^{m} G_l^+(t, x(t), c(t), u) + \sum_{\tau=1}^{q}\sum_{\alpha=1}^{p}\Big|H_\alpha^\tau(t, x(t), c(t), v) - \frac{\partial x^\tau}{\partial t^\alpha}\Big|\Big]e\Big\}dt\Big\},$$

where $e = (1, \ldots, 1) \in \mathbb{R}^s$, $u \in \mathcal{U}$ and $v \in \mathcal{V}$ are the uncertain parameters, $\rho(t)$ is a non-zero non-negative penalty parameter and the penalty function $G_l^+(t, x(t), c(t), u)$ is defined as

$$G_l^+(t, x(t), c(t), u) = \begin{cases} 0, & \text{if } G_l(t, x(t), c(t), u) \leqq 0, \\ G_l(t, x(t), c(t), u), & \text{if } G_l(t, x(t), c(t), u) > 0. \end{cases} \tag{3.4.1}$$

Thus, the associated robust counterpart to the problem $(\text{MMCOPCU})_\rho$ is given by

$$(\text{RMMCOPCU})_\rho \quad \min_{(x(\cdot),c(\cdot))} \Big\{ \mathscr{P}(t,x(t),c(t),u,v,\rho(t)) = \int_\Gamma \Big\{ f(t,x(t),c(t)) + \rho(t)\Big[\sum_{l=1}^{m} G_l^{+}(t,x(t),c(t),u) + \sum_{\tau=1}^{q}\sum_{\alpha=1}^{p} \Big| H_\alpha^\tau(t,x(t),c(t),v) - \frac{\partial x^\tau}{\partial t^\alpha} \Big| \Big] e \Big\} dt \Big\},$$

for all $u \in \mathcal{U}$ and $v \in \mathcal{V}$.

Definition 3.4.0.1 A point $(\bar{x}, \bar{c}) \in \mathscr{X} \times \mathscr{C}$ is said to be a weak robust efficient solution to the problem $(\text{MMCOPCU})_\rho$, if there does not exist another point $(x, c) \in \mathscr{X} \times \mathscr{C}$ such that

$$\mathscr{P}(\pi, \bar{u}, \bar{v}, \rho) < \mathscr{P}(\bar{\pi}, \bar{u}, \bar{v}, \rho),$$

where $\bar{u} \in \mathcal{U}$ and $\bar{v} \in \mathcal{V}$.

Definition 3.4.0.2 The Lagrange functional $L((x, c), \nu, \gamma, u, v) : \mathscr{X} \times \mathscr{C} \times \mathbb{R}_+^m \times \mathbb{R}^{pq} \times \mathcal{U} \times \mathcal{V} \to \mathbb{R}^s$ associated with the multi-dimensional multi-objective optimization problem (MMCOPCU) is defined as

$$L((x,c),\nu,\gamma,u,v) = \int_\Gamma \Big\{ f(t,x,c) + \nu^T G(t,x,c,u)e + \gamma^T (H(t,x,c,v) - x_\alpha)e \Big\} dt.$$

Definition 3.4.0.3 A point $((\bar{x}, \bar{c}), \bar{\nu}, \bar{\gamma}, \bar{u}, \bar{v}) \in \mathcal{D} \times \mathbb{R}_+^m \times \mathbb{R}^{pq} \times \mathcal{U} \times \mathcal{V}$ is said to be a robust saddle-point in the multi-dimensional multi-objective optimization problem (MMCOPCU), iff

(i) $L((\bar{x}, \bar{c}), \nu, \gamma, u, v) \leqq L((\bar{x}, \bar{c}), \bar{\nu}, \bar{\gamma}, \bar{u}, \bar{v}), \quad \forall \nu \in \mathbb{R}_+^m, \quad \forall \gamma \in \mathbb{R}^{pq}, \quad \forall u \in \mathcal{U}, \forall v \in \mathcal{V}$,
(ii) $L((x, c), \bar{\nu}, \bar{\gamma}, \bar{u}, \bar{v}) \geqq L((\bar{x}, \bar{c}), \bar{\nu}, \bar{\gamma}, \bar{u}, \bar{v}), \ \forall (x, c) \in \mathscr{X} \times \mathscr{C}$.

Note: The exactness of the penalization is an important property of each exact penalty function method in which there exists a threshold value $\bar{\rho} \geq 0$ of the penalty parameter ρ such that, for every $\rho \geq \bar{\rho}$,

$$\arg\min \Big\{ \int_\Gamma f(t,x,c)dt \Big| (x,c) \in \mathcal{D} \Big\} = \arg\min \Big\{ \int_\Gamma \mathscr{P}(t,x,c,u,v,\rho)dt \ \Big| (x,c) \in \mathscr{X} \times \mathscr{C} \Big\}.$$

In the following theorem, we shall show that the robust saddle-point of the multi-dimensional multi-objective optimization problem (MMCOPCU) and a weak robust efficient solution of its associated unconstrained multi-dimensional multi-objective optimization problem $(\text{MMCOPCU})_\rho$ are equivalent if the penalty parameter is assumed to be sufficiently large.

Theorem 3.4.0.1 *Let* $((\bar{x}, \bar{c}), \bar{\nu}, \bar{\gamma}, \bar{u}, \bar{v})$ *be a robust saddle-point to the multi-dimensional multi-objective optimization problem* (MMCOPCU). *Then,* $(\bar{x}, \bar{c})$ *is* *a weak robust efficient solution to* (MMCOPCU)$_\rho$, *if the penalty parameter* ρ *is assumed to be sufficiently large, that is,* $\rho \geq \max\{\bar{\nu}, |\bar{\gamma}|\}$.

Proof We proceed by contradiction and consider $(\bar{x}, \bar{c})$ is not a weak robust efficient solution to (MMCOPCU)$_\rho$. Then, there exists $(\hat{x}, \hat{c}) \in \mathscr{X} \times \mathscr{C}$ such that

$$\int_\Gamma \mathscr{P}(\hat{\pi}, \bar{u}, \bar{v}, \rho)dt < \int_\Gamma \mathscr{P}(\bar{\pi}, \bar{u}, \bar{v}, \rho)dt.$$

By using the definition of the penalized multi-dimensional multi-objective optimization problem (MMCOPCU)$_\rho$, we get

$$\int_\Gamma \Big\{f(\hat{\pi}) + \rho\Big[\sum_{l=1}^{m} G_l^+(\hat{\pi}, \bar{u}) + \sum_{\tau=1}^{q}\sum_{\alpha=1}^{p} \Big|H_\alpha^\tau(\hat{\pi}, \bar{v}) - \frac{\partial x^\tau}{\partial t^\alpha}\Big|\Big]e\Big\}dt$$
$$< \int_\Gamma \Big\{f(\bar{\pi}) + \rho\Big[\sum_{l=1}^{m} G_l^+(\bar{\pi}, \bar{u}) + \sum_{\tau=1}^{q}\sum_{\alpha=1}^{p} \Big|H_\alpha^\tau(\bar{\pi}, \bar{v}) - \frac{\partial x^\tau}{\partial t^\alpha}\Big|\Big]e\Big\}dt.$$

The above inequality together with the robust feasibility of $(\bar{x}, \bar{c})$ in the multi-dimensional multi-objective optimization problem (MMCOPCU) and (3.4.1) yields

$$\int_\Gamma \Big\{f(\hat{\pi}) + \rho\Big[\sum_{l=1}^{m} G_l^+(\hat{\pi}, \bar{u}) + \sum_{\tau=1}^{q}\sum_{\alpha=1}^{p} \Big|H_\alpha^\tau(\hat{\pi}, \bar{v}) - \frac{\partial x^\tau}{\partial t^\alpha}\Big|\Big]e\Big\}dt < \int_\Gamma f(\bar{\pi})dt.$$

Since $\rho \geq \max\{\bar{\nu}, |\bar{\gamma}|\}$, we have

$$\int_\Gamma \Big\{f(\hat{\pi}) + \sum_{l=1}^{m} \bar{\nu}_l G_l^+(\hat{\pi}, \bar{u})e + \sum_{\tau=1}^{q}\sum_{\alpha=1}^{p} \Big|\bar{\gamma}_\alpha^\tau\Big(H_\alpha^\tau(\hat{\pi}, \bar{v}) - \frac{\partial x^\tau}{\partial t^\alpha}\Big)\Big|e\Big\}dt < \int_\Gamma f(\bar{\pi})dt.$$

Again using (3.4.1), we get

$$\int_\Gamma \Big\{f(\hat{\pi}) + \sum_{l=1}^{m} \bar{\nu}_l G_l(\hat{\pi}, \bar{u})e + \sum_{\tau=1}^{q}\sum_{\alpha=1}^{p} \bar{\gamma}_\alpha^\tau\Big(H_\alpha^\tau(\hat{\pi}, \bar{v}) - \frac{\partial x^\tau}{\partial t^\alpha}\Big)e\Big\}dt < \int_\Gamma f(\bar{\pi})dt,$$

equivalently,

$$\int_\Gamma \Big\{f(\hat{\pi}) + \bar{\nu}^T G(\hat{\pi}, \bar{u})e + \bar{\gamma}^T\Big(H(\hat{\pi}, \bar{v}) - x_\alpha\Big)e\Big\}dt < \int_\Gamma f(\bar{\pi})dt. \tag{3.4.2}$$

On the other hand, since $((\bar{x}, \bar{c}), \bar{\nu}, \bar{\gamma}, \bar{u}, \bar{v})$ is a robust saddle-point to the multi-dimensional multi-objective optimization problem (MMCOPCU), by Definition 3.4.0.3 (i), we have

$$L((\bar{x},\bar{c}),\nu,\gamma,u,\upsilon) \leqq L((\bar{x},\bar{c}),\bar{\nu},\bar{\gamma},\bar{u},\bar{\upsilon}),\ \forall \nu \in \mathbb{R}^m_+, \gamma \in \mathbb{R}^{pq}, u \in \mathcal{U}, \upsilon \in \mathcal{V}.$$

By considering the definition of the Lagrange functional, we get

$$\int_\Gamma \Big\{ f(\bar{\pi}) + \nu^T G(\bar{\pi},u)e + \gamma^T \Big(H(\bar{\pi},\upsilon) - x_\alpha \Big) e \Big\} dt$$
$$\leqq \int_\Gamma \Big\{ f(\bar{\pi}) + \bar{\nu}^T G(\bar{\pi},\bar{u})e + \bar{\gamma}^T \Big(H(\bar{\pi},\bar{\upsilon}) - x_\alpha \Big) e \Big\} dt.$$

If we set $\nu = 0$, $\gamma = 0$, we obtain

$$\int_\Gamma \bar{\nu}^T G(\bar{\pi},\bar{u})\, dt \geqq 0.$$

Also, the robust feasibility of $(\bar{x},\bar{c})$ in the problem (MMCOPCU) yields

$$\int_\Gamma \bar{\nu}^T G(\bar{\pi},\bar{u})\, dt \leqq 0.$$

On combining the above two inequalities, we get $\int_\Gamma \bar{\nu}^T G(\bar{\pi},\bar{u})\, dt = 0$. Further, the robust feasibility of $(\bar{x},\bar{c})$ in the problem (MMCOPCU) also admits that $\int_\Gamma \Big(H(\bar{\pi},\bar{\upsilon}) - x_\alpha \Big) dt = 0$. Therefore, the inequality (3.4.2) can be rewritten as

$$\int_\Gamma \Big\{ f(\hat{\pi}) + \bar{\nu}^T G(\hat{\pi},\bar{u})e + \bar{\gamma}^T \Big(H(\hat{\pi},\bar{\upsilon}) - x_\alpha \Big) e \Big\} dt$$
$$< \int_\Gamma \Big\{ f(\bar{\pi}) + \bar{\nu}^T G(\bar{\pi},\bar{u})e + \bar{\gamma}^T \Big(H(\bar{\pi},\bar{\upsilon}) - x_\alpha \Big) e \Big\} dt.$$

By using the definition of the Lagrange functional, we get

$$L((\hat{x},\hat{c}),\bar{\nu},\bar{\gamma},\bar{u},\bar{\upsilon}) < L((\bar{x},\bar{c}),\bar{\nu},\bar{\gamma},\bar{u},\bar{\upsilon}),\ \forall (\hat{x},\hat{c}) \in \mathscr{X} \times \mathscr{C},$$

which contradicts Definition 3.4.0.3 (ii). Hence, the proof is complete. □

Let us consider Example 3.3.0.1 to show the relationship between the solution set of multi-dimensional multi-objective optimization problem and its associated penalized problem via saddle-point.

Example 3.4.0.1 The Lagrange functional for the problem (MMCOPCU1) is given by

$$L((x,c),\nu,\gamma,u,\upsilon) = \int_{\Gamma_{0,1}} \Big\{ (c^2 + x\tan^{-1}(x)) - \nu\frac{1}{3}ux + \gamma_1^1\Big(x_{t^1} - 1 + c\upsilon_1^1\Big)$$
$$+ \gamma_2^1\Big(x_{t^2} - 1 + c\upsilon_2^1\Big),\ (x + c) - \nu\frac{1}{3}ux + \gamma_1^1\Big(x_{t^1} - 1 + c\upsilon_1^1\Big)$$

$$+ \gamma_2^1\left(x_{t^2} - 1 + cv_2^1\right)\Big\}dt^1dt^2.$$

One can be easily seen that the following inequalities:

$$L((\bar{x}, \bar{c}), \nu, \gamma, u, v) - L((\bar{x}, \bar{c}), \bar{\nu}, \bar{\gamma}, \bar{u}, \bar{v}) \leqq (0, 0),$$
$$\forall \nu \in \mathbb{R}, \gamma = (\gamma_1^1, \gamma_2^1) \in \mathbb{R}^2, u \in \mathcal{U}, v_1^1 \in \mathcal{V}_1^1, v_2^1 \in \mathcal{V}_2^1,$$
$$\text{and } L((x, c), \bar{\nu}, \bar{\gamma}, \bar{u}, \bar{v}) - L((\bar{x}), \bar{c}, \bar{\nu}, \bar{\gamma}, \bar{u}, \bar{v}) \geqq (0, 0) \forall (x, c) \in \mathscr{X} \times \mathscr{C} \text{ and } v_1^1 = v_2^1 \in [0, 1]$$

are satisfied at $((\bar{x}, \bar{c}), \bar{\nu}, \bar{\gamma}, \bar{u}, \bar{v}) = ((0, \frac{3}{4}), 0, (1, \frac{1}{2}), \frac{1}{2}, (1, 1))$. This shows that $((\bar{x}, \bar{c}), \bar{\nu}, \bar{\gamma}, \bar{u}, \bar{v}) = ((0, \frac{3}{4}), 0, (1, \frac{1}{2}), \frac{1}{2}, (1, 1))$ is a saddle-point for the problem (MMCOPCU1).

Now, the penalized multi-dimensional multi-objective optimization problem in the face of data uncertainty $(\text{MMCOPCU1})_\rho$ associated with the multi-dimensional multi-objective optimization problem (MMCOPCU1) is defined as follows:

$$(\text{MMCOPCU1})_\rho \quad \min_{(x(\cdot), c(\cdot))} \int_{\Gamma_{0,1}} \Big\{(c^2 + x\tan^{-1}(x)) + \rho\Big[\max\{0, -\frac{1}{3}ux\}$$
$$+ \left|x_{t^1} - 1 + cv_1^1\right| + \left|x_{t^2} - 1 + cv_2^1\right|\Big], (x + c) + \rho\Big[\max\{0, -\frac{1}{3}ux\}$$
$$+ \left|x_{t^1} - 1 + cv_1^1\right| + \left|x_{t^2} - 1 + cv_2^1\right|\Big]\Big\}dt^1dt^2.$$

Thus, the robust counterpart for the problem (MMCOPCU1) is given by

$$(\text{RMMCOPCU1})_\rho \quad \min_{(x(\cdot), c(\cdot))} \int_{\Gamma_{0,1}} \Big\{(c^2 + x\tan^{-1}(x)) + \rho\Big[\max\{0, -\frac{1}{3}ux\}$$
$$+ \left|x_{t^1} - 1 + cv_1^1\right| + \left|x_{t^2} - 1 + cv_2^1\right|\Big], (x + c) + \rho\Big[\max\{0, -\frac{1}{3}ux\}$$
$$+ \left|x_{t^1} - 1 + cv_1^1\right| + \left|x_{t^2} - 1 + cv_2^1\right|\Big]\Big\}dt^1dt^2,$$

for all $u \in \mathcal{U}$, $v_1^1 \in \mathcal{V}_1^1$ and $v_2^1 \in \mathcal{V}_2^1$.

In virtue of Definition 3.4.0.1, it can be observed that the inequality $\mathscr{P}(\pi, \bar{u}, \bar{v}, \rho) - \mathscr{P}(\bar{\pi}, \bar{u}, \bar{v}, \rho) \geq 0$ is satisfied at $(\bar{x}, \bar{c}) = (0, \frac{3}{4})$. This shows that $(0, \frac{3}{4})$ is a weak robust efficient solution to $(\text{MMCOPCU1})_\rho$ with $\rho \geq 1$. Hence, Theorem 3.4.0.1 is verified.

Proposition 3.4.0.1 *Let* $(\bar{x}, \bar{c}) \in \mathscr{X} \times \mathscr{C}$ *be a weak robust efficient solution to the problem* $(\text{MMCOPCU})_{\bar{\rho}}$. *Then, there does not exist another point* $(x, c) \in \mathcal{D}$ *such that*

$$\int_\Gamma f(\pi)dt \leq \int_\Gamma f(\bar{\pi})dt.$$

Proof Since $(\bar{x}, \bar{c})$ is a weak robust efficient solution to the problem $(\text{MMCOPCU})_{\bar{\rho}}$, then there does not exist another point $(x, c) \in \mathscr{X} \times \mathscr{C}$ such that

$$\mathscr{P}(\pi, \bar{u}, \bar{v}, \bar{\rho}) < \mathscr{P}(\bar{\pi}, \bar{u}, \bar{v}, \bar{\rho}).$$

By using the definition of the penalized semi-infinite multi-dimensional multi-objective optimization problem $(\text{MMCOPCU})_\rho$, we get

$$\int_\Gamma \Big\{f(\pi) + \bar{\rho}\Big[\sum_{l=1}^{m} G_l^+(\pi, \bar{u}) + \sum_{\tau=1}^{q}\sum_{\alpha=1}^{p}\Big|H_\alpha^\tau(\pi, \bar{v}) - x_\alpha\Big|\Big]e\Big\}dt$$
$$< \int_\Gamma \Big\{f(\bar{\pi}) + \bar{\rho}\Big[\sum_{l=1}^{m} G_l^+(\bar{\pi}, \bar{u}) + \sum_{\tau=1}^{q}\sum_{\alpha=1}^{p}\Big|H_\alpha^\tau(\bar{\pi}, \bar{v}) - x_\alpha\Big|\Big]e\Big\}dt.$$

Since $\mathcal{D} \subseteq \mathscr{X} \times \mathscr{C}$, therefore, it follows that there does not exist another point $(x, c) \in \mathcal{D}$ such that

$$\int_\Gamma \Big\{f(\pi) + \bar{\rho}\Big[\sum_{l=1}^{m} G_l^+(\pi, \bar{u}) + \sum_{\tau=1}^{q}\sum_{\alpha=1}^{p}\Big|H_\alpha^\tau(\pi, \bar{v}) - x_\alpha\Big|\Big]e\Big\}dt$$
$$< \int_\Gamma \Big\{f(\bar{\pi}) + \bar{\rho}\Big[\sum_{l=1}^{m} G_l^+(\bar{\pi}, \bar{u}) + \sum_{\tau=1}^{q}\sum_{\alpha=1}^{p}\Big|H_\alpha^\tau(\bar{\pi}, \bar{v}) - x_\alpha\Big|\Big]e\Big\}dt.$$

Equivalently, one can say that the following relation holds for all $(x, c) \in \mathcal{D}$:

$$\int_\Gamma \Big\{f_k(\pi) + \bar{\rho}\Big[\sum_{l=1}^{m} G_l^+(\pi, \bar{u}) + \sum_{\tau=1}^{q}\sum_{\alpha=1}^{p}\Big|H_\alpha^\tau(\pi, \bar{v}) - \frac{\partial x^\tau}{\partial t^\alpha}\Big|\Big]\Big\}dt$$
$$\geqq \int_\Gamma \Big\{f_k(\bar{\pi}) + \bar{\rho}\Big[\sum_{l=1}^{m} G_l^+(\bar{\pi}, \bar{u}) + \sum_{\tau=1}^{q}\sum_{\alpha=1}^{p}\Big|H_\alpha^\tau(\bar{\pi}, \bar{v}) - \frac{\partial x^\tau}{\partial t^\alpha}\Big|\Big]\Big\}dt.$$

By virtue of the relation (3.4.1), we get

$$\int_\Gamma f_k(\pi)dt \geqq \int_\Gamma f_k(\bar{\pi})dt.$$

Therefore, we conclude that there does not exist another point $(x, c) \in \mathcal{D}$ such that

$$\int_\Gamma f(\pi)dt < \int_\Gamma f(\bar{\pi})dt.$$

This completes the proof. □

Theorem 3.4.0.2 *Let* $(\bar{x}, \bar{c})$ *be a weak robust efficient solution to the multi-dimensional multi-objective optimization problem* $(\text{MMCOPCU})_{\bar{\rho}}$. *Assume that* $\mathcal{D}$ *is a compact subset of* $\mathscr{X} \times \mathscr{C}$ *and the inequality* $\int_\Gamma \mathscr{P}(\pi, \bar{u}, \bar{v}, \rho)dt \geq \int_\Gamma \mathscr{P}(\bar{\pi}, \bar{u}, \bar{v}, \rho)dt$ *holds for any* $\rho \geq \bar{\rho}$ *and for all* $(x, c) \in \mathcal{D}$. *Further, if* $\int_\Gamma \mu^T f(\pi)dt$, $\int_\Gamma \nu^T G(\pi, u)dt$ *and* $\int_\Gamma \gamma^T (H(\pi, v) - x_\alpha)dt$ *are convex at* $(\bar{x}, \bar{c})$ *on*

$\mathscr{X} \times \mathscr{C}$, *then* $((\bar{x}, \bar{c}), \bar{v}, \bar{\gamma}, \bar{u}, \bar{v})$ *is a robust saddle-point to the multi-dimensional multi-objective optimization problem* (MMCOPCU).

Proof Since $(\bar{x}, \bar{c})$ is a weak robust efficient solution to the problem $(\text{MMCOPCU})_{\bar{\rho}}$, then, by Proposition 3.4.0.1, there does not exist another point $(x, c) \in \mathcal{D}$ such that

$$\int_{\Gamma} f(\pi)dt < \int_{\Gamma} f(\bar{\pi})dt,$$

equivalently,

$$\int_{\Gamma} f_k(\pi)dt < \int_{\Gamma} f_k(\bar{\pi})dt, \forall k = \overline{1, s}, \tag{3.4.3}$$

which specify that the objective functional $\int_{\Gamma} f_k(\pi)dt$ is bounded below in the compact subset $\mathcal{D}$.

Now, to show that $(\bar{x}, \bar{c})$ is a robust feasible solution to the multi-dimensional multi-objective optimization problem (MMCOPCU), we proceed by contradiction and assume that $(\bar{x}, \bar{c})$ is not a robust feasible solution to the problem (MMCOPCU). Then, by virtue of (3.4.1), we have

$$\int_{\Gamma} \Big\{ \sum_{l=1}^{m} G_l^{+}(\bar{\pi}, \bar{u}) + \sum_{\tau=1}^{q} \sum_{\alpha=1}^{p} \Big| H_{\alpha}^{\tau}(\bar{\pi}, \bar{v}) - \frac{\partial x^{\tau}}{\partial t^{\alpha}} \Big| \Big\} dt > 0. \tag{3.4.4}$$

Let $(\hat{x}, \hat{c})$ be any robust feasible solution to the problem (MMCOPCU). By assumptions, the following inequality:

$$\int_{\Gamma} \mathscr{P}(\hat{\pi}, \bar{u}, \bar{v}, \rho)dt \geq \int_{\Gamma} \mathscr{P}(\bar{\pi}, \bar{u}, \bar{v}, \rho)dt$$

holds for any $\rho > \bar{\rho}$ and for all $(\hat{x}, \hat{c}) \in \mathcal{D}$.

By using the definition of the penalized multi-dimensional multi-objective optimization problem (MMCOPCU), we get

$$\int_{\Gamma} \Big\{ f_k(\hat{\pi}) + \rho \Big[\sum_{l=1}^{m} G_l^{+}(\hat{\pi}, \bar{u}) + \sum_{\tau=1}^{q} \sum_{\alpha=1}^{p} \Big| H_{\alpha}^{\tau}(\hat{\pi}, \bar{v}) - \frac{\partial x^{\tau}}{\partial t^{\alpha}} \Big| \Big] \Big\} dt$$
$$\geq \int_{\Gamma} \Big\{ f_k(\bar{\pi}) + \rho \Big[\sum_{l=1}^{m} G_l^{+}(\bar{\pi}, \bar{u}) + \sum_{\tau=1}^{q} \sum_{\alpha=1}^{p} \Big| H_{\alpha}^{\tau}(\bar{\pi}, \bar{v}) - \frac{\partial x^{\tau}}{\partial t^{\alpha}} \Big| \Big] \Big\} dt. \tag{3.4.5}$$

On the other hand, if we set

$$\rho > \max\left\{ \frac{\int_\Gamma f_k(\hat{\pi})dt - \int_\Gamma f_k(\bar{\pi})dt}{\int_\Gamma \left\{ \sum_{l=1}^m G_l^+(\bar{\pi}, \bar{u}) + \sum_{\tau=1}^q \sum_{\alpha=1}^p \left| H_\alpha^\tau(\bar{\pi}, \bar{v}) - \frac{\partial x^\tau}{\partial t^\alpha} \right| \right\} dt}, \bar{\rho}; \forall (x, c) \in \mathcal{D} \right\},$$

the inequalities (3.4.3) and (3.4.4) imply that ρ is a positive real number. Thus, we have

$$\int_\Gamma \left\{ f_k(\bar{\pi}) + \rho \Big[\sum_{l=1}^m G_l^+(\bar{\pi}, \bar{u}) + \sum_{\tau=1}^q \sum_{\alpha=1}^p \left| H_\alpha^\tau(\bar{\pi}, \bar{v}) - \frac{\partial x^\tau}{\partial t^\alpha} \right| \Big] \right\} dt > \int_\Gamma f_k(\hat{\pi}) dt.$$

Further, the robust feasibility of $(\hat{x}, \hat{c})$ to the problem (MMCOPCU) and in virtue of (3.4.1), the above inequality can be written as

$$\int_\Gamma \left\{ f_k(\bar{\pi}) + \rho \Big[\sum_{l=1}^m G_l^+(\bar{\pi}, \bar{u}) + \sum_{\tau=1}^q \sum_{\alpha=1}^p \left| H_\alpha^\tau(\bar{\pi}, \bar{v}) - \frac{\partial x^\tau}{\partial t^\alpha} \right| \Big] \right\} dt$$
$$> \int_\Gamma \left\{ f_k(\hat{\pi}) + \rho \Big[\sum_{l=1}^m G_l^+(\hat{\pi}, u) + \sum_{\tau=1}^q \sum_{\alpha=1}^p \left| H_\alpha^\tau(\hat{\pi}, v) - \frac{\partial x^\tau}{\partial t^\alpha} \right| \Big] \right\} dt,$$

which contradicts the inequality (3.4.5). Therefore, $(\bar{x}, \bar{c})$ is a robust feasible solution to the problem (MMCOPCU). Hence, the inequality (3.4.3) directly concludes that $(\bar{x}, \bar{c})$ is a weak robust efficient solution to the problem (MMCOPCU).

We shall now show that $((\bar{x}, \bar{c}), \bar{\nu}, \bar{\gamma}, \bar{u}, \bar{v})$ is a robust saddle-point to the problem (MMCOPCU). Since $(\bar{x}, \bar{c})$ is a weak robust efficient solution to the problem (MMCOPCU), then, by Theorem 2.1, $(\bar{x}, \bar{c})$ is a robust KT point. Therefore, there exist $\bar{\mu} \in \mathbb{R}^s$, $\bar{\nu} \in \mathbb{R}_+^m$ and $\bar{\gamma} \in \mathbb{R}^{pq}$ and the uncertain parameters $\bar{u} \in \mathcal{U}$, $\bar{v} \in \mathcal{V}$ such that the robust necessary efficiency conditions (3.2.1)–(3.2.4) are satisfied at $(\bar{x}, \bar{c})$ on $\mathscr{X} \times \mathscr{C}$.

From the robust feasibility of $(\bar{x}, \bar{c})$ to the problem (MMCOPCU), together with the necessary efficiency condition (3.2.3), we have

$$\int_\Gamma \nu^T G(\bar{\pi}, u) dt \leqq \int_\Gamma \bar{\nu}^T G(\bar{\pi}, \bar{u}) dt, \ \forall \nu \in \mathbb{R}_+^m, \forall u \in \mathcal{U}.$$

It follows that

$$\int_\Gamma \left\{ f(\bar{\pi}) + \nu^T G(\bar{\pi}, u) + \gamma^T \Big(H(\bar{\pi}, v) - x_\alpha \Big) \right\} dt$$
$$\leqq \int_\Gamma \left\{ f(\bar{\pi}) + \bar{\nu}^T G(\bar{\pi}, \bar{u}) + \bar{\gamma}^T \Big(H(\bar{\pi}, \bar{v}) - x_\alpha \Big) \right\} dt.$$

By using the definition of Lagrange functional, we get

$$L(\bar{x}, \bar{c}, \nu, \gamma, u, v) \leqq L(\bar{x}, \bar{c}, \bar{\nu}, \bar{\gamma}, \bar{u}, \bar{v}).$$

To prove the second condition of Definition 3.4.0.3, we proceed by contradiction and assume that

$$L(x, c, \bar{\nu}, \bar{\gamma}, \bar{u}, \bar{v}) < L(\bar{x}, \bar{c}, \bar{\nu}, \bar{\gamma}, \bar{u}, \bar{v}), \ \forall (x, c) \in \mathscr{X} \times \mathscr{C}.$$

Again, by considering the definition of Lagrange functional, we get

$$\int_{\Gamma} \Big\{ f(\pi) + \bar{\nu}^T G(\pi, \bar{u}) e + \bar{\gamma}^T \Big(H(\pi, \bar{v}) - x_\alpha \Big) e \Big\} dt < \int_{\Gamma} \Big\{ f(\bar{\pi}) + \bar{\nu}^T G(\bar{\pi}, \bar{u}) e + \bar{\gamma}^T \Big(H(\bar{\pi}, \bar{v}) - x_\alpha \Big) e \Big\} dt. \quad (3.4.6)$$

On the other hand, from the hypotheses $\int_{\Gamma} \mu^T f(\pi) dt$, $\int_{\Gamma} \nu^T G(\pi, u) dt$ and $\int_{\Gamma} \gamma^T (H(\pi, v) - x_\alpha) dt$ are convex at $(\bar{x}, \bar{c})$ on $\mathscr{X} \times \mathscr{C}$, we get

$$\int_{\Gamma} \Big\{ \bar{\mu}^T f(\pi) - \bar{\mu}^T f(\bar{\pi}) \Big\} dt \geqq \int_{\Gamma} (x - \bar{x}) \bar{\mu}^T f_x(\bar{\pi}) dt + \int_{\Gamma} (c - \bar{c}) \bar{\mu}^T f_c(\bar{\pi}) dt, \quad (3.4.7)$$

$$\int_{\Gamma} \Big\{ \bar{\nu}^T G(\pi, \bar{u}) - \bar{\nu}^T G(\bar{\pi}, \bar{u}) \Big\} dt \geqq \int_{\Gamma} (x - \bar{x}) \bar{\nu}^T G_x(\bar{\pi}, \bar{u}) dt + \int_{\Gamma} (c - \bar{c}) \bar{\nu}^T G_c(\bar{\pi}, \bar{u}) dt, \quad (3.4.8)$$

and

$$\int_{\Gamma} \Big\{ \bar{\gamma}^T \Big(H(\pi, \bar{v}) - x_\alpha \Big) - \bar{\gamma}^T \Big(H(\bar{\pi}, \bar{v}) - x_\alpha \Big) \Big\} dt \geqq \int_{\Gamma} (x - \bar{x}) \Big\{ \bar{\gamma}^T H_x(\bar{\pi}, \bar{v}) + \bar{\gamma}_t \Big\} dt + \int_{\Gamma} (c - \bar{c}) \bar{\gamma}^T H_c(\bar{\pi}, \bar{v}) dt. \quad (3.4.9)$$

On combining the inequalities (3.4.7)–(3.4.9), together with the robust necessary efficiency conditions (3.2.1) and (3.2.2), we have

$$\int_{\Gamma} \Big\{ \bar{\mu}^T f(\pi) + \bar{\nu}^T G(\pi, \bar{u}) + \bar{\gamma}^T \Big(H(\pi, \bar{v}) - x_\alpha \Big) \Big\} dt \geqq \int_{\Gamma} \Big\{ \bar{\mu}^T f(\bar{\pi}) + \bar{\nu}^T G(\bar{\pi}, \bar{u}) + \bar{\gamma}^T \Big(H(\bar{\pi}, \bar{v}) - x_\alpha \Big) \Big\} dt,$$

equivalently,

$$\int_{\Gamma} \Big\{ f(\pi) + \bar{\nu}^T G(\pi, \bar{u}) e + \bar{\gamma}^T \Big(H(\pi, \bar{v}) - x_\alpha \Big) e \Big\} dt \geqq \int_{\Gamma} \Big\{ f(\bar{\pi}) + \bar{\nu}^T G(\bar{\pi}, \bar{u}) e + \bar{\gamma}^T \Big(H(\bar{\pi}, \bar{v}) - x_\alpha \Big) e \Big\} dt,$$

which contradicts the inequality (3.4.6). Hence, this completes the proof. □

Example 3.4.0.2 Let $p = 2, q = r = 1, \mathscr{U} = [1, 2], \mathscr{V}_1^1 = \mathscr{V}_2^1 = [0, \frac{1}{2}]$. Γ_{t_0, t_1} is a rectangle (in particular cases, a square) fixed by the diagonally opposite points $t_0 = (t_0^1, t_0^2)$ and $t_1 = (t_1^1, t_1^2)$ in $\mathbb{R}^2$.

We consider the following multi-dimensional multi-objective optimization problem in the face of data uncertainty as follows:

$$\text{(MMCOPCU2)} \quad \min_{(x(\cdot),c(\cdot))} \int_{\Gamma_{0,\frac{1}{2}}} (x^2 + c^2, \frac{1}{2}c + e^x)\, dt^1 dt^2$$

$$\text{subject to} \quad 1 - ux \leqq 0, \tag{3.4.10}$$

$$x_{t^1} = \frac{1}{2} - v_1^1 c, \tag{3.4.11}$$

$$x_{t^2} = \frac{1}{2} - v_2^1 c, \tag{3.4.12}$$

$$x(0,0) = 0, \ x(\frac{1}{2}, \frac{1}{2}) = \frac{1}{3}, \tag{3.4.13}$$

where $t = (t^1, t^2) \in \Gamma_{0,\frac{1}{2}}$.
Thus, the robust counterpart for the problem (MMCOPCU2) is

$$\text{(RMMCOPCU2)} \quad \min_{(x(\cdot),c(\cdot))} \int_{\Gamma_{0,\frac{1}{2}}} (x^2 + c^2, \frac{1}{2}c + e^x)\, dt^1 dt^2$$

$$\text{subject to} \quad 1 - ux \leqq 0, \ \forall u \in [1, 2],$$

$$x_{t^1} = \frac{1}{2} - v_1^1 c, \ \forall v_1^1 \in [0, \frac{1}{2}],$$

$$x_{t^2} = \frac{1}{2} - v_2^1 c, \ \forall v_2^1 \in [0, \frac{1}{2}],$$

$$x(0,0) = 0, \ x(\frac{1}{2}, \frac{1}{2}) = \frac{1}{3},$$

where $t = (t^1, t^2) \in \Gamma_{0,\frac{1}{2}}$. Clearly, $\mathcal{D} = \{(x, c) \in \mathscr{X} \times \mathscr{C} : x \geqq 1 \wedge x_{t^1} = \frac{1}{2} - v_1^1 c \wedge x_{t^2} = \frac{1}{2} - v_2^1 c, x(0, 0) = 0, x(\frac{1}{2}, \frac{1}{2}) = \frac{1}{3}\}$ is a robust feasible solution set to (MMCOPCU2).

Now, we find a weak robust efficient solution to the problem (MMCOPCU2), i.e. find the control function $\bar{c} : \Gamma_{0,\frac{1}{2}} \to \mathbb{R}$ (that determines the state function $\bar{x} : \Gamma_{0,\frac{1}{2}} \to \mathbb{R}$) satisfying the dynamical systems (3.4.11) and (3.4.12) with respect to boundary conditions (3.4.13). Additionally, we assume that the state function is affine.

Let $(\bar{x}, \bar{c})$ be a robust feasible solution to the problem (MMCOPCU2). The robust necessary efficiency conditions (3.2.1)–(3.2.3) at $(\bar{x}, \bar{c})$ with the Lagrange multipliers $\bar{\mu}_1, \bar{\mu}_2, \bar{\nu}, \bar{\gamma}_1^1, \bar{\gamma}_2^1$ and the uncertain parameters $\bar{u}, \bar{v}_1^1, \bar{v}_2^1$ are as follows:

$$2\bar{\mu}_1 \bar{x} - \bar{\mu}_2 e^x - \bar{\nu}\bar{u} + (\bar{\gamma}_1^1)_{t^1} + (\bar{\gamma}_2^1)_{t^2} = 0, \tag{3.4.14}$$

$$2\bar{\mu}_1 \bar{c} + \bar{\mu}_2 \frac{1}{2} - \bar{\gamma}_1^1 \bar{v}_1^1 - \bar{\gamma}_2^1 \bar{v}_2^1 = 0, \tag{3.4.15}$$

$$- \bar{\nu}(1 - \bar{u}\bar{x}) = 0, \tag{3.4.16}$$

where relation (3.4.16) is satisfied if either $\bar{\nu} = 0$ or $1 - \bar{u}\bar{x} = 0$, for all $t = (t^1, t^2) \in \text{int}\Gamma_{0,\frac{1}{2}}$ (interior of hyperparallelepiped Γ). Let $\bar{\nu} \neq 0$, then $1 - \bar{u}\bar{x} = 0$ implies that $\bar{x} = \frac{1}{\bar{u}}$. Clearly, the boundary conditions (3.4.13) are failing at $\bar{x} = \frac{1}{\bar{u}}$, $\forall \bar{u} \in \mathcal{U}$. Consequently, $\bar{\nu} = 0$.

Case 1. Assume that the Lagrange multipliers $\bar{\gamma}_\alpha^\tau$, $\tau = \overline{1,q}, \alpha = \overline{1,p}$ are constant functions, say $\bar{\gamma}_1^1 = h_1$ and $\bar{\gamma}_2^1 = h_2$, where $h_1, h_2 \in \mathbb{R}_+$. Therefore, Eq. (3.4.15) gives

$$\bar{c} = \frac{2h_1\bar{v}_1^1 + 2h_2\bar{v}_2^1 - \bar{\mu}_2}{4\bar{\mu}_1}, \ \forall t \in \Gamma_{0,\frac{1}{2}},$$

which shows that $\bar{c}$ is a constant function on $\Gamma_{0,\frac{1}{2}}$. Taking into account the equality constraints (3.4.11) and (3.4.12), it follows that

$$\bar{x} = \Big(\frac{1}{2} - \bar{v}_1^1\frac{2h_1\bar{v}_1^1 + 2h_2\bar{v}_2^1 - \bar{\mu}_2}{4\bar{\mu}_1}\Big)t^1 + \Big(\frac{1}{2} - \bar{v}_2^1\frac{2h_1\bar{v}_1^1 + 2h_2\bar{v}_2^1 - \bar{\mu}_2}{4\bar{\mu}_1}\Big)t^2 + h_3, \ h_3 \in \mathbb{R}_+.$$

By using the boundary condition (3.4.13) along with the Lagrange multiplier $\bar{\mu}_1 = 1$, $\bar{\mu}_2 = 0$ and the uncertain parameters $\bar{v}_1^1 = \frac{1}{2}$ and $\bar{v}_2^1 = \frac{1}{2}$, we get $h_3 = 0$ and $h_1 + h_2 = \frac{4}{3}$, and therefore

$$\bar{x} = \frac{1}{3}(t^1 + t^2), \ \bar{c} = \frac{1}{3}, \ \forall t \in \Gamma_{0,\frac{1}{2}}.$$

Now, we shall show that $(\bar{x}, \bar{c}) = (\frac{1}{3}(t^1 + t^2), \frac{1}{3})$ at $t^1 = t^2 = 0$, i.e. $(\bar{x}, \bar{c}) = (0, \frac{1}{3})$ is a weak robust efficient solution to the problem (MMCOPCU2). In virtue of Definition 3.2.0.1, for all $(x, c) \in \mathcal{D}$, we have

$$\begin{aligned}\int_{\Gamma_{0,\frac{1}{2}}} f(\pi)dt - \int_{\Gamma_{0,\frac{1}{2}}} f(\bar{\pi})dt &= \int_{\Gamma_{0,\frac{1}{2}}} \Big(x^2 + c^2, \frac{1}{2}c + e^x\Big)dt^1dt^2 - \int_{\Gamma_{0,\frac{1}{2}}} \Big(\bar{x}^2 + \bar{c}^2, \frac{1}{2}\bar{c} + e^{\bar{x}}\Big)dt^1dt^2 \\ &= \int_{\Gamma_{0,\frac{1}{2}}} \Big(x^2 + c^2, \frac{1}{2}c + e^x\Big)dt^1dt^2 - \int_{\Gamma_{0,\frac{1}{2}}} \Big(\frac{1}{9}, \frac{7}{6}\Big)dt^1dt^2 \\ &= \int_{\Gamma_{0,\frac{1}{2}}} \Big\{\Big(x^2 + c^2 - \frac{1}{9}\Big), \Big(\frac{1}{2}c + e^x - \frac{7}{6}\Big)\Big\}dt^1dt^2. \end{aligned} \quad (3.4.17)$$

On solving Eq. (3.4.11), we have $c = \frac{1}{v_1^1}\Big(\frac{1}{2} - x_{t^1}\Big)$. Therefore, the above Eq. (3.4.17) can be written as

$$\begin{aligned}\int_{\Gamma_{0,\frac{1}{2}}} f(\pi)dt - \int_{\Gamma_{0,\frac{1}{2}}} f(\bar{\pi})dt = \int_{\Gamma_{0,\frac{1}{2}}} \Big\{\Big(x^2 + \Big[\frac{1}{v_1^1}\Big(\frac{1}{2} - x_{t^1}\Big)\Big]^2 - \frac{1}{9}\Big), \\ \times \Big(\frac{1}{2v_1^1}\Big(\frac{1}{2} - x_{t^1}\Big) + e^x - \frac{7}{6}\Big)\Big\}dt^1dt^2. \end{aligned} \quad (3.4.18)$$

Now, it remains to find the state function x which satisfies the boundary conditions (3.4.13).

The equality constraints (3.4.11) and (3.4.12) follow that

$$x_{t^1} = x_{t^2}, \ \forall t \in \Gamma_{0,\frac{1}{2}}.$$

Applying the necessary Euler-Lagrange PDE (see, for instance, [10]) for Eq. (3.4.18), it follows that

$$x = \phi(t^1 + t^2) + \psi, \ \phi, \psi \in \mathbb{R}, \ \forall t \in \Gamma_{0,\frac{1}{2}}.$$

The above solution together with the boundary conditions (3.4.13) imply that $\phi = \frac{1}{3}$ and $\psi = 0$. Therefore, $x = \frac{1}{3}(t^1 + t^2)$. Hence, Eq. (3.4.18) yields

$$\begin{aligned}\int_{\Gamma_{0,\frac{1}{2}}} f(\pi)dt - \int_{\Gamma_{0,\frac{1}{2}}} f(\bar{\pi})dt &= \int_{\Gamma_{0,\frac{1}{2}}} \Big\{\Big(\Big(\tfrac{1}{3}(t^1+t^2)\Big)^2 + \Big[\tfrac{1}{v_1^1}\Big(\tfrac{1}{2}-\tfrac{1}{3}\Big)\Big]^2 - \tfrac{1}{9}\Big), \\ &\quad \times \Big(\frac{1}{2v_1^1}\Big(\frac{1}{2}-\frac{1}{3}\Big) + e^{\frac{1}{3}(t^1+t^2)} - \frac{7}{6}\Big)\Big\}dt^1dt^2 \\ &= \Big(\frac{1}{144(v_1^1)^2} - \frac{1}{288}, \frac{1}{48v_1^1} + 0.00436\Big) \geq 0, \ \forall v_1^1 \in [0, \frac{1}{2}].\end{aligned}$$

Similarly, on solving Eq. (3.4.12), we have $c = \frac{1}{v_2^1}\Big(\frac{1}{2} - x_{t^2}\Big)$ and the inequality (3.4.17) concludes that $\int_{\Gamma_{0,\frac{1}{2}}} f(\pi)dt - \int_{\Gamma_{0,\frac{1}{2}}} f(\bar{\pi})dt \geq 0$. In both the cases, it results that $(0, \frac{1}{3})$ is a weak robust efficient solution to the problem (MMCOPCU2) for which the necessary efficiency conditions (3.4.14)–(3.4.16) are satisfied with the Lagrange multipliers $\bar{\mu}_1 = 1, \bar{\mu}_2 = 0, \bar{\nu} = 0$ and $\bar{\gamma}_1^1 = 1, \bar{\gamma}_2^1 = \frac{1}{3}$ and the uncertain parameters $\bar{u} = 2, \bar{v}_1^1 = \bar{v}_2^1 = \frac{1}{2}$.

Now, we formulate the penalized multi-dimensional multi-objective optimization problem $(\text{MMCOPCU2})_\rho$ associated to the multi-dimensional multi-objective optimization problem (MMCOPCU2) as follows:

$$\begin{aligned}(\text{MMCOPCU2})_\rho \quad \min_{(x(\cdot),c(\cdot))} \int_{\Gamma_{0,\frac{1}{2}}} &\Big\{(x^2+c^2) + \rho\Big[\max\{0, 1-ux\} + \Big|x_{t^1} - \frac{1}{2} + v_1^1c\Big| \\ &+ \Big|x_{t^2} - \frac{1}{2} + v_2^1c\Big|\Big], (\frac{1}{2}c + e^x) + \rho\Big[\max\{0, 1-ux\} \\ &+ \Big|x_{t^1} - \frac{1}{2} + v_1^1c\Big| + \Big|x_{t^2} - \frac{1}{2} + v_2^1c\Big|\Big]\Big\}dt^1dt^2.\end{aligned}$$

Thus, the robust counterpart for $(\text{MMCOPCU2})_\rho$ is given as follows:

$$(\text{RMMCOPCU2})_\rho \quad \min_{(x(\cdot),c(\cdot))} \int_{\Gamma_{0,\frac{1}{2}}} \Big\{(x^2+c^2)+\rho\Big[\max\{0,1-ux\}+\Big|x_{t^1}-\frac{1}{2}+v_1^1c\Big| + \Big|x_{t^2}-\frac{1}{2}+v_2^1c\Big|\Big], (\frac{1}{2}c+e^x)+\rho\Big[\max\{0,1-ux\} + \Big|x_{t^1}-\frac{1}{2}+v_1^1c\Big|+\Big|x_{t^2}-\frac{1}{2}+v_2^1c\Big|\Big]\Big\}dt^1dt^2,$$

for all $u \in \mathcal{U}$, $v_1^1 \in \mathcal{V}_1^1$, $v_2^1 \in \mathcal{V}_2^1$.

In virtue of Definition (3.4.0.1), it can be observed that the inequality $\mathscr{P}(\pi,\bar{u},\bar{v},\rho)-\mathscr{P}(\bar{\pi},\bar{u},\bar{v},\rho) \geq 0$ is satisfied at $(\bar{x},\bar{c})=(0,\frac{1}{3})$ for all $(x,c) \in \mathscr{X}\times\mathscr{C}$. This shows that $(0,\frac{1}{3})$ is a weak robust efficient solution of the penalized multi-dimensional optimization problem $(\text{MMCOPCU2})_\rho$ with $\rho \geq 1$. Further, it can be seen that the objective and constraint functionals are convex at $(0,\frac{1}{3})$ on $\mathscr{X}\times\mathscr{C}$.

The uncertain Lagrange functional is given as follows:

$$L((x,c),\nu,\gamma,u,v)=\int_{\Gamma_{0,\frac{1}{2}}} \Big\{(x^2+c^2)+\nu(1-ux)+\gamma_1^1\Big(x_{t^1}-\frac{1}{2}+v_1^1c\Big) + \gamma_2^1\Big(x_{t^2}-\frac{1}{2}+v_2^1c\Big), (\frac{1}{2}c+e^x)+\nu(1-ux) + \gamma_1^1\Big(x_{t^1}-\frac{1}{2}+v_1^1c\Big)+\gamma_2^1\Big(x_{t^2}-\frac{1}{2}+v_2^1c\Big)\Big\}dt^1dt^2.$$

Now, the following inequalities:

$$L((\bar{x},\bar{c}),\nu,\gamma,\bar{u},\bar{v})-L((\bar{x},\bar{c}),\bar{\nu},\bar{\gamma},\bar{u},\bar{v}) \leqq (0,0),$$
$$\forall \nu \in \mathbb{R}, \gamma=(\gamma_1^1,\gamma_2^1) \in \mathbb{R}^2, u \in \mathcal{U}, v_1^1 \in \mathcal{V}_1^1, v_2^1 \in \mathcal{V}_2^1,$$
$$\text{and } L((x,c),\bar{\nu},\bar{\gamma},\bar{u},\bar{v})-L((\bar{x},\bar{c}),\bar{\nu},\bar{\gamma},\bar{u},\bar{v}) \geqq (0,0), \ \forall (x,c) \in \mathscr{X}\times\mathscr{C} \text{ and } v_1^1=v_2^1 \in [0,\frac{1}{2}],$$

are satisfied at $((\bar{x},\bar{c}),\bar{\nu},\bar{\gamma},\bar{u},\bar{v})=((0,\frac{1}{3}),0,(\frac{1}{3},1),2,(\frac{1}{2},\frac{1}{2}))$. Thus, $((0,\frac{1}{3}),0,(\frac{1}{3},1),2,(\frac{1}{2},\frac{1}{2}))$ is a robust saddle-point of (MMCOPCU2). Hence, Theorem 3.4.0.2 is verified.

Case 2. Assume that the Lagrange multipliers $\bar{\gamma}_\alpha^\tau$ are functions of t, say $\bar{\gamma}_1^1 = t^2$, $\bar{\gamma}_2^1=t^1$. Thus, by Eq. (3.4.15), we have

$$\bar{c}=\frac{2t^2\bar{v}_1^1+2t^1\bar{v}_2^1-\bar{\mu}_2}{4\bar{\mu}_1}, \quad \forall t \in \Gamma_{0,\frac{1}{2}}. \tag{3.4.19}$$

On solving Eq. (3.4.11) and (3.4.12) with the help of Eq. (3.4.19), we have

$$\bar{x}=\frac{1}{4\mu_1}\Big\{\frac{1}{2}(t^2+t^1)-\frac{v_1^1}{4}\Big[\frac{(t^1)^2}{2}+t^1t^2\Big]-\frac{v_2^1}{4}\Big[\frac{(t^2)^2}{2}+t^1t^2\Big]\Big\}+h_4, \ h_4 \in \mathbb{R}_+.$$

On simplify the above equation with boundary conditions (3.4.13), we get

$$\bar{x} = \frac{1}{4\mu_1}\left\{\frac{1}{2}(t^2 + t^1) - \frac{v_1^1}{4}\left[\frac{(t^1)^2}{2} + t^1 t^2\right] - \frac{v_2^1}{4}\left[\frac{(t^2)^2}{2} + t^1 t^2\right]\right\},$$

which shows that the state function is not an affine function. Consequently, we get a contradiction with the hypothesis.

Theorem 3.4.0.3 *Let $(\bar{x}, \bar{c})$ be a robust feasible solution to the multi-dimensional multi-objective optimization problem* (MMCOPCU) *for which the necessary efficiency conditions (3.2.1)–(3.2.4) are satisfied. Assume that $\int_\Gamma \mu^T f(\pi)dt$, $\int_\Gamma \nu^T G(\pi, \bar{u})dt$ and $\int_\Gamma \gamma^T (H(\pi, v) - x_\alpha)dt$ are convex at $(\bar{x}, \bar{c})$ on $\mathscr{X} \times \mathscr{C}$. Further, if the penalty parameter ρ is assumed to be sufficiently large, (it is sufficient to set $\rho \geq \max\{\bar{\nu}, |\bar{\gamma}|\}$), then $(\bar{x}, \bar{c})$ is a weak robust efficient solution to the multi-dimensional multi-objective optimization problem* $(\text{MMCOPCU})_\rho$.

Proof To prove $(\bar{x}, \bar{c})$ is a weak robust efficient solution to the multi-dimensional multi-objective optimization problem $(\text{MMCOPCU})_\rho$, it is sufficient to prove that $((\bar{x}, \bar{c}), \bar{\nu}, \bar{\gamma}, \bar{u}, \bar{v})$ is a robust saddle-point to the problem (MMCOPCU).

From the robust feasibility of $(\bar{x}, \bar{c})$ to the problem (MMCOPCU) together with the necessary efficiency condition (3.2.3), we have

$$\int_\Gamma \nu^T G(\bar{\pi}, u)dt \leq \int_\Gamma \bar{\nu}^T G(\bar{\pi}, \bar{u})dt, \ \forall \nu \in \mathbb{R}^m_+,$$

equivalently,

$$\int_\Gamma \left\{f(\bar{\pi}) + \nu^T G(\bar{\pi}, u)e + \gamma^T \Big(H(\bar{\pi}, v) - x_\alpha\Big)e\right\}dt$$
$$\leq \int_\Gamma \left\{f(\bar{\pi}) + \bar{\nu}^T G(\bar{\pi}, \bar{u})e + \bar{\gamma}^T \Big(H(\bar{\pi}, \bar{v}) - x_\alpha\Big)e\right\}dt.$$

From the definition of Lagrange functional, we get

$$L((\bar{x}, \bar{c}), \nu, \gamma, u, v) \leqq L((\bar{x}, \bar{c}), \bar{\nu}, \bar{\gamma}, \bar{u}, \bar{v}). \tag{3.4.20}$$

Since $\int_\Gamma \mu^T f(\pi)dt$, $\int_\Gamma \nu^T G(\pi, u)dt$ and $\int_\Gamma \gamma^T (H(\pi, v) - x_\alpha)dt$ are convex at $(\bar{x}, \bar{c})$ on $\mathscr{X} \times \mathscr{C}$, on the line of Theorem 3.4.0.2, we get

$$\int_\Gamma \left\{ f(\pi) + \bar{\nu}^T G(\pi, \bar{u})e + \bar{\gamma}^T \Big(H(\pi, \bar{v}) - x_\alpha\Big)e\right\}dt$$
$$\geqq \int_\Gamma \left\{ f(\bar{\pi}) + \bar{\nu}^T G(\bar{\pi}, \bar{u})e + \bar{\gamma}^T \Big(H(\bar{\pi}, \bar{v}) - x_\alpha\Big)e\right\}dt.$$

Again, from the definition of Lagrange functional, we get

$$L((x, c), \bar{\nu}, \bar{\gamma}, \bar{u}, \bar{v}) \geqq L((\bar{x}, \bar{c}), \bar{\nu}, \bar{\gamma}, \bar{u}, \bar{v}), \ \forall (x, c) \in \mathscr{X} \times \mathscr{C}. \tag{3.4.21}$$

The inequalities (3.4.20) and (3.4.21) ensure that $((\bar{x}, \bar{c}), \bar{\nu}, \bar{\gamma}, \bar{u}, \bar{v})$ is a robust saddle-point to the problem (MMCOPCU). Hence, by Theorem 3.4.0.1, we conclude that $(\bar{x}, \bar{c})$ is a weak robust efficient solution to the multi-dimensional multi-objective optimization problem $(\text{MMCOPCU})_\rho$. □

References

1. W.I. Zangwill, Nonlinear programming via penalty functions. Manag. Sci. **13**, 344–358 (1967)
2. T. Antczak, The vector exact l_1 penalty method for nondifferentiable convex multi-objective programming problems. Appl. Math. Comput. **218**, 9095–9106 (2012)
3. F. Alvarez, Absolute minimizer in convex programming by exponential penalty. J. Convex Anal. **7**, 197–202 (2000)
4. T. Antczak, Vector exponential penalty function method for nondifferentiable multi-objective programming problems. Bull. Malays. Math. Sci. Soc. **41**, 657–686 (2016)
5. S. Liu, E. Feng, The exponential penalty function method for multi-objective programming problems. Optim. Meth. Softw. **25**, 667–675 (2010)
6. B. Li, C.J. Yu, K.L. Teo, G.R. Duan, An exact penalty function method for continuous inequality constrained optimal control problem. J. Optim. Theory Appl. **151**, 260–291 (2011)
7. A.J. Preeti, An exact l_1 penalty function method for multi-dimensional first-order PDE constrained control optimisation problem. Eur. J. Control **52**, 34–41 (2020)
8. A.J. Preeti, M. Arana-Jiménez, An exact l_1 penalty function method for a multi-dimensional optimization problem with data uncertainty. Optim. Control Appl. Meth. **41**, 1705–1717 (2020)
9. Şt. Mititelu, Optimality and duality for invex multi-dimensional control problems with mixed constraints. J. Adv. Math. Stud. **2**, 25–34 (2009)
10. S. Treanţă, Higher-order Hamilton dynamics and Hamilton-Jacobi divergence PDE. Comput. Math. Appl. **75**, 547–560 (2018)

Chapter 4
Multi-dimensional Variational Control Problem with Data Uncertainty in Objective and Constraint Functionals

4.1 Introduction

Insufficient information, outdated sources, a big amount of data, sampling discrepancy, etc. arise the uncertainty in the data, which generates the barrier to find the exact solution to the problem. In this way, the robust approach is growing rapidly for finding the solution to the problem in the face of data uncertainty. The purpose of this method is to minimize the maximum uncertainty involved in the problem. This fact motivated the researchers to study various optimization problems in the face of data uncertainty and developed new results on this topic. Liu and Yuan [1] generated a robust algorithm that converges to a point that satisfies a certain first-order necessary efficiency condition when the considered problem is itself infeasible. Jeyakumar [2] characterized the robust approach when the uncertainty consists in the constraints function. Whereas, Wei *et al.* [3] examined the robust optimization problems under the strictly robust counterpart via the image space analysis. More precisely, a relationship between the uncertain optimization problem and its image problem was proved, which provides an approach to tackle minimax problems. Sun *et al.* [4] studied the robust convex optimization problem and developed the associated saddle-point theory. Besides this, Preeti *et al.* [5] extended the robust approach over multi-dimensional control problem in the face of data uncertainty and used the penalty function method to simplify the considered constrained problem.

4.2 Problem Description

Throughout the chapter, we are considering the following notations and working hypotheses:

- $\mathbb{R}^p$, $\mathbb{R}^q$, $\mathbb{R}^r$ and $\mathbb{R}^n$ are Euclidean spaces of dimensions p, q, r and n, respectively.
- $\Gamma_{t_0,t_1} \subset \mathbb{R}^p$ is a hyperparallelepiped fixed by the diagonally opposite points $t_0 = (t_0^\alpha)$ and $t_1 = (t_1^\alpha)$, $\alpha = \overline{1,p}$ and the point $t = ((t^\alpha), \alpha = \overline{1,p}) \in \Gamma_{t_0,t_1} \subset \mathbb{R}^p$.

A. Jayswal et al., *Multi-dimensional Control Problems*, Industrial and Applied Mathematics, https://doi.org/10.1007/978-981-19-6561-6_4

- $\mathscr{X}$ is the space of state functions (piecewise smooth) $x = (x^\tau) : \Gamma_{t_0,t_1} \subset \mathbb{R}^p \to \mathbb{R}^q$ and $\frac{\partial x}{\partial t^\alpha} = x_\alpha$ denotes the partial derivative of x with respect to t^α.
- $\mathscr{C}$ is the space of control functions (piecewise continuous) $c = (c^j) : \Gamma_{t_0,t_1} \subset \mathbb{R}^p \to \mathbb{R}^r$.
- $dt = dt^1 \dots dt^p$ is the volume element on $\mathbb{R}^p \supset \Gamma_{t_0,t_1}$.
- T denotes the transpose of a vector.
- We use the following convention for inequalities and equalities: for $x, y \in \mathbb{R}^n$, we have

$$
\begin{aligned}
(i)\ & x < y \Leftrightarrow x_i < y_i, \forall i = \overline{1,n},\\
(ii)\ & x = y \Leftrightarrow x_i = y_i, \ \forall i = \overline{1,n},\\
(iii)\ & x \leqq y \Leftrightarrow x_i \leq y_i, \ \forall i = \overline{1,n},\\
(iv)\ & x \leq y \Leftrightarrow x_i \leq y_i, \ \forall i = \overline{1,n} \text{ and } x_i < y_i \text{ for some } i.
\end{aligned}
$$

The multi-dimensional multi-objective optimization problem with data uncertainty in the objective and constraint functionals is defined as follows:

$$
\begin{aligned}
\text{(MMCOPOCU)} \quad & \min_{(x(\cdot),c(\cdot))} \left\{ \int_{\Gamma_{t_0,t_1}} f(t, x(t), c(t), w)dt \right.\\
& = \left. \left(\int_{\Gamma_{t_0,t_1}} f_1(t, x(t), c(t), w_1)dt, \dots, \int_{\Gamma_{t_0,t_1}} f_s(t, x(t), c(t), w_s)dt \right) \right\}\\
\text{subject to} \quad & G(t, x(t), c(t), u) \leqq 0,\\
& x_\alpha = H(t, x(t), c(t), v),\\
& x(t_0) = x_0, \ x(t_1) = x_1,
\end{aligned}
$$

where $f = (f_1, \dots, f_s); (f_k) : \Gamma_{t_0,t_1} \times \mathscr{X} \times \mathscr{C} \times \mathcal{W}_k \to \mathbb{R}^s, k = \overline{1,s}$, $G = (G_1, \dots, G_m); (G_l) : \Gamma_{t_0,t_1} \times \mathscr{X} \times \mathscr{C} \times \mathcal{U}_l \to \mathbb{R}^m$, $l = \overline{1,m}$, $H = (H_\alpha^\tau) : \Gamma_{t_0,t_1} \times \mathscr{X} \times \mathscr{C} \times \mathcal{V}_\alpha^\tau \to \mathbb{R}^{pq}, \alpha = \overline{1,p}, \tau = \overline{1,q}$ are C^∞-class functionals. Consider $w = (w_k)$, $u = (u_l)$ and $v = (v_\alpha^\tau)$ are the uncertain parameters for some convex compact subsets $\mathcal{W} = (\mathcal{W}_k) \subset \mathbb{R}^k$, $\mathcal{U} = (\mathcal{U}_l) \subset \mathbb{R}^m$ and $\mathcal{V} = (\mathcal{V}_\alpha^\tau) \subset \mathbb{R}^{pq}$, respectively. The functions H_α satisfy the closeness conditions (complete integrability conditions) $D_\beta H_\alpha = D_\alpha H_\beta, \alpha, \beta = \overline{1,p}, \alpha \neq \beta$, where the total derivative is denoted by D_β.

The associated robust counterpart of the multi-dimensional multi-objective optimization problem (MMCOPOCU) is defined as follows:

$$
\begin{aligned}
\text{(RMMCOPOCU)} \quad & \min_{(x(\cdot),c(\cdot))} \left\{ \int_{\Gamma_{t_0,t_1}} \max_{w \in \mathcal{W}} f(t, x(t), c(t), w)dt \right.\\
& = \left. \left(\int_{\Gamma_{t_0,t_1}} \max_{w_1 \in \mathcal{W}_1} f_1(t, x(t), c(t), w_1)dt, \dots, \int_{\Gamma_{t_0,t_1}} \max_{w_s \in \mathcal{W}_s} f_s(t, x(t), c(t), w_s)dt \right) \right\}\\
\text{subject to} \quad & G(t, x(t), c(t), u) \leqq 0, \forall u \in \mathcal{U},\\
& x_\alpha = H(t, x(t), c(t), v), \forall v \in \mathcal{V},\\
& x(t_0) = x_0, \ x(t_1) = x_1,
\end{aligned}
$$

where f, G and H are defined as above in the multi-dimensional multi-objective optimization problem (MMCOPOCU).

We denote $\mathcal{D} = \{(x, c) \in \mathcal{X} \times \mathcal{C} : \ G(t, x(t), c(t), u) \leqq 0, x_\alpha = H(t, x(t), c(t), v), x(t_0) = x_0, x(t_1) = x_1\}$ the set of all feasible solutions in (RMMCOPOCU) and we say that it is the robust feasible solution set to the problem (MMCOPOCU).

From now on, to simplify our presentation, we introduce some notations as follows: $x = x(t), \hat{x} = \hat{x}(t), \bar{x} = \bar{x}(t), c = c(t), \hat{c} = \hat{c}(t), \bar{c} = \bar{c}(t), \pi = (t, x(t), c(t)), \bar{\pi} = (t, \bar{x}(t), \bar{c}(t)), \hat{\pi} = (t, \hat{x}(t), \hat{c}(t)), \rho = \rho(t), \bar{\rho} = \bar{\rho}(t)$ and $\Gamma = \Gamma_{t_0,t_1}$.

The partial derivatives of Lagrange multiplier γ and the partial derivatives associated with f are defined as

$$\gamma_t = \begin{bmatrix} \frac{\partial \gamma_1^\tau}{\partial t^1} & \cdots & \frac{\partial \gamma_1^\tau}{\partial t^p} \\ \vdots & \ddots & \vdots \\ \frac{\partial \gamma_p^\tau}{\partial t^1} & \cdots & \frac{\partial \gamma_p^\tau}{\partial t^p} \end{bmatrix}, \quad f_x = \begin{bmatrix} \frac{\partial f_1}{\partial x^1} & \cdots & \frac{\partial f_1}{\partial x^q} \\ \vdots & \ddots & \vdots \\ \frac{\partial f_s}{\partial x^1} & \cdots & \frac{\partial f_s}{\partial x^q} \end{bmatrix}, \quad f_c = \begin{bmatrix} \frac{\partial f_1}{\partial c^1} & \cdots & \frac{\partial f_1}{\partial c^r} \\ \vdots & \ddots & \vdots \\ \frac{\partial f_s}{\partial c^1} & \cdots & \frac{\partial f_s}{\partial c^r} \end{bmatrix}.$$

Similarly, we have G_x and G_c using matrices with m rows and H_x, H_c, x_α using matrices with p rows.

In general, for a vector optimization problem, it is difficult to find a point which optimizes more than one objective simultaneously. Therefore, the notion of an efficient solution is used to determine an optimality in vector optimization problems.

Definition 4.2.0.1 A point $(\bar{x}, \bar{c}) \in \mathcal{D}$ is said to be weak robust efficient solution to the multi-dimensional multi-objective optimization problem (MMCOPOCU), if there does not exist another point $(x, c) \in \mathcal{D}$ such that

$$\int_\Gamma \max_{w \in \mathcal{W}} f(\pi, w)dt < \int_\Gamma \max_{w \in \mathcal{W}} f(\bar{\pi}, w)dt.$$

Definition 4.2.0.2 A functional $F(x, c, \bar{w}) = \int_\Gamma f(t, x(t), x_\sigma(t), c(t), \bar{w})dt$ is said to be convex at $(\bar{x}, \bar{c})$, if the following inequality:

$$\begin{aligned} F(x, c, \bar{w}) - F(\bar{x}, \bar{c}, \bar{w}) \geqq & \int_\Gamma (x(t) - \bar{x}(t)) f_x(t, \bar{x}(t), \bar{x}_\sigma(t), \bar{c}(t), \bar{w})dt \\ & + \int_\Gamma (x_\sigma(t) - \bar{x}_\sigma(t)) f_{x_\sigma}(t, \bar{x}(t), \bar{x}_\sigma(t), \bar{c}(t), \bar{w})dt \\ & + \int_\Gamma (c(t) - \bar{c}(t)) f_c(t, \bar{x}(t), \bar{x}_\sigma(t), \bar{c}(t), \bar{w})dt \end{aligned}$$

holds for all $(x(t), c(t))$ on $\mathcal{X} \times \mathcal{C}$.

Following the same lines formulated in Mititelu [6], we establish the following result. Thus, the proof is omitted. The proof can be read (it is free) at the following link: https://www.journal.fairpartners.ro/volume-22009-no-1_3.html

Theorem 4.2.0.1 *Let $(\bar{x}, \bar{c}) \in \mathcal{D}$ be a weak robust efficient solution to the problem (MMOPOCU). Further assume that* $\max_{w \in \mathcal{W}} f(\pi, w) = f(\pi, \bar{w})$. *If the constraint conditions (for the existence of the multipliers) hold, then there exist the scalar vector $\bar{\mu} \in \mathbb{R}^s$, the piecewise smooth Lagrange multipliers $\bar{\nu} = (\bar{\nu}_l(t)) \in \mathbb{R}^{\Lambda}_{+}, l \in \mathbb{R}^{\Lambda}, \bar{\gamma} = (\bar{\gamma}^{\tau}_{\alpha}(t)) \in \mathbb{R}^{pq}, \tau = \overline{1, q}, \alpha = \overline{1, p}$, and the uncertain parameters $\bar{u} \in \mathcal{U}, \bar{v} \in \mathcal{V}$ such that $(\bar{x}, \bar{c})$ satisfies the following conditions:*

$$\bar{\mu}^T f_x(\bar{\pi}, \bar{w}) + \bar{\nu}^T G_x(\bar{\pi}, \bar{u}) + \bar{\gamma}^T H_x(\bar{\pi}, \bar{v}) + \bar{\gamma}_t = 0, \tag{4.2.1}$$

$$\bar{\mu}^T f_c(\bar{\pi}, \bar{w}) + \bar{\nu}^T G_c(\bar{\pi}, \bar{v}) + \bar{\gamma}^T H_c(\bar{\pi}, \bar{v}) = 0, \tag{4.2.2}$$

$$\bar{\nu}^T G(\bar{\pi}, \bar{u}) = 0, \tag{4.2.3}$$

$$\bar{\mu} \geq 0, \bar{\nu} \geqq 0 \tag{4.2.4}$$

hold for all $t \in \Gamma$, except at discontinuities.

Remark 4.2.0.1 The conditions (4.2.1)–(4.2.4) are known as robust necessary efficiency conditions for the multi-dimensional multi-objective optimization problem (MMCOPOCU).

4.3 Robust Sufficient Efficiency Conditions

In this section, we will derive the robust sufficient efficiency conditions for the multi-dimensional multi-objective optimization problem (MMCOPOCU) which states that any robust feasible solution to the problem (MMCOPOCU) will be a weak robust efficient solution to the considered problem (MMCOPOCU) if the involved functionals are convex.

Theorem 4.3.0.1 *Let $(\bar{x}, \bar{c})$ be a robust feasible solution to the multi-dimensional multi-objective optimization problem (MMCOPOCU) and the robust necessary efficiency conditions (4.2.1)–(4.2.4) be satisfied at $(\bar{x}, \bar{c})$ and* $\max_{w \in \mathcal{W}} f(\pi, w) = f(\pi, \bar{w})$. *Further, assume that $\int_{\Gamma} \bar{\mu}^T f(\pi, \bar{w}) dt, \int_{\Gamma} \bar{\nu}^T G(\pi, \bar{u}) dt$ and $\int_{\Gamma} \bar{\gamma}^T (H(\pi, \bar{v}) - x_{\alpha}) dt$ are convex at $(\bar{x}, \bar{c})$. Then, $(\bar{x}, \bar{c})$ is a weak robust efficient solution to the multi-dimensional multi-objective optimization problem (MMCOPOCU).*

Proof Contrary to the results, we assume that $(\bar{x}, \bar{c})$ is not a weak robust efficient solution to the multi-dimensional multi-objective optimization problem (MMCOPOCU). Then, there exists $(\hat{x}, \hat{c}) \in \mathcal{D}$ such that

$$\int_{\Gamma} \max_{w \in \mathcal{W}} f(\hat{\pi}, w) dt < \int_{\Gamma} \max_{w \in \mathcal{W}} f(\bar{\pi}, w) dt.$$

By assumption, $\max_{w \in \mathcal{W}} f(\pi, w) = f(\pi, \bar{w})$, we get

$$\int_{\Gamma} f(\hat{\pi}, \bar{w})dt < \int_{\Gamma} f(\bar{\pi}, \bar{w})dt. \tag{4.3.1}$$

Since $(\bar{x}, \bar{c})$ satisfies the conditions (4.2.1)–(4.2.4), therefore, multiplying Eq. (4.2.1) and (4.2.2) by $(\hat{x} - \bar{x})$ and $(\hat{c} - \bar{c})$, respectively, and integrating them, we get

$$\int_{\Gamma} (\hat{x} - \bar{x})\left\{\bar{\mu}^T f_x(\bar{\pi}, \bar{w}) + \bar{\upsilon}^T G_x(\bar{\pi}, \bar{u}) + \bar{\gamma}^T H_x(\bar{\pi}, \bar{v}) + \bar{\gamma}_t\right\}dt$$
$$+ \int_{\Gamma} (\hat{c} - \bar{c})\left\{\bar{\mu}^T f_c(\bar{\pi}, \bar{w}) + \bar{\upsilon}^T G_c(\bar{\pi}, \bar{u}) + \bar{\gamma}^T H_c(\bar{\pi}, \bar{v})\right\}dt = 0. \tag{4.3.2}$$

On the other hand, from assumption $\int_{\Gamma} \bar{\mu}^T f(\pi, \bar{w})dt$ is convex at $(\bar{x}, \bar{c})$, we have

$$\int_{\Gamma} \left\{\bar{\mu}^T f(\hat{\pi}, \bar{w}) - \bar{\mu}^T f(\bar{\pi}, \bar{w})\right\}dt \geqq \int_{\Gamma} (\hat{x} - \bar{x})\bar{\mu}^T f_x(\bar{\pi}, \bar{w})dt + \int_{\Gamma} (\hat{c} - \bar{c})\bar{\mu}^T f_c(\bar{\pi}, \bar{w})dt,$$

which along with the inequality (4.3.1) gives

$$\int_{\Gamma} (\hat{x} - \bar{x})\bar{\mu}^T f_x(\bar{\pi}, \bar{w})dt + \int_{\Gamma} (\hat{c} - \bar{c})\bar{\mu}^T f_c(\bar{\pi}, \bar{w})dt < 0. \tag{4.3.3}$$

Again, by assumption $\int_{\Gamma} \bar{\upsilon}^T G(\pi, u)dt$ is convex at $(\bar{x}, \bar{c})$, we obtain

$$\int_{\Gamma} \left\{\bar{\upsilon}^T G(\hat{\pi}, \bar{u}) - \bar{\upsilon}^T G(\bar{\pi}, \bar{u})\right\}dt \geqq \int_{\Gamma} (\hat{x} - \bar{x})\bar{\upsilon}^T G_x(\bar{\pi}, \bar{u})dt + \int_{\Gamma} (\hat{c} - \bar{c})\bar{\upsilon}^T G_c(\bar{\pi}, \bar{u})dt.$$

The above inequality together with the robust feasibility of $(\hat{x}, \hat{c})$ in the multi-dimensional multi-objective optimization problem (MMCOPOCU) and the condition (4.2.3) yields

$$\int_{\Gamma} (\hat{x} - \bar{x})\bar{\upsilon}^T G_x(\bar{\pi}, \bar{u})dt + \int_{\Gamma} (\hat{c} - \bar{c})\bar{\upsilon}^T G_c(\bar{\pi}, \bar{u})dt \leqq 0. \tag{4.3.4}$$

Further, from the assumption $\int_{\Gamma} \bar{\gamma}^T (H(\pi, v) - x_{\alpha})dt$ is convex at $(\bar{x}, \bar{c})$ and the robust feasibility of $(\hat{x}, \hat{c})$ in the multi-dimensional multi-objective optimization problem (MMCOPOCU), we obtain

$$\int_{\Gamma} (\hat{x} - \bar{x})\left\{\bar{\gamma}^T H_x(\bar{\pi}, \bar{v}) + \bar{\gamma}_t\right\}dt + \int_{\Gamma} (\hat{c} - \bar{c})\bar{\gamma}^T H_c(\bar{\pi}, \bar{v})dt \leqq 0. \tag{4.3.5}$$

On adding inequalities (4.3.3), (4.3.4) and (4.3.5), we have

$$\int_{\Gamma} (\hat{x} - \bar{x})\Big\{\bar{\mu}^T f_x(\bar{\pi}, \bar{w}) + \bar{\nu}^T G_x(\bar{\pi}, \bar{u}) + \bar{\gamma}^T H_x(\bar{\pi}, \bar{v}) + \bar{\gamma}_t\Big\}dt$$
$$+ \int_{\Gamma} (\hat{c} - \bar{c})\Big\{\bar{\mu}^T f_c(\bar{\pi}, \bar{w}) + \bar{\nu}^T G_c(\bar{\pi}, \bar{u}) + \bar{\gamma}^T H_c(\bar{\pi}, \bar{v})\Big\}dt < 0,$$

which contradicts the relation (4.3.2). Hence, this completes the proof. □

The following example verifies the robust sufficient efficiency conditions for a weak robust efficient solution to the multi-dimensional multi-objective optimization problem (MMCOPOCU).

Example 4.3.0.1 Let $p = 2, q = r = 1, k = 2, \mathcal{W}_1 = [1, 2], \mathcal{W}_2 = [2, 3], \mathcal{U} = [\frac{1}{2}, 1], \mathcal{V}_1^1 = \mathcal{V}_2^1 = [\frac{1}{2}, \frac{3}{2}]$ and Γ_{t_0,t_1} is fixed by the diagonally opposite points $t_0 = (t_0^1, t_0^2)$ and $t_1 = (t_1^1, t_1^2)$ in $\mathbb{R}^2$.

We consider the following multi-dimensional multi-objective optimization problem in the face of data uncertainty:

$$\text{(MMCOPOCU1)} \quad \min_{(x(\cdot),c(\cdot))} \int_{\Gamma_{0,\frac{1}{3}}} (e^x + w_1,\ c^2 + (w_2 x)^2)\, dt^1 dt^2$$

$$\text{subject to} \quad u(x-2)(x+1) \leqq 0, \tag{4.3.6}$$
$$x_{t^1} = v_1^1 - c, \tag{4.3.7}$$
$$x_{t^2} = v_2^1 - c, \tag{4.3.8}$$
$$x(0,0) = 0,\ x\left(\frac{1}{3}, \frac{1}{3}\right) = \frac{1}{2}, \tag{4.3.9}$$

where $t = (t^1, t^2) \in \Gamma_{0,\frac{1}{3}}$.

The associated robust counterpart for the multi-dimensional multi-objective optimization problem (MMCOPOCU1) is defined as follows:

$$\text{(RMMCOPOCU1)} \quad \min_{(x(\cdot),c(\cdot))} \int_{\Gamma_{0,\frac{1}{3}}} \Big(\max_{w_1 \in \mathcal{W}_1} (e^x + w_1),\ \max_{w_2 \in \mathcal{W}_2} (c^2 + (w_2 x)^2)\Big) dt^1 dt^2,$$

$$\text{subject to} \quad u(x-2)(x+1) \leqq 0,\ \forall\, u \in \mathcal{U}$$
$$x_{t^1} = v_1^1 - c,\ \forall v_1^1 \in \mathcal{V}_1^1,$$
$$x_{t^2} = v_2^1 - c,\ \forall v_2^1 \in \mathcal{V}_2^1,$$
$$x(0,0) = 0,\ x\left(\frac{1}{3}, \frac{1}{3}\right) = \frac{1}{2},$$

where $t = (t^1, t^2) \in \Gamma_{0,\frac{1}{3}}$. Clearly $\mathcal{D} = \{(x, c) \in \mathscr{X} \times \mathscr{C} : -1 \leq x \leq 2 \wedge x_{t^1} = v_1^1 - c \wedge x_{t^2} = v_2^1 - c, x(0,0) = 0, x\left(\frac{1}{3}, \frac{1}{3}\right) = \frac{1}{2}\}$ is a robust feasible solution set to (MMCOPOCU1).

Now, we are interested to find a weak robust efficient solution to the problem (MMCOPOCU1). It means that we have to find the control function $\bar{c} : \Gamma_{0,\frac{1}{3}} \to \mathbb{R}$

(that determines the state function $\bar{x} : \Gamma_{0,\frac{1}{3}} \to \mathbb{R}$) which satisfies the dynamical systems (4.3.7) and (4.3.8) with respect to the boundary conditions (4.3.9). Additionally, we assume that the state function is affine.

Let $(\bar{x}, \bar{c})$ be a robust feasible solution to the problem (MMCOPOCU1). The necessary efficiency conditions (4.2.1)–(4.2.4) at $(\bar{x}, \bar{c})$ with the Lagrange multipliers $\bar{\mu}_1, \bar{\mu}_2, \bar{\nu}, \bar{\gamma}_1^1, \bar{\gamma}_2^1$ and the uncertain parameters $\max_{w \in \mathcal{W}} f(\pi, w) = f(\pi, \bar{w}), \bar{u}, \bar{v}_1^1, \bar{v}_2^1$ are as follows:

$$\bar{\mu}_1 e^{\bar{x}} + 2\bar{\mu}_2 \bar{w}_2^2 x + \bar{\nu}(2\bar{x} - 1) + (\bar{\gamma}_1^1)_{t^1} + (\bar{\gamma}_2^1)_{t^2} = 0, \tag{4.3.10}$$

$$\bar{\mu}_2 2\bar{c} - (\bar{\gamma}_1^1 + \bar{\gamma}_2^1) = 0, \tag{4.3.11}$$

$$\bar{\nu}\bar{u}(\bar{x} - 2)(\bar{x} + 1) = 0, \tag{4.3.12}$$

where relation (4.3.12) is satisfied if either $\bar{\nu} = 0$ or $\bar{u}(\bar{x} - 2)(\bar{x} + 1) = 0$, for all $t = (t^1, t^2) \in \text{int}\Gamma_{0,\frac{1}{3}}$. Let $\bar{\nu} \neq 0$, then $\bar{u}(\bar{x} - 2)(\bar{x} + 1) = 0$ implies that $\bar{x} = -1, 2$. Clearly, the boundary conditions (4.3.9) are failing at $\bar{x} = -1, 2$. Consequently, $\bar{\nu} = 0$.

Case 1. Assume that the Lagrange multipliers $\bar{\gamma}_\alpha^\tau$, $\tau = 1, \alpha = 1, 2$ are constant functions, say $\bar{\gamma}_1^1 = d_1$ and $\bar{\gamma}_2^1 = d_2$, where $d_1, d_2 \in \mathbb{R}_+$. Therefore, Eq. (4.3.11) gives

$$\bar{c} = \frac{d_1 + d_2}{2\bar{\mu}_1}, \quad \forall t \in \Gamma_{0,\frac{1}{3}},$$

which shows that $\bar{c}$ is a constant function on $\Gamma_{0,\frac{1}{3}}$. Taking into account the equality constraints (4.3.7) and (4.3.8), it follows that

$$\bar{x} = \left(\bar{v}_1^1 - \frac{d_1 + d_2}{2\bar{\mu}_1}\right)t^1 + \left(\bar{v}_2^1 - \frac{d_1 + d_2}{2\bar{\mu}_1}\right)t^2 + d_3, \ d_3 \in \mathbb{R}_+.$$

By using the boundary condition (4.3.9) along with Lagrange multiplier $\bar{\mu}_1 = 1, \bar{\mu}_2 = 0$ and the uncertain parameters $\bar{w}_1 = 2, \bar{w}_2 = 3$, we get $d_3 = 0$ and $d_1 + d_2 = \frac{3}{4}$, and therefore

$$\bar{x} = \frac{3}{4}(t^1 + t^2), \ \bar{c} = \frac{3}{4}, \ \forall t \in \Gamma_{0,\frac{1}{3}}.$$

One can easily verify that the necessary efficiency conditions (4.3.10), (4.3.11) and (4.3.12) are satisfied at $(\bar{x}, \bar{c}) = (\frac{3}{4}(t^1 + t^2), \frac{3}{4})$ at $t^1 = t^2 = 0$, i.e. $(0, \frac{1}{3})$ with the Lagrange multipliers $\bar{\mu}_1 = 1, \bar{\mu}_2 = 0, \bar{\nu} = 0, \bar{\gamma}_1^1 + \bar{\gamma}_2^1 = \frac{3}{4}$ and the uncertain parameters $\bar{w}_1 = 2, \bar{w}_2 = 3, \bar{u} = 1, \bar{v}_1^1 = \frac{3}{2}, \bar{v}_2^1 = \frac{3}{2}$. Further, it can also be easily verified that the involved functionals are convex at $(\bar{x}, \bar{c}) = (0, \frac{3}{4})$. Hence, all the conditions of Theorem 4.3.0.1 are satisfied, which ensure that $(0, \frac{3}{4})$ is also a weak robust efficient solution to the problem (MMCOPOCU1).

Case 2. Assume that the Lagrange multipliers $\bar{\gamma}_{\alpha}^{\tau}, \tau = 1, \alpha = 1, 2$ are functions of t, say $\bar{\gamma}_1^1 = t^2, \bar{\gamma}_2^1 = t^1$. So, from Eq. (4.3.11), we have

$$\bar{c} = \frac{t^2 + t^1}{2\bar{\mu}_1}, \ \forall t \in \Gamma_{0,\frac{1}{3}}. \tag{4.3.13}$$

On solving Eq. (4.3.7) and (4.3.8) with the help of Eq. (4.3.13) together with the Lagrange multiplier $\bar{\mu}_1 = 1$, we have

$$\bar{x} = \frac{3}{2}(t^1 + t^2) - \frac{1}{2}\Big[\frac{(t^1)^2}{2} + \frac{(t^2)^2}{2} + 2t^1t^2\Big] + d_3, \ d_3 \in \mathbb{R}_+,$$

and, by considering the boundary conditions (4.3.9), we get

$$\bar{x} = \frac{3}{2}(t^1 + t^2) - \frac{1}{2}\Big[\frac{(t^1)^2}{2} + \frac{(t^2)^2}{2} + 2t^1t^2\Big].$$

This shows that the state function is not an affine function. Consequently, we get a contradiction with the hypothesis.

4.4 The Exact l_1 Penalty Function Method

In this section, we use the exact l_1 penalty function method to construct a penalized multi-dimensional multi-objective optimization problem associated with the multi-dimensional multi-objective optimization problem (MMCOPOCU). This is defined as

$$(\text{MMCOPOCU})_{\rho} \quad \min_{(x(\cdot),c(\cdot))} \Big\{\mathscr{P}(t, x(t), c(t), w, u, v, \rho(t)) = \int_{\Gamma} \Big\{f(t, x(t), c(t), w)$$
$$+ \rho(t)\Big[\sum_{l=1}^{m} G_l^{+}(t, x(t), c(t), u) + \sum_{\tau=1}^{q}\sum_{\alpha=1}^{p}\Big|H_{\alpha}^{\tau}(t, x(t), c(t), v) - \frac{\partial x^{\tau}}{\partial t^{\alpha}}\Big|\Big]e\Big\}dt\Big\},$$

where $e = (1, \ldots, 1) \in \mathbb{R}^s$, $w \in \mathcal{W}$, $u \in \mathcal{U}$ and $v \in \mathcal{V}$ are the uncertain parameters, $\rho(t)$ is a non-zero non-negative penalty parameter and the penalty function $G_l^{+}(t, x(t), c(t), u)$ is defined as

$$G_l^{+}(t, x(t), c(t), u) = \begin{cases} 0, & \text{if } G_l(t, x(t), c(t), u) \leqq 0, \\ G_l(t, x(t), c(t), u), & \text{if } G_l(t, x(t), c(t), u) > 0. \end{cases} \tag{4.4.1}$$

The associated robust counterpart to the problem $(\text{MMCOPOCU})_{\rho}$ is given by

$$(\text{RMMCOPOCU})_\rho \quad \min_{(x(\cdot),c(\cdot))} \Big\{ \max_{w\in\mathcal{W}} \mathscr{P}(t,x(t),c(t),w,u,v,\rho(t)) = \int_\Gamma \Big\{ \max_{w\in\mathcal{W}} f(t,x(t),c(t),w)$$
$$+\rho(t)\Big[\sum_{l=1}^{m} G_l^+(t,x(t),c(t),u) + \sum_{\tau=1}^{q}\sum_{\alpha=1}^{p}\Big|H_\alpha^\tau(t,x(t),c(t),v) - \frac{\partial x^\tau}{\partial t^\alpha}\Big|\Big]e\Big\}dt\Big\},$$

for all $u \in \mathcal{U}$ and $v \in \mathcal{V}$.

Definition 4.4.0.1 A point $(\bar{x},\bar{c}) \in \mathscr{X}\times\mathscr{C}$ is said to be a weak robust efficient solution to the problem $(\text{MMCOPOCU})_\rho$, if there does not exist another point $(x,c) \in \mathscr{X}\times\mathscr{C}$ such that

$$\max_{w\in\mathcal{W}} \mathscr{P}(\pi,w,\bar{u},\bar{v},\rho) < \max_{w\in\mathcal{W}} \mathscr{P}(\bar{\pi},w,\bar{u},\bar{v},\rho),$$

where $\bar{u} \in \mathcal{U}$ and $\bar{v} \in \mathcal{V}$.

Theorem 4.4.0.1 *Let $(\bar{x},\bar{c})$ be a robust feasible solution to the multi-dimensional multi-objective optimization problem (MMCOPOCU) and the necessary efficient conditions (4.2.1)–(4.2.4) be satisfied at $(\bar{x},\bar{c})$ with respect to the considered Lagrange multipliers and $\max_{w\in\mathcal{W}} f(\pi,w) = f(\pi,\bar{w})$. Further, assume that $\int_\Gamma \bar{\mu}^T f(\pi,\bar{w})dt$, $\int_\Gamma \bar{v}^T G(\pi,u)dt$ and $\int_\Gamma \bar{\gamma}^T(H(\pi,v)-x_\alpha)dt$ are convex at $(\bar{x},\bar{c})$ on $\mathscr{X}\times\mathscr{C}$. If $\rho \geq \max\{\bar{v},|\bar{\gamma}|\}$, then $(\bar{x},\bar{c})$ is a weak robust efficient solution to the penalized multi-dimensional multi-objective optimization problem (MMCOPOCU)ρ.*

Proof Contrary to the results, we assume that $(\bar{x},\bar{c})$ is not a weak robust efficient solution to the problem $(\text{MMCOPOCU})_\rho$. Then, there exists another point $(\hat{x},\hat{c}) \in \mathscr{X}\times\mathscr{C}$ such that

$$\max_{w\in\mathcal{W}} \mathscr{P}(\hat{\pi},w,\bar{u},\bar{v},\rho) < \max_{w\in\mathcal{W}} \mathscr{P}(\bar{\pi},w,\bar{u},\bar{v},\rho).$$

By using the definition of the penalized multi-dimensional multi-objective optimization problem $(\text{MMCOPOCU})_\rho$, we get

$$\int_\Gamma \Big\{ \max_{w\in\mathcal{W}} f(\hat{\pi},w) + \rho\Big[\sum_{l=1}^{m} G_l^+(\hat{\pi},\bar{u}) + \sum_{\tau=1}^{q}\sum_{\alpha=1}^{p}\Big|H_\alpha^\tau(\hat{\pi},\bar{v}) - \frac{\partial x^\tau}{\partial t^\alpha}\Big|\Big]e\Big\}dt$$
$$< \int_\Gamma \Big\{ \max_{w\in\mathcal{W}} f(\bar{\pi},w) + \rho\Big[\sum_{l=1}^{m} G_l^+(\bar{\pi},\bar{u}) + \sum_{\tau=1}^{q}\sum_{\alpha=1}^{p}\Big|H_\alpha^\tau(\bar{\pi},\bar{v}) - \frac{\partial x^\tau}{\partial t^\alpha}\Big|\Big]e\Big\}dt.$$

By assumption $\max_{p\in\mathcal{W}} f(\pi,w) = f(\pi,\bar{w})$, we get

$$\int_{\Gamma}\Big\{f(\hat{\pi},\bar{w})+\rho\Big[\sum_{l=1}^{m}G_l^{+}(\hat{\pi},\bar{u})+\sum_{\tau=1}^{q}\sum_{\alpha=1}^{p}\Big|H_{\alpha}^{\tau}(\hat{\pi},\bar{v})-\frac{\partial x^{\tau}}{\partial t^{\alpha}}\Big|\Big]e\Big\}dt$$
$$<\int_{\Gamma}\Big\{f(\bar{\pi},\bar{w})+\rho\Big[\sum_{l=1}^{m}G_l^{+}(\bar{\pi},\bar{u})+\sum_{\tau=1}^{q}\sum_{\alpha=1}^{p}\Big|H_{\alpha}^{\tau}(\bar{\pi},\bar{v})-\frac{\partial x^{\tau}}{\partial t^{\alpha}}\Big|\Big]e\Big\}dt.$$

From the above inequality together with the robust feasibility of $(\bar{x},\bar{c})$ to the problem (MMCOPOCU) and (4.4.1), we have

$$\int_{\Gamma}\Big\{f(\hat{\pi},\bar{w})+\rho\Big[\sum_{l=1}^{m}G_l^{+}(\hat{\pi},\bar{u})+\sum_{\tau=1}^{q}\sum_{\alpha=1}^{p}\Big|H_{\alpha}^{\tau}(\hat{\pi},\bar{v})-\frac{\partial x^{\tau}}{\partial t^{\alpha}}\Big|\Big]e\Big\}dt<\int_{\Gamma}f(\bar{\pi},\bar{w})dt.$$

On multiplying the above inequality by scalar vector $\bar{\mu}$ on the both sides, we get

$$\int_{\Gamma}\Big\{\bar{\mu}^Tf(\hat{\pi},\bar{w})+\bar{\mu}^T\rho\Big[\sum_{l=1}^{m}G_l^{+}(\hat{\pi},\bar{u})+\sum_{\tau=1}^{q}\sum_{\alpha=1}^{p}\Big|H_{\alpha}^{\tau}(\hat{\pi},\bar{v})-\frac{\partial x^{\tau}}{\partial t^{\alpha}}\Big|\Big]e\Big\}dt$$
$$<\int_{\Gamma}\bar{\mu}^Tf(\bar{\pi},\bar{w})dt.$$

As $\bar{\mu}^Te=1$, it follows that

$$\int_{\Gamma}\Big\{\bar{\mu}^Tf(\hat{\pi},\bar{w})+\rho\Big[\sum_{l=1}^{m}G_l^{+}(\hat{\pi},\bar{u})+\sum_{\tau=1}^{q}\sum_{\alpha=1}^{p}\Big|H_{\alpha}^{\tau}(\hat{\pi},\bar{v})-\frac{\partial x^{\tau}}{\partial t^{\alpha}}\Big|\Big]\Big\}dt$$
$$<\int_{\Gamma}\bar{\mu}^Tf(\bar{\pi},\bar{w})dt.$$

Since $\rho\geq\max\{\bar{\nu},|\bar{\gamma}|\}$, the above inequality yields

$$\int_{\Gamma}\Big\{\bar{\mu}^Tf(\hat{\pi},\bar{w})+\sum_{l=1}^{m}\bar{\nu}_lG_l^{+}(\hat{\pi},\bar{u})+\sum_{\tau=1}^{q}\sum_{\alpha=1}^{p}\bar{\gamma}_{\alpha}^{\tau}\Big|H_{\alpha}^{\tau}(\hat{\pi},\bar{v})-\frac{\partial x^{\tau}}{\partial t^{\alpha}}\Big|\Big\}dt$$
$$<\int_{\Gamma}\bar{\mu}^Tf(\bar{\pi},\bar{w})dt. \tag{4.4.2}$$

From the condition (4.2.3) along with the robust feasibility of $(\bar{x},\bar{c})$ to the problem (MMCOPOCU), the inequality (4.4.2) can be written as

$$\int_{\Gamma}\Big\{\bar{\mu}^Tf(\hat{\pi},\bar{w})+\sum_{l=1}^{m}\bar{\nu}_lG_l(\hat{\pi},\bar{u})+\sum_{\tau=1}^{q}\sum_{\alpha=1}^{p}\bar{\gamma}_{\alpha}^{\tau}\Big(H_{\alpha}^{\tau}(\hat{\pi},\bar{v})-\frac{\partial x^{\tau}}{\partial t^{\alpha}}\Big)\Big\}dt$$
$$<\int_{\Gamma}\Big\{\bar{\mu}^Tf(\bar{\pi},\bar{w})+\sum_{l=1}^{m}\bar{\nu}_lG_l(\bar{\pi},\bar{u})+\sum_{\tau=1}^{q}\sum_{\alpha=1}^{p}\bar{\gamma}_{\alpha}^{\tau}\Big(H_{\alpha}^{\tau}(\bar{\pi},\bar{v})-\frac{\partial x^{\tau}}{\partial t^{\alpha}}\Big)\Big\}dt,$$

equivalently,

$$\int_{\Gamma}\Big\{\bar{\mu}^T f(\hat{\pi},\bar{w})+\bar{\nu}^T G(\hat{\pi},\bar{u})+\bar{\gamma}^T(H(\hat{\pi},\bar{v})-x_\alpha)\Big\}dt < \int_{\Gamma}\Big\{\bar{\mu}^T f(\bar{\pi},\bar{w})+\bar{\nu}^T G(\bar{\pi},\bar{u})+\bar{\gamma}^T(H(\bar{\pi},\bar{v})-x_\alpha)\Big\}dt. \tag{4.4.3}$$

On the other hand, by assumption $\int_\Gamma \bar{\mu}^T f(\pi,\bar{w})dt$, $\int_\Gamma \bar{\nu}^T G(\pi,u)dt$ and $\int_\Gamma \bar{\gamma}^T(H(\pi,v)-x_\alpha)dt$ are convex at $(\bar{x},\bar{c})$ on $\mathscr{X}\times\mathscr{C}$, we have

$$\int_{\Gamma}\Big\{\bar{\mu}^T f(\hat{\pi},\bar{w})-\bar{\mu}^T f(\bar{\pi},\bar{w})\Big\}dt \geqq \int_{\Gamma}(\hat{x}-\bar{x})\bar{\mu}^T f_x(\bar{\pi},\bar{w})dt+\int_{\Gamma}(\hat{c}-\bar{c})\bar{\mu}^T f_c(\bar{\pi},\bar{w})dt, \tag{4.4.4}$$

$$\int_{\Gamma}\Big\{\bar{\nu}^T G(\hat{\pi},\bar{u})-\bar{\nu}^T G(\bar{\pi},\bar{u})\Big\}dt \geqq \int_{\Gamma}(\hat{x}-\bar{x})\bar{\nu}^T G_x(\bar{\pi},\bar{u})dt+\int_{\Gamma}(\hat{c}-\bar{c})\bar{\nu}^T G_c(\bar{\pi},\bar{u})dt, \tag{4.4.5}$$

and
$$\int_{\Gamma}\Big\{\bar{\gamma}^T(H(\hat{\pi},\bar{v})-x_\alpha)-\bar{\gamma}^T(H(\bar{\pi},\bar{v})-x_\alpha)\Big\}dt \geqq \int_{\Gamma}(\hat{x}-\bar{x})\Big\{\bar{\gamma}^T H_x(\bar{\pi},\bar{v})+\bar{\gamma}_t\Big\}dt+\int_{\Gamma}(\hat{c}-\bar{c})\bar{\gamma}^T H_c(\bar{\pi},\bar{v})dt. \tag{4.4.6}$$

On combining the inequalities (4.4.4)–(4.4.6) together with the conditions (4.2.1)–(4.2.2), we have

$$\int_{\Gamma}\Big\{\bar{\mu}^T f(\hat{\pi},\bar{w})+\bar{\nu}^T G(\hat{\pi},\bar{u})+\bar{\gamma}^T(H(\hat{\pi},\bar{v})-x_\alpha)\Big\}dt \geqq \int_{\Gamma}\Big\{\bar{\mu}^T f(\bar{\pi},\bar{w})+\bar{\nu}^T G(\bar{\pi},\bar{u})+\bar{\gamma}^T(H(\bar{\pi},\bar{v})-x_\alpha)\Big\}dt,$$

which contradicts the inequality (4.4.3). Hence, this completes the proof. □

Let us consider Example 4.3.0.1 to show the relationship between the solution set of multi-dimensional multi-objective optimization problem and its associated penalized problem.

Example 4.4.0.1 Now, the penalized multi-dimensional multi-objective optimization problem in the face of data uncertainty $(\text{MMCOPOCU1})_\rho$ associated with the multi-dimensional multi-objective optimization problem (MMCOPOCU1) is defined as follows:

$$(\text{MMCOPOCU1})_\rho \quad \min_{(x(\cdot),c(\cdot))} \Big(\int_{\Gamma_{0,\frac{1}{3}}} \Big\{ (e^x + w_1) + \rho \Big[\max\{0, u(x-2)(x+1)\} + \Big| x_{t^1} - v_1^1 + c \Big| + \Big| x_{t^2} - v_2^1 + c \Big| \Big] \Big\} dt^1 dt^2, \int_{\Gamma_{0,\frac{1}{3}}} \Big\{ (c^2 + (w_2 x)^2) + \rho \Big[\max\{0, u(x-2)(x+1)\} + \Big| x_{t^1} - v_1^1 + c \Big| + \Big| x_{t^2} - v_2^1 + c \Big| \Big] \Big\} dt^1 dt^2 \Big).$$

Therefore, the robust counterpart for $(\text{MMCOPOCU1})_\rho$ is formulated as follows:

$$(\text{MMCOPOCU1})_\rho \quad \min_{(x(\cdot),c(\cdot))} \Big(\int_{\Gamma_{0,\frac{1}{3}}} \Big\{ \max_{w_1 \in \mathcal{W}_1} (e^x + w_1) + \rho \Big[\max\{0, u(x-2)(x+1)\} + \Big| x_{t^1} - v_1^1 + c \Big| + \Big| x_{t^2} - v_2^1 + c \Big| \Big] \Big\} dt^1 dt^2, \int_{\Gamma_{0,\frac{1}{3}}} \Big\{ \max_{w_2 \in \mathcal{W}_2} (c^2 + (w_2 x)^2) + \rho \Big[\max\{0, u(x-2)(x+1)\} + \Big| x_{t^1} - v_1^1 + c \Big| + \Big| x_{t^2} - v_2^1 + c \Big| \Big] \Big\} dt^1 dt^2 \Big).$$

In virtue of Definition 4.4.0.1, it can be observed that the inequality $\max_{w \in \mathcal{W}} \mathscr{P}(\pi, w, \bar{u}, \bar{v}, \rho) - \max_{w \in \mathcal{W}} \mathscr{P}(\bar{\pi}, w, \bar{u}, \bar{v}, \rho) \geq 0$ at $(\bar{x}, \bar{c}) = (0, \frac{3}{4})$. This shows that $(0, \frac{3}{4})$ is a weak robust efficient solution to $(\text{MMCOPOCU1})_\rho$ with $\rho \geq 1$. Hence, Theorem 4.4.0.1 is verified.

Proposition 4.4.0.1 *Let $(\bar{x}, \bar{c}) \in \mathscr{X} \times \mathscr{C}$ be a weak robust efficient solution to the problem $(MMCOPOCU)_{\bar{\rho}}$ and $\max_{w \in \mathcal{W}} f(\pi, w) = f(\pi, \bar{w})$. Then, there does not exist another point $(x, c) \in \mathcal{D}$ such that*

$$\int_\Gamma f(\pi, \bar{w}) dt < \int_\Gamma f(\bar{\pi}, \bar{w}) dt.$$

Proof Since $(\bar{x}, \bar{c})$ is a weak robust efficient solution to the problem $(\text{MMCOPOCU})_{\bar{\rho}}$, then there does not exist another point $(x, c) \in \mathscr{X} \times \mathscr{C}$ such that

$$\max_{w \in \mathcal{W}} \mathscr{P}(\pi, w, \bar{u}, \bar{v}, \bar{\rho}) < \max_{w \in \mathcal{W}} \mathscr{P}(\bar{\pi}, w, \bar{u}, \bar{v}, \bar{\rho}).$$

By using the definition of the penalized semi-infinite multi-dimensional multi-objective optimization problem $(\text{MMCOPOCU})_\rho$, we get

$$\int_\Gamma \Big\{ \max_{w \in \mathcal{W}} f(\pi, w) + \bar{\rho} \Big[\sum_{l=1}^m G_l^+(\pi, \bar{u}) + \sum_{\tau=1}^q \sum_{\alpha=1}^p \Big| H_\alpha^\tau(\pi, \bar{v}) - x_\alpha \Big| \Big] e \Big\} dt < \int_\Gamma \Big\{ \max_{w \in \mathcal{W}} f(\bar{\pi}, w) + \bar{\rho} \Big[\sum_{l=1}^m G_l^+(\bar{\pi}, \bar{u}) + \sum_{\tau=1}^q \sum_{\alpha=1}^p \Big| H_\alpha^\tau(\bar{\pi}, \bar{v}) - x_\alpha \Big| \Big] e \Big\} dt.$$

Since $\max_{w \in \mathcal{W}} f(\pi, w) = f(\pi, \bar{w})$, we have

$$\int_{\Gamma}\Big\{f(\pi,\bar{w})+\bar{\rho}\Big[\sum_{l=1}^{m}G_l^{+}(\pi,\bar{u})+\sum_{\tau=1}^{q}\sum_{\alpha=1}^{p}\Big|H_{\alpha}^{\tau}(\pi,\bar{v})-x_{\alpha}\Big|\Big]e\Big\}dt$$
$$<\int_{\Gamma}\Big\{f(\bar{\pi},\bar{w})+\bar{\rho}\Big[\sum_{l=1}^{m}G_l^{+}(\bar{\pi},\bar{u})+\sum_{\tau=1}^{q}\sum_{\alpha=1}^{p}\Big|H_{\alpha}^{\tau}(\bar{\pi},\bar{v})-x_{\alpha}\Big|\Big]e\Big\}dt.$$

Since $\mathcal{D}\subseteq\mathscr{X}\times\mathscr{C}$, therefore, it follows that there does not exist another point $(x,c)\in\mathcal{D}$ such that

$$\int_{\Gamma}\Big\{f(\pi,\bar{w})+\bar{\rho}\Big[\sum_{l=1}^{m}G_l^{+}(\pi,\bar{u})+\sum_{\tau=1}^{q}\sum_{\alpha=1}^{p}\Big|H_{\alpha}^{\tau}(\pi,\bar{v})-x_{\alpha}\Big|\Big]e\Big\}dt$$
$$<\int_{\Gamma}\Big\{f(\bar{\pi},\bar{w})+\bar{\rho}\Big[\sum_{l=1}^{m}G_l^{+}(\bar{\pi},\bar{u})+\sum_{\tau=1}^{q}\sum_{\alpha=1}^{p}\Big|H_{\alpha}^{\tau}(\bar{\pi},\bar{v})-x_{\alpha}\Big|\Big]e\Big\}dt.$$

Equivalently, one can say that the following relation holds for all $(x,c)\in\mathcal{D}$:

$$\int_{\Gamma}\Big\{f_k(\pi,\bar{w})+\bar{\rho}\Big[\sum_{l=1}^{m}G_l^{+}(\pi,\bar{u})+\sum_{\tau=1}^{q}\sum_{\alpha=1}^{p}\Big|H_{\alpha}^{\tau}(\pi,\bar{v})-\frac{\partial x^{\tau}}{\partial t^{\alpha}}\Big|\Big]\Big\}dt$$
$$\geqq\int_{\Gamma}\Big\{f_k(\bar{\pi},\bar{w})+\bar{\rho}\Big[\sum_{l=1}^{m}G_l^{+}(\bar{\pi},\bar{u})+\sum_{\tau=1}^{q}\sum_{\alpha=1}^{p}\Big|H_{\alpha}^{\tau}(\bar{\pi},\bar{v})-\frac{\partial x^{\tau}}{\partial t^{\alpha}}\Big|\Big]\Big\}dt.$$

By virtue of the relation (4.4.1), we get

$$\int_{\Gamma}f_k(\pi,\bar{w})dt\geqq\int_{\Gamma}f_k(\bar{\pi},\bar{w})dt.$$

Therefore, we conclude that there does not exist another point $(x,c)\in\mathcal{D}$ such that

$$\int_{\Gamma}f(\pi,\bar{w})dt<\int_{\Gamma}f(\bar{\pi},\bar{w})dt.$$

This completes the proof. □

Now, we prove the converse result of Theorem 4.4.0.1 without imposing any convexity assumption on the involved functionals.

Theorem 4.4.0.2 *Let* $(\bar{x},\bar{c})$ *be a weak robust efficient solution to the multi-dimensional multi-objective optimization problem* $(MMCOPOCU)_{\bar{\rho}}$. *Assume that* $\max_{w\in\mathcal{W}}\mathscr{P}(\pi,w,\bar{u},\bar{v},\rho)\geqq\max_{w\in\mathcal{W}}\mathscr{P}(\bar{\pi},w,\bar{u},\bar{v},\rho)$ *is satisfied for any* $\rho\geq\bar{\rho}$ *and for all* $(x,c)\in\mathcal{D}$ *and* $\max_{w\in\mathcal{W}}f(\pi,w)=f(\pi,\bar{w})$. *If* $\rho\geq\max\{\bar{v},|\bar{\gamma}|\}$, *then* $(\bar{x},\bar{c})$ *is a weak robust efficient solution to the multi-dimensional multi-objective optimization problem (MMCOPOCU).*

Proof Since $(\bar{x}, \bar{c})$ is a weak robust efficient solution to the problem $(\text{MMCOPOCU})_{\bar{\rho}}$, by Proposition 4.4.0.1, there does not exist another point $(x, c) \in \mathcal{D}$ such that

$$\int_{\Gamma} f(\pi, \bar{w})dt < \int_{\Gamma} f(\bar{\pi}, \bar{w})dt,$$

equivalently,

$$\int_{\Gamma} f_k(\pi, \bar{w})dt < \int_{\Gamma} f_k(\bar{\pi}, \bar{w})dt, \ \forall k = \overline{1, s}, \tag{4.4.7}$$

which specify that the objective functional $\int_{\Gamma} f_k(\pi, \bar{w})dt$ is bounded below in the compact subset $\mathcal{D}$.

Now, to show that $(\bar{x}, \bar{c})$ is a weak robust efficient solution to the problem (MMCOPOCU), firstly, we shall prove that $(\bar{x}, \bar{c}) \in \mathcal{D}$. We proceed by contradiction and assume that $(\bar{x}, \bar{c}) \notin \mathcal{D}$. Therefore, by virtue of (4.4.1), we have

$$\int_{\Gamma} \Big\{ \sum_{l=1}^{m} G_l^{+}(\bar{\pi}, \bar{u}) + \sum_{\tau=1}^{q} \sum_{\alpha=1}^{p} \Big| H_{\alpha}^{\tau}(\bar{\pi}, \bar{v}) - \frac{\partial x^{\tau}}{\partial t^{\alpha}} \Big| \Big\} e \, dt > 0. \tag{4.4.8}$$

Let (x, c) be any robust feasible solution to the problem (MMCOPOCU). By assumption, the following inequality:

$$\max_{w \in \mathcal{W}} \mathscr{P}(\pi, w, \bar{u}, \bar{v}, \rho)dt \geqq \max_{w \in \mathcal{W}} \mathscr{P}(\bar{\pi}, w, \bar{u}, \bar{v}, \rho)dt$$

holds for any $(x, c) \in \mathcal{D}$ and for all $\rho \geq \bar{\rho}$.
By considering the definition of the penalized multi-dimensional multi-objective optimization problem (MMCOPOCU), we get

$$\int_{\Gamma} \Big\{ \max_{w \in \mathcal{W}} f(\pi, w) + \rho \Big[\sum_{l=1}^{m} G_l^{+}(\pi, \bar{u}) + \sum_{\tau=1}^{q} \sum_{\alpha=1}^{p} \Big| H_{\alpha}^{\tau}(\pi, \bar{v}) - \frac{\partial x^{\tau}}{\partial t^{\alpha}} \Big| \Big] e \Big\} dt$$
$$\geqq \int_{\Gamma} \Big\{ \max_{w \in \mathcal{W}} f(\bar{\pi}, w) + \rho \Big[\sum_{l=1}^{m} G_l^{+}(\bar{\pi}, \bar{u}) + \sum_{\tau=1}^{q} \sum_{\alpha=1}^{p} \Big| H_{\alpha}^{\tau}(\bar{\pi}, \bar{v}) - \frac{\partial x^{\tau}}{\partial t^{\alpha}} \Big| \Big] e \Big\} dt.$$

Since $\max_{w \in \mathcal{W}} f(\pi, w) = f(\pi, \bar{w})$, we get

$$\int_\Gamma \Big\{f(\pi, \bar{w}) + \rho\Big[\sum_{l=1}^{m} G_l^+(\pi, \bar{u}) + \sum_{\tau=1}^{q}\sum_{\alpha=1}^{p}\Big|H_\alpha^\tau(\pi, \bar{v}) - \frac{\partial x^\tau}{\partial t^\alpha}\Big|\Big]e\Big\}dt$$
$$\geqq \int_\Gamma \Big\{f(\bar{\pi}, \bar{w}) + \rho\Big[\sum_{l=1}^{m} G_l^+(\bar{\pi}, \bar{u}) + \sum_{\tau=1}^{q}\sum_{\alpha=1}^{p}\Big|H_\alpha^\tau(\bar{\pi}, \bar{v}) - \frac{\partial x^\tau}{\partial t^\alpha}\Big|\Big]e\Big\}dt. \tag{4.4.9}$$

On the other hand, if we set

$$\rho > \max\left\{\frac{\int_\Gamma\{f(\pi, \bar{w}) - f(\bar{\pi}, \bar{w})\}dt}{\int_\Gamma\Big\{\sum_{l=1}^{m} G_l^+(\bar{\pi}, \bar{u}) + \sum_{\tau=1}^{q}\sum_{\alpha=1}^{p}\Big|H_\alpha^\tau(\bar{\pi}, \bar{v}) - \frac{\partial x^\tau}{\partial t^\alpha}\Big|\Big\}e\, dt}, \bar{\rho};\ (x, c) \in \mathcal{D}\right\},$$

the inequalities (4.4.7) and (4.4.8) imply that ρ is a positive real number. Then, we have

$$\int_\Gamma \Big\{f(\bar{\pi}, \bar{w}) + \rho\Big[\sum_{l=1}^{m} G_l^+(\bar{\pi}, \bar{u}) + \sum_{\tau=1}^{q}\sum_{\alpha=1}^{p}\Big|H_\alpha^\tau(\bar{\pi}, \bar{v}) - \frac{\partial x^\tau}{\partial t^\alpha}\Big|\Big]e\Big\}dt > \int_\Gamma f(\pi, \bar{w})dt.$$

Further, from the robust feasibility of (x, c) to the problem (MMCOPOCU) and in virtue of (4.4.1), the above inequality can be written as

$$\int_\Gamma \Big\{f(\bar{\pi}, \bar{w}) + \rho\Big[\sum_{l=1}^{m} G_l^+(\bar{\pi}, \bar{u}) + \sum_{\tau=1}^{q}\sum_{\alpha=1}^{p}\Big|H_\alpha^\tau(\bar{\pi}, \bar{v}) - \frac{\partial x^\tau}{\partial t^\alpha}\Big|\Big]e\Big\}dt$$
$$> \int_\Gamma \Big\{f(\pi, \bar{w}) + \rho\Big[\sum_{l=1}^{m} G_l^+(\pi, \bar{u}) + \sum_{\tau=1}^{q}\sum_{\alpha=1}^{p}\Big|H_\alpha^\tau(\pi, \bar{v}) - \frac{\partial x^\tau}{\partial t^\alpha}\Big|\Big]e\Big\}dt,$$

which contradicts the inequality (4.4.9). Therefore, $(\bar{x}, \bar{c}) \in \mathcal{D}$, i.e. $(\bar{x}, \bar{c})$ is a robust feasible solution to the problem (MMCOPOCU). Hence, the inequality (4.4.7) directly concludes that $(\bar{x}, \bar{c})$ is a weak robust efficient solution to the problem (MMCOPOCU). □

4.5 Uncertain Lagrange Functional

In this section, we introduce the uncertain Lagrange functional for (MMCOPOCU) and prove the relationship between a weak robust efficient solution to the problem (MMCOPOCU) and (MMCOPOCU)$_\rho$ under convexity assumption of the uncertain Lagrange functional.

Definition 4.5.0.1 The uncertain Lagrange functional $L(x, c, \mu, \nu, \gamma, w, u, v) : \mathscr{X} \times \mathscr{C} \times \mathbb{R}^s \times \mathbb{R}^m_+ \times \mathbb{R}^{pq} \times \mathcal{W} \times \mathcal{U} \times \mathcal{V} \to \mathbb{R}$ to the problem (MMCOPOCU) is defined as follows:

$$L(x, c, \mu, \nu, \gamma, w, u, v) = \int_\Gamma \left\{\mu^T f(\pi, w) + \nu^T G(\pi, u) + \gamma^T (H(\pi, v) - x_\alpha)\right\} dt.$$

Now, we show that (MMCOPOCU) and (MMCOPOCU)$_\rho$ consist of the same weak robust efficient solution set under the convexity assumption of the uncertain Lagrange functional.

Theorem 4.5.0.1 *Let $(\bar{x}, \bar{c})$ be a robust feasible solution to the multi-dimensional multi-objective optimization problem (MMCOPOCU) at which the robust necessary efficiency conditions (4.2.1)–(4.2.4) hold and* $\max_{w \in \mathcal{W}} f(\pi, w) = f(\pi, \bar{w})$. *Further, assume that if the Lagrange functional $L(x, c, \mu, \nu, \gamma, w, u, v)$ is convex at $(\bar{x}, \bar{c})$ on $\mathscr{X} \times \mathscr{C}$ and $\rho \geq \max\{\bar{\nu}, |\bar{\gamma}|\}$, then $(\bar{x}, \bar{c})$ is also a weak robust efficient solution to the multi-dimensional multi-objective optimization problem (MMCOPOCU)$_\rho$.*

Proof We proceed by contradiction and assume that $(\bar{x}, \bar{c})$ is not a weak robust efficient solution to the problem (MMCOPOCU)$_\rho$. Then, there exists another point $(\hat{x}, \hat{c}) \in \mathscr{X} \times \mathscr{C}$ such that

$$\max_{w \in \mathcal{W}} \mathscr{P}(\hat{\pi}, w, \bar{u}, \bar{v}, \rho) < \max_{w \in \mathcal{W}} \mathscr{P}(\bar{\pi}, w, \bar{u}, \bar{v}, \rho).$$

Then, on the line of Theorem 4.4.0.1, we get

$$\int_\Gamma \Big\{\sum_{k=1}^{s} \bar{\mu}_k f_k(\hat{\pi}, \bar{w}) + \sum_{l=1}^{m} \bar{\nu}_l G_l(\hat{\pi}, \bar{u}) + \sum_{\tau=1}^{q}\sum_{\alpha=1}^{p} \bar{\gamma}_\alpha^\tau \Big(H_\alpha^\tau(\hat{\pi}, \bar{v}) - \frac{\partial x^\tau}{\partial t^\alpha}\Big)\Big\} dt$$
$$< \int_\Gamma \Big\{\sum_{k=1}^{s} \bar{\mu}_k f_k(\bar{\pi}, \bar{w}) + \sum_{l=1}^{m} \bar{\nu}_l G_l(\bar{\pi}, \bar{u}) + \sum_{\tau=1}^{q}\sum_{\alpha=1}^{p} \bar{\gamma}_\alpha^\tau \Big(H_\alpha^\tau(\bar{\pi}, \bar{v}) - \frac{\partial x^\tau}{\partial t^\alpha}\Big)\Big\} dt,$$

equivalently,

$$\int_\Gamma \left\{\bar{\mu}^T f(\hat{\pi}, \bar{w}) + \bar{\nu}^T G(\hat{\pi}, \bar{u}) + \bar{\gamma}^T (H(\hat{\pi}, \bar{v}) - x_\alpha)\right\} dt$$
$$< \int_\Gamma \left\{\bar{\mu}^T f(\bar{\pi}, \bar{w}) + \bar{\nu}^T G(\bar{\pi}, \bar{u}) + \bar{\gamma}^T (H(\bar{\pi}, \bar{v}) - x_\alpha)\right\} dt. \tag{4.5.1}$$

On the other hand, by assumption that the uncertain Lagrange functional is convex at $(\bar{x}, \bar{c})$ on $\mathscr{X} \times \mathscr{C}$, we have

$$L(\hat{x}, \hat{c}, \bar{\mu}, \bar{\nu}, \bar{\gamma}, \bar{w}, \bar{u}, \bar{v}) - L(\bar{x}, \bar{c}, \bar{\mu}, \bar{\nu}, \bar{\gamma}, \bar{w}, \bar{u}, \bar{v})$$
$$\geqq (\hat{x} - \bar{x}) L_x(\bar{x}, \bar{c}, \bar{\mu}, \bar{\nu}, \bar{\gamma}, \bar{w}, \bar{u}, \bar{v})$$
$$+ (\hat{c} - \bar{c}) L_c(\bar{x}, \bar{c}, \bar{\mu}, \bar{\nu}, \bar{\gamma}, \bar{w}, \bar{u}, \bar{v}).$$

By using the definition of the Lagrange functional, we get

$$\int_{\Gamma} \Big\{\bar{\mu}^T f(\hat{\pi}, \bar{w}) + \bar{v}^T G(\hat{\pi}, \bar{u}) + \bar{\gamma}^T (H(\hat{\pi}, \bar{v}) - x_\alpha)\Big\} dt - \int_{\Gamma} \Big\{\bar{\mu}^T f(\bar{\pi}, \bar{w}) + \bar{v}^T G(\bar{\pi}, \bar{u})$$
$$+ \bar{\gamma}^T (H(\bar{\pi}, \bar{v}) - x_\alpha)\Big\} dt \geqq \int_{\Gamma} (\hat{x} - \bar{x}) \Big\{\bar{\mu}^T f_x(\bar{\pi}, \bar{w}) + \bar{v}^T G_x(\bar{\pi}, \bar{u}) + \bar{\gamma}^T H_x(\bar{\pi}, \bar{v}) + \bar{\gamma}_t \Big\} dt$$
$$+ \int_{\Gamma} (\hat{c} - \bar{c}) \Big\{\bar{\mu}^T f_c(\bar{\pi}, \bar{w}) + \bar{v}^T G_c(\bar{\pi}, \bar{u}) + \bar{\gamma}^T H_c(\bar{\pi}, \bar{v})\Big\} dt.$$

Using the robust necessary efficiency conditions (4.2.1) and (4.2.2), the above inequality can be written as

$$\int_{\Gamma} \Big\{\bar{\mu}^T f(\hat{\pi}, \bar{w}) + \bar{v}^T G(\hat{\pi}, \bar{u}) + \bar{\gamma}^T (H(\hat{\pi}, \bar{v}) - x_\alpha)\Big\} dt$$
$$\geqq \int_{\Gamma} \Big\{\bar{\mu}^T f(\bar{\pi}, \bar{w}) + \bar{v}^T G(\bar{\pi}, \bar{u}) + \bar{\gamma}^T (H(\bar{\pi}, \bar{v}) - x_\alpha)\Big\} dt,$$

which contradicts the inequality (4.5.1). Hence, this completes the proof. □

Now, we give the following example to validate Theorem 4.5.0.1.

Example 4.5.0.1 Let $p = 2, q = r = 1, k = 2, \mathcal{W}_1 = [0, \frac{1}{2}], \mathcal{W}_2 = [0, 1], \mathcal{U} = [0, 1], \mathcal{V}_1^1 = \mathcal{V}_2^1 = [0, \frac{1}{2}]$ and Γ_{t_0, t_1} is fixed by the diagonally opposite points $t_0 = (t_0^1, t_0^2)$ and $t_1 = (t_1^1, t_1^2)$ in $\mathbb{R}^2$.

We consider the following multi-dimensional multi-objective optimization problem as follows:

$$\text{(MMCOPOCU2)} \quad \min_{(x(\cdot), c(\cdot))} \int_{\Gamma_{0, \frac{2}{3}}} (x^2 - \frac{1}{2} w_1, \; \frac{1}{2} w_2 c^2 + x^2) \, dt^1 dt^2$$

$$\text{subject to} \quad u(x + 2)(x - 1) \leqq 0, \tag{4.5.2}$$

$$x_{t^1} = \frac{3}{2} - v_1^1 c, \tag{4.5.3}$$

$$x_{t^2} = \frac{3}{2} - v_2^1 c, \tag{4.5.4}$$

$$x(0, 0) = 0, \; x\left(\frac{2}{3}, \frac{2}{3}\right) = 1, \tag{4.5.5}$$

where $t = (t^1, t^2) \in \Gamma_{0, \frac{2}{3}}$.

Therefore, the robust counterpart for the problem (MMCOPOCU2) is defined as follows:

$$\text{(RMMCOPOCU2)} \quad \min_{(x(\cdot),c(\cdot))} \int_{\Gamma_{0,\frac{2}{3}}} \Big(\max_{w_1 \in \mathcal{W}_1} (x^2 - \frac{1}{2} w_1), \max_{w_2 \in \mathcal{W}_2} (\frac{1}{2} w_2 c^2 + x^2) \Big) \, dt^1 dt^2$$

$$\text{subject to} \quad u(x+2)(x-1) \leqq 0, \ \forall u \in \mathcal{U},$$

$$x_{t^1} = \frac{3}{2} - v_1^1 c, \ \forall v_1^1 \in \mathcal{V}_1^1,$$

$$x_{t^2} = \frac{3}{2} - v_2^1 c, \ \forall v_2^1 \in \mathcal{V}_2^1,$$

$$x(0,0) = 0, \ x\Big(\frac{2}{3}, \frac{2}{3}\Big) = 1,$$

where $t = (t^1, t^2) \in \Gamma_{0,\frac{2}{3}}$. Clearly $\mathcal{D} = \{(x,c) \in \mathscr{X} \times \mathscr{C} : -2 \leq x \leq 1 \wedge x_{t^1} = \frac{3}{2} - v_1^1 c \wedge x_{t^2} = \frac{3}{2} - v_2^1 c, \ x(0,0) = 0, x\big(\frac{2}{3}, \frac{2}{3}\big) = 1\}$ is a robust feasible solution set to (MMCOPOCU2).

Now, we find a weak robust efficient solution to the problem (MMCOPOCU2), i.e. find the control function $\bar{c} : \Gamma_{0,\frac{2}{3}} \to \mathbb{R}$ (that determines the state function $\bar{x} : \Gamma_{0,\frac{2}{3}} \to \mathbb{R}$) satisfying the dynamical systems (4.5.3) and (4.5.4) with respect to boundary conditions (4.5.5). Additionally, we assume that the state function is affine.

Let $(\bar{x}, \bar{c})$ be a robust feasible solution to the problem (MMCOPOCU2) and there exist the Lagrange multipliers $\bar{\mu}_1, \bar{\mu}_2, \bar{\nu}, \bar{\gamma}_1^1, \bar{\gamma}_2^1$ and the uncertain parameters $\max_{w \in \mathcal{W}} f(\pi, w) = f(\pi, \bar{w}), \bar{u}, \bar{v}_1^1, \bar{v}_2^1$ such that the robust necessary efficiency conditions are as follows:

$$2\bar{\mu}_1 \bar{x} + 2\bar{\mu}_2 \bar{x} + \bar{\nu}\bar{u}(2\bar{x} + 1) + (\bar{\gamma}_1^1)_{t^1} + (\bar{\gamma}_2^1)_{t^2} = 0, \tag{4.5.6}$$

$$\bar{\mu}_2 \bar{c} \bar{w}_2 - \bar{\gamma}_1^1 \bar{v}_1^1 - \bar{\gamma}_2^1 \bar{v}_2^1 = 0, \tag{4.5.7}$$

$$\bar{\nu}\bar{u}(\bar{x} + 2)(\bar{x} - 1) = 0, \tag{4.5.8}$$

where relation (4.5.8) is satisfied if either $\bar{\nu} = 0$ or $\bar{u}(\bar{x} + 2)(\bar{x} - 1) = 0$, for all $t = (t^1, t^2) \in \text{int}\Gamma_{0,\frac{2}{3}}$ (interior of hyperparallelepiped Γ). Let $\bar{\nu} \neq 0$, then $\bar{u}(\bar{x} + 2)(\bar{x} - 1) = 0$ implies that $\bar{x} = -2, 1$. Clearly, the boundary conditions (4.5.5) are failing at $\bar{x} = -2, 1$. Consequently, $\bar{\nu} = 0$.

Case 1. Assume that the Lagrange multipliers $\bar{\gamma}_\alpha^\tau, \tau = \overline{1,q}, \alpha = \overline{1,p}$ are constant functions, say $\bar{\gamma}_1^1 = h_1$ and $\bar{\gamma}_2^1 = h_2$, where $h_1, h_2 \in \mathbb{R}_+$. Therefore, Eq. (4.5.7) gives

$$\bar{c} = \frac{\bar{v}_1^1 h_1 + \bar{v}_2^1 h_2}{2\bar{\mu}_2 \bar{w}_2}, \ \forall t \in \Gamma_{0,\frac{2}{3}},$$

which shows that $\bar{c}$ is a constant function on $\Gamma_{0,\frac{2}{3}}$. Taking into account of equality constraints (4.5.3) and (4.5.4), it follows that

$$\bar{x} = \Big(\frac{3}{2} - \bar{v}_1^1 \frac{\bar{v}_1^1 h_1 + \bar{v}_2^1 h_2}{2\bar{\mu}_2 \bar{w}_2}\Big) t^1 + \Big(\frac{3}{2} - \bar{v}_2^1 \frac{\bar{v}_1^1 h_1 + \bar{v}_2^1 h_2}{2\bar{\mu}_2 \bar{w}_2}\Big) t^2 + h_3, \ h_3 \in \mathbb{R}_+.$$

By using the boundary condition (4.5.5) along with the Lagrange multiplier $\bar{\mu}_2 = 1$ and the uncertain parameters $\bar{w}_2 = 1$, $\bar{v}_1^1 = \bar{v}_2^1 = \frac{1}{2}$, we get $h_3 = 0$ and $h_1 + h_2 = 1$, and therefore,

$$\bar{x} = \frac{3}{4}(t^1 + t^2),\ \bar{c} = \frac{3}{2},\ \forall t \in \Gamma_{0,\frac{2}{3}}.$$

Now, we shall show that $(\bar{x}, \bar{c}) = (\frac{3}{4}(t^1 + t^2), \frac{3}{2})$ at $t^1 = t^2 = 0$, i.e. $(\bar{x}, \bar{c}) = (0, \frac{3}{2})$ is a weak robust efficient solution to the problem (MMCOPOCU2). In virtue of Definition 4.2.0.1, for all $(x, c) \in \mathcal{D}$, we have

$$\begin{aligned}
\int_{\Gamma_{0,\frac{2}{3}}} \max_{w \in W} f(\pi, w)dt - \int_{\Gamma_{0,\frac{2}{3}}} \max_{w \in W} f(\bar{\pi}, w)dt
&= \int_{\Gamma_{0,\frac{2}{3}}} f(\pi, \bar{w})dt - \int_{\Gamma_{0,\frac{2}{3}}} f(\bar{\pi}, \bar{w})dt\\
&= \int_{\Gamma_{0,\frac{2}{3}}} \left(x^2 - \frac{1}{2}\bar{w}_1,\ \frac{1}{2}\bar{w}_2 c^2 + x^2\right)dt^1dt^2\\
&\quad - \int_{\Gamma_{0,\frac{2}{3}}} \left(\bar{x}^2 - \frac{1}{2}\bar{w}_1,\ \frac{1}{2}\bar{w}_2 \bar{c}^2 + \bar{x}^2\right)dt^1dt^2\\
&= \int_{\Gamma_{0,\frac{2}{3}}} \left(x^2 - \frac{1}{4},\ \frac{1}{2}c^2 + x^2\right)dt^1dt^2 - \int_{\Gamma_{0,\frac{2}{3}}} \left(-\frac{1}{4},\ \frac{9}{8}\right)dt^1dt^2\\
&= \int_{\Gamma_{0,\frac{2}{3}}} \left\{x^2, \left(\frac{1}{2}c^2 + x^2 - \frac{9}{8}\right)\right\}dt^1dt^2.
\end{aligned} \tag{4.5.9}$$

On solving Eq. (4.5.3), we have $c = \frac{1}{v_1^1}(\frac{3}{2} - x_{t^1})$. Therefore, the above Eq. (4.5.9) can be written as

$$\begin{aligned}
\int_{\Gamma_{0,\frac{2}{3}}} \max_{w \in W} f(\pi, w)dt - \int_{\Gamma_{0,\frac{2}{3}}} \max_{w \in W} f(\bar{\pi}, w)dt
&= \int_{\Gamma_{0,\frac{2}{3}}} \left\{x^2, \left(\frac{1}{2(v_1^1)^2}\left(\frac{3}{2} - x_{t^1}\right)^2 + x^2 - \frac{9}{8}\right)\right\}dt^1dt^2.
\end{aligned} \tag{4.5.10}$$

Now, it remains to find the state function x which satisfies the boundary conditions (4.5.5).

The equality constraints (4.5.3) and (4.5.4) follow that

$$x_{t^1} = x_{t^2},\ \forall t \in \Gamma_{0,\frac{2}{3}}.$$

Applying the necessary Euler-Lagrange PDE (see, for instance, [7]) for Eq. (4.5.10), it follows

$$x = \phi(t^1 + t^2) + \psi,\ \phi, \psi \in \mathbb{R},\ \forall t \in \Gamma_{0,\frac{2}{3}}.$$

The above solution together with the boundary conditions (4.5.5) imply that $\phi = \frac{3}{4}$ and $\psi = 0$. Therefore, $x = \frac{3}{4}(t^1 + t^2)$. Hence, Eq. (4.5.10) yields

$$\begin{aligned}
&\int_{\Gamma_{0,\frac{2}{3}}} \max_{w \in \mathcal{W}} f(\pi, w)dt - \int_{\Gamma_{0,\frac{2}{3}}} \max_{w \in \mathcal{W}} f(\bar{\pi}, w)dt \\
&= \int_{\Gamma_{0,\frac{2}{3}}} \Big\{\frac{9}{16}(t^1+t^2)^2, \Big(\frac{1}{2(v_1^1)^2}\Big(\frac{3}{2} - \frac{3}{4}\Big)^2 + \frac{9}{16}(t^1+t^2)^2 - \frac{9}{8}\Big)\Big\}dt^1dt^2 \\
&= \Big(\frac{7}{54}, \frac{1}{8(v_1^1)^2} - \frac{10}{27}\Big) \geq 0,\ \forall v_1^1 \in \mathcal{V}_1^1.
\end{aligned}$$

Similarly, on solving Eq. (4.5.4), we have $c = \frac{1}{v_2^1}(\frac{3}{2} - x_{t^2})$ and the inequality (4.5.9) concludes that $\int_{\Gamma_{0,\frac{2}{3}}} \max_{w \in \mathcal{W}} f(\pi, w)dt - \int_{\Gamma_{0,\frac{2}{3}}} \max_{w \in \mathcal{W}} f(\bar{\pi}, w)dt \geq 0$. In both the cases, it shows that $(0, \frac{3}{2})$ is a weak robust efficient solution to the problem (MMCOPOCU2) at which the necessary efficiency conditions (4.5.6)–(4.5.8) are satisfied with the Lagrange multipliers $\bar{\mu}_1 = 1, \bar{\mu}_2 = 0, \bar{\nu} = 0$ and $\bar{\gamma}_1^1 + \bar{\gamma}_2^1 = 3$ and the uncertain parameters $\bar{w}_1 = \frac{1}{2}, \bar{w}_2 = 1, \bar{u} = 1, \bar{v}_1^1 = \bar{v}_2^1 = \frac{1}{2}$. Further, it can be seen that the Lagrange functional is also convex at $(\bar{x}, \bar{c}) = (0, \frac{3}{2})$ on $\mathscr{X} \times \mathscr{C}$.

Now, we formulate the penalized multi-dimensional multi-objective optimization problem in the face of data uncertainty $(\text{MMCOPOCU2})_\rho$ associated to the multi-dimensional multi-objective optimization problem (MMCOPOCU2) as follows:

$$\begin{aligned}
&(\text{MMCOPOCU2})_\rho \min_{(x(\cdot),c(\cdot))} \Big(\int_{\Gamma_{0,\frac{2}{3}}} \Big\{(x^2 - \frac{1}{2}w_1) + \rho\Big[\max\{0, u(x+2)(x-1)\} \\
&+ |x_{t^1} - \frac{3}{2} + v_1^1 c| + |x_{t^2} - \frac{3}{2} + v_2^1 c|\Big]\Big\}dt^1dt^2, \int_{\Gamma_{0,\frac{2}{3}}} \Big\{(\frac{1}{2}w2c^2 + x^2) \\
&+ \rho\Big[\max\{0, u(x+2)(x-1)\} + |x_{t^1} - \frac{3}{2} + v_1^1 c| + |x_{t^2} - \frac{3}{2} + v_2^1 c|\Big]\Big\}dt^1dt^2\Big).
\end{aligned}$$

Therefore, the robust counterpart for $(\text{MMCOPOCU2})_\rho$ is defined as follows:

$$\begin{aligned}
&(\text{RMMCOPOCU2})_\rho \min_{(x(\cdot),c(\cdot))} \Big(\int_{\Gamma_{0,\frac{2}{3}}} \Big\{\max_{w_1 \in \mathcal{W}_1}(x^2 - \frac{1}{2}w_1) + \rho\Big[\max\{0, u(x+2)(x-1)\} \\
&+ |x_{t^1} - \frac{3}{2} + v_1^1 c| + |x_{t^2} - \frac{3}{2} + v_2^1 c|\Big]\Big\}dt^1dt^2, \int_{\Gamma_{0,\frac{2}{3}}} \Big\{\max_{w_2 \in \mathcal{W}_2}(\frac{1}{2}w2c^2 + x^2) \\
&+ \rho\Big[\max\{0, u(x+2)(x-1)\} + |x_{t^1} - \frac{3}{2} + v_1^1 c| + |x_{t^2} - \frac{3}{2} + v_2^1 c|\Big]\Big\}dt^1dt^2\Big).
\end{aligned}$$

In virtue of Definition 4.4.0.1, it can be observed that the inequality $\mathscr{P}(\pi, w, \bar{u}, \bar{v}, \rho) - \mathscr{P}(\bar{\pi}, w, \bar{u}, \bar{v}, \rho) \geq 0$ at $(\bar{x}, \bar{c}) = (0, \frac{3}{2})$ for all $(x, c) \in \mathscr{X} \times \mathscr{C}$. This shows that $(0, \frac{3}{2})$ is a weak robust efficient solution of the penalized multi-dimensional optimization problem $(\text{MMCOPOCU2})_\rho$ with $\rho \geq 1$. Hence, Theorem 4.5.0.1 is verified.

Case 2. Assume that the Lagrange multipliers $\bar{\gamma}_\alpha^\tau$, $\tau = 1, \alpha = 1, 2$ are functions of t, say $\bar{\gamma}_1^1 = t^2$, $\bar{\gamma}_2^1 = t^1$. Thus, by Eq. (4.5.7), we have

$$\bar{c} = \frac{t^2 + t^1}{2\bar{\mu}_1 \bar{w}_1}, \ \forall t \in \Gamma_{0, \frac{2}{3}}. \tag{4.5.11}$$

On solving Eq. (4.5.3) and (4.5.4) with the help of Eq. (4.5.11), we have

$$\bar{x} = \left\{\frac{3}{2}(t^1 + t^2) - \frac{1}{2}v_1^1\left(t^1t^2 + \frac{(t^1)^2}{2}\right) - \frac{1}{2}v_2^1\left(t^1t^2 + \frac{(t^2)^2}{2}\right)\right\} + h_4, \ h_4 \in \mathbb{R}_+,$$

and, by considering the boundary conditions (4.5.5), we get

$$\bar{x} = \left\{\frac{3}{2}(t^1 + t^2) - \frac{1}{2}v_1^1\left(t^1t^2 + \frac{(t^1)^2}{2}\right) - \frac{1}{2}v_2^1\left(t^1t^2 + \frac{(t^2)^2}{2}\right)\right\},$$

which shows that the state function is not affine function. Consequently, we get a contradiction with the hypothesis.

References

1. X.W. Liu, Y.X. Yuan, A robust algorithm for optimization with general equality and inequality constraints. SIAM J. Sci. Comput. **22**, 517–534 (2000)
2. V. Jeyakumar, J.H. Wang, G. Li, Lagrange multiplier characterizations of robust best approximations under constraint data uncertainty. J. Math. Anal. Appl. **393**, 285–297 (2012)
3. H.Z. Wei, C.R. Chen, S.J. Li, Characterizations for optimality conditions of general robust optimization problems. J. Optim. Theory Appl. **177**, 835–856 (2018)
4. X.K. Sun, K.L. Teo, J. Zeng, X.L. Guo, On approximate solutions and saddle point theorems for robust convex optimization. Optim. Lett. **14**, 1711–1730 (2020)
5. P.A. Jayswal, M. Arana-Jiménez, Robust saddle-point criteria for multi-dimensional optimization problems with data uncertainty. Int. J. Control (2020). https://doi.org/10.1080/00207179.2020.1759827
6. Şt. Mititelu, Optimality and duality for invex multi-dimensional control problems with mixed constraints, J. Adv. Math. Stud. **2**, 25–34 (2009)
7. S. Treanţă, Higher-order Hamilton dynamics and Hamilton-Jacobi divergence PDE. Comput. Math. Appl. **75**, 547–560 (2018)

Chapter 5
The Modified Approach for Multi-dimensional Optimization Problem with Data Uncertainty

5.1 Introduction

Optimal control theory is a mature mathematical discipline and has yet to be an active field of research due to extensive application in various areas, such as economics, engineering, ecology, medicine, management sciences and many more. This is a consequence of the fact that a variety of real-life problems from the aforesaid areas can be modeled as optimal control optimization problems.

Many works on optimal control optimization problems have been centered on finding the associated solutions through the study of the involved functional or by using appropriate optimality conditions or to make an equivalence with an another mathematical programming problem associated to optimal control optimization problem (see, for example, [1–9]).

In this chapter, we present an effective method to solve the multi-dimensional optimal control optimization problems, which is known as the modified objective function method. Antczak [10] introduced this approach to find an efficient solution for a class of differentiable multi-objective optimization problems. Thereafter, this approach was used for solving various mathematical programming problems involving complex objective functions (see, for example, [11–17]) and other.

5.2 Problem Description

Throughout the chapter, we are considering the following notations and working hypotheses:

- $\mathbb{R}^p$, $\mathbb{R}^q$, $\mathbb{R}^r$ and $\mathbb{R}^n$ are Euclidean spaces of dimensions p, q, r and n, respectively.
- $\Gamma_{t_0,t_1} \subset \mathbb{R}^p$ is a hyperparallelepiped fixed by the diagonally opposite points $t_0 = (t_0^\alpha)$ and $t_1 = (t_1^\alpha)$, $\alpha = \overline{1, p}$ and the point $t = ((t^\alpha), \alpha = \overline{1, p}) \in \Gamma_{t_0,t_1} \subset \mathbb{R}^p$.

A. Jayswal et al., *Multi-dimensional Control Problems*, Industrial and Applied Mathematics, https://doi.org/10.1007/978-981-19-6561-6_5

- $\mathscr{X}$ is the space of state functions (piecewise smooth) $x = (x^\tau) : \Gamma_{t_0,t_1} \subset \mathbb{R}^p \to \mathbb{R}^q$ and $\frac{\partial x}{\partial t^\alpha} = x_\alpha$ denotes the partial derivative of x with respect to t^α.
- $\mathscr{C}$ is the space of control functions (piecewise continuous) $c = (c^j) : \Gamma_{t_0,t_1} \subset \mathbb{R}^p \to \mathbb{R}^r$.
- $dt = dt^1 \dots dt^p$ is the volume element on $\mathbb{R}^p \supset \Gamma_{t_0,t_1}$.
- T denotes the transpose of a vector.
- We use the following convention for inequalities and equalities: for $x, y \in \mathbb{R}^n$, we have

$$\begin{aligned}
(i)\ & x < y \Leftrightarrow x_i < y_i, \forall i = \overline{1,n},\\
(ii)\ & x = y \Leftrightarrow x_i = y_i,\ \forall i = \overline{1,n},\\
(iii)\ & x \leqq y \Leftrightarrow x_i \leq y_i,\ \forall i = \overline{1,n},\\
(iv)\ & x \leq y \Leftrightarrow x_i \leq y_i,\ \forall i = \overline{1,n} \text{ and } x_i < y_i \text{ for some } i.
\end{aligned}$$

The multi-dimensional multi-objective optimization problem with data uncertainty in the objective and constraint functionals is defined as follows:

$$\begin{aligned}
\text{(MMOPU)} \quad & \min_{(x(\cdot),c(\cdot))} \Big\{ \int_{\Gamma_{t_0,t_1}} f(t, x(t), c(t), w)dt \\
& = \Big(\int_{\Gamma_{t_0,t_1}} f_1(t, x(t), c(t), w_1)dt, \dots, \int_{\Gamma_{t_0,t_1}} f_s(t, x(t), c(t), w_s)dt \Big) \Big\} \\
\text{subject to} \quad & G(t, x(t), c(t), u) \leqq 0, \\
& x_\alpha = H(t, x(t), c(t), v), \\
& x(t_0) = x_0,\ x(t_1) = x_1,
\end{aligned}$$

where $f = (f_1, \dots, f_s)$; $(f_k) : \Gamma_{t_0,t_1} \times \mathscr{X} \times \mathscr{C} \times \mathcal{W}_k \to \mathbb{R}^s, k = \overline{1,s}$, $G = (G_1, \dots, G_m)$; $(G_l) : \Gamma_{t_0,t_1} \times \mathscr{X} \times \mathscr{C} \times \mathcal{U}_l \to \mathbb{R}^m$, $l = \overline{1,m}, H = (H^\tau_\alpha) : \Gamma_{t_0,t_1} \times \mathscr{X} \times \mathscr{C} \times \mathcal{V}^\tau_\alpha \to \mathbb{R}^{pq}, \alpha = \overline{1,p}, \tau = \overline{1,q}$ are C^∞-class functionals. Consider $w = (w_k)$, $u = (u_l)$ and $v = (v^\tau_\alpha)$ are the uncertain parameters for some convex compact subsets $\mathcal{W} = (\mathcal{W}_k) \subset \mathbb{R}^s$, $\mathcal{U} = (\mathcal{U}_l) \subset \mathbb{R}^m$ and $\mathcal{V} = (\mathcal{V}^\tau_\alpha) \subset \mathbb{R}^{pq}$, respectively. The functions H_α satisfy the closeness conditions (complete integrability conditions) $D_\beta H_\alpha = D_\alpha H_\beta, \alpha, \beta = \overline{1,p}, \alpha \neq \beta$, where the total derivative is denoted by D_β.

The associated robust counterpart of the multi-dimensional multi-objective optimization problem (MMOPU) is defined as follows:

$$\begin{aligned}
\text{(RMMOPU)} \quad & \min_{(x(\cdot),c(\cdot))} \Big\{ \int_{\Gamma_{t_0,t_1}} \max_{w \in \mathcal{W}} f(t, x(t), c(t), w)dt \\
& = \Big(\int_{\Gamma_{t_0,t_1}} \max_{w_1 \in \mathcal{W}_1} f_1(t, x(t), c(t), w_1)dt, \dots, \int_{\Gamma_{t_0,t_1}} \max_{w_s \in \mathcal{W}_s} f_s(t, x(t), c(t), w_s)dt \Big) \Big\} \\
\text{subject to} \quad & G(t, x(t), c(t), u) \leqq 0, \forall u \in \mathcal{U}, \\
& x_\alpha = H(t, x(t), c(t), v), \forall v \in \mathcal{V}, \\
& x(t_0) = x_0,\ x(t_1) = x_1,
\end{aligned}$$

where f, G and H are defined as above in the multi-dimensional multi-objective optimization problem (MMOPU).

We denote $\mathcal{D} = \{(x, c) \in \mathcal{X} \times \mathcal{C} : \ G(t, x(t), c(t), u) \leqq 0, x_\alpha = H(t, x(t), c(t), v), x(t_0) = x_0, x(t_1) = x_1\}$ the set of all feasible solutions in (RMMOPU) and we say that it is the robust feasible solution set to the problem (MMOPU).

From now on, to simplify our presentation, we introduce some notations as follows: $x = x(t), \hat{x} = \hat{x}(t), \bar{x} = \bar{x}(t), \tilde{x} = \tilde{x}(t), c = c(t), \hat{c} = \hat{c}(t), \bar{c} = \bar{c}(t), \tilde{c} = \tilde{c}(t), \pi = (t, x(t), c(t)), \bar{\pi} = (t, \bar{x}(t), \bar{c}(t)), \hat{\pi} = (t, \hat{x}(t), \hat{c}(t)), \tilde{\pi} = (t, \tilde{x}(t), \tilde{c}(t)), \rho = \rho(t), \bar{\rho} = \bar{\rho}(t)$ and $\Gamma = \Gamma_{t_0, t_1}$.

The partial derivatives of Lagrange multiplier γ and the partial derivatives associated with f are defined as

$$\gamma_t = \begin{bmatrix} \frac{\partial \gamma_1^\tau}{\partial t^1} & \cdots & \frac{\partial \gamma_1^\tau}{\partial t^p} \\ \vdots & \ddots & \vdots \\ \frac{\partial \gamma_p^\tau}{\partial t^1} & \cdots & \frac{\partial \gamma_p^\tau}{\partial t^p} \end{bmatrix}, \quad f_x = \begin{bmatrix} \frac{\partial f_1}{\partial x^1} & \cdots & \frac{\partial f_1}{\partial x^q} \\ \vdots & \ddots & \vdots \\ \frac{\partial f_s}{\partial x^1} & \cdots & \frac{\partial f_s}{\partial x^q} \end{bmatrix}, \quad f_c = \begin{bmatrix} \frac{\partial f_1}{\partial c^1} & \cdots & \frac{\partial f_1}{\partial c^r} \\ \vdots & \ddots & \vdots \\ \frac{\partial f_s}{\partial c^1} & \cdots & \frac{\partial f_s}{\partial c^r} \end{bmatrix}.$$

Similarly, we have G_x and G_c using matrices with m rows and H_x, H_c, x_α using matrices with p rows.

In general, for a vector optimization problem, it is not so easy to find a point that optimizes more than one objective simultaneously. Therefore, the notion of an efficient solution is used to determine an optimality in vector optimization problems.

Definition 5.2.0.1 A point $(\bar{x}, \bar{c}) \in \mathcal{D}$ is said to be weak robust efficient solution to the multi-dimensional multi-objective optimization problem (MMOPU), if there does not exist another point $(x, c) \in \mathcal{D}$ such that

$$\int_\Gamma \max_{w \in \mathcal{W}} f(\pi, w) dt < \int_\Gamma \max_{w \in \mathcal{W}} f(\bar{\pi}, w) dt.$$

Definition 5.2.0.2 A functional $F(x, c, \bar{w}) = \int_\Gamma f(t, x(t), x_\sigma(t), c(t), \bar{w}) dt$ is said to be a convex at $(\bar{x}(t), \bar{c}(t))$, if the following inequality

$$\begin{aligned} F(x, c, \bar{w}) - F(\bar{x}, \bar{c}, \bar{w}) \geqq & \int_\Gamma (x(t) - \bar{x}(t)) f_x(t, \bar{x}(t), \bar{x}_\sigma(t), \bar{c}(t), \bar{w}) dt \\ & + \int_\Gamma (x_\sigma(t) - \bar{x}_\sigma(t)) f_{x_\sigma}(t, \bar{x}(t), \bar{x}_\sigma(t), \bar{c}(t), \bar{w}) dt \\ & + \int_\Gamma (c(t) - \bar{c}(t)) f_c(t, \bar{x}(t), \bar{x}_\sigma(t), \bar{c}(t), \bar{w}) dt \end{aligned}$$

holds for all $(x(t), c(t))$ on $\mathcal{X} \times \mathcal{C}$.

Following the same lines formulated in Mititelu [5], we establish the following result. Thus, the proof is omitted. The proof can be read (it is free) at the following link: https://www.journal.fairpartners.ro/volume-22009-no-1_3.html.

Theorem 5.2.0.1 *Let $(\bar{x}, \bar{c}) \in \mathcal{D}$ be a weak robust efficient solution to the problem (MMOPU). Further assume that* $\max_{w \in \mathcal{W}} f(\pi, w) = f(\pi, \bar{w})$. *If the constraint conditions (for the existence of the multipliers) hold, then there exist the scalar vector $\bar{\mu} \in \mathbb{R}^s$, the piecewise smooth Lagrange multipliers $\bar{\nu} = (\bar{\nu}_l(t)) \in \mathbb{R}^m_+, l \in \mathbb{R}^m, \bar{\gamma} = (\bar{\gamma}^\tau_\alpha(t)) \in \mathbb{R}^{pq}, \tau = \overline{1, q}, \alpha = \overline{1, p}$, and the uncertain parameters $\bar{u} \in \mathcal{U}, \bar{v} \in \mathcal{V}$ such that $(\bar{x}, \bar{c})$ satisfies the following conditions:*

$$\bar{\mu}^T f_x(\bar{\pi}, \bar{w}) + \bar{\nu}^T G_x(\bar{\pi}, \bar{u}) + \bar{\gamma}^T H_x(\bar{\pi}, \bar{v}) + \bar{\gamma}_t = 0, \tag{5.2.1}$$

$$\bar{\mu}^T f_c(\bar{\pi}, \bar{w}) + \bar{\nu}^T G_c(\bar{\pi}, \bar{v}) + \bar{\gamma}^T H_c(\bar{\pi}, \bar{v}) = 0, \tag{5.2.2}$$

$$\bar{\nu}^T G(\bar{\pi}, \bar{u}) = 0, \tag{5.2.3}$$

$$\bar{\mu} \geq 0, \bar{\nu} \geqq 0 \tag{5.2.4}$$

hold for all $t \in \Gamma$, except at discontinuities.

Remark 5.2.0.1 The conditions (5.2.1)–(5.2.4) are known as the robust necessary efficiency conditions for the multi-dimensional multi-objective optimization problem (MMOPU).

5.3 Modified Multi-dimensional Multi-objective Optimization Problem

In this section, we use the modified objective function method to reduce the complexity in the considered multi-dimensional multi-objective optimization problem (MMOPU).

Let $(\bar{x}, \bar{c})$ be a given arbitrary robust feasible solution to the multi-dimensional multi-objective optimization problem (MMOPU). The modified multi-dimensional multi-objective optimization problem associated with the original problem (MMOPU) is defined as

$$(\text{MMOPU})_\Omega \quad \min_{(x(\cdot), c(\cdot))} \int_{\Gamma_{t_0, t_1}} \Big\{ (x(t) - \bar{x}(t)) f_x(t, \bar{x}(t), \bar{c}(t), w) + (c(t) - \bar{c}(t)) f_c(t, \bar{x}(t), \bar{c}(t), w) \Big\} dt$$

$$\text{subject to} \quad G(t, x(t), c(t), u) \leq 0,$$

$$x_\alpha = H(t, x(t), c(t), v),$$

$$x(t_0) = x_0, \; x(t_1) = x_1,$$

where $t \in \Gamma_{t_0, t_1}$, and the functionals f, G and H are given as in the original problem (MMOPU).

The associated robust counterpart of the multi-dimensional multi-objective optimization problem $(\text{MMOPU})_\Omega$ is defined as

$$\begin{aligned}
(\text{RMMOPU})_{\Omega} \quad & \min_{(x(\cdot),c(\cdot))} \left\{ \int_{\Gamma_{t_0,t_1}} \max_{w \in \mathcal{W}} \left\{ (x(t) - \bar{x}(t)) f_x(t, \bar{x}(t), \bar{c}(t), w) \right. \right. \\
& \qquad \left. \left. + (c(t) - \bar{c}(t)) f_c(t, \bar{x}(t), \bar{c}(t), w) \right\} dt \right. \\
\text{subject to} \quad & G(t, x(t), c(t), u) \leqq 0, \ \forall u \in \mathcal{U}, \\
& x_\alpha = H(t, x(t), c(t), v), \ \forall v \in \mathcal{V}, \\
& x(t_0) = x_0, \ x(t_1) = x_1,
\end{aligned}$$

where $t \in \Gamma_{t_0,t_1}$.

Remark 5.3.0.1 Note that the robust feasible solution set to the problem $(\text{MMOPU})_{\Omega}$ is the same as in the problem (MMOPU), and, thus, it is also denoted by $\mathcal{D}$.

Definition 5.3.0.1 A point $(\hat{x}, \hat{c}) \in \mathcal{D}$ is said to be a weak robust efficient solution to the modified multi-dimensional multi-objective optimization problem $(\text{MMOPU})_{\Omega}$, if there does not exist another point $(x, c) \in \mathcal{D}$ such that

$$\begin{aligned}
& \int_{\Gamma} \max_{w \in \mathcal{W}} \Big[(x - \bar{x}) f_x(\bar{\pi}, w) + (c - \bar{c}) f_c(\bar{\pi}, w) \Big] dt \\
& \qquad < \int_{\Gamma} \max_{w \in \mathcal{W}} \Big[(\hat{x} - \bar{x}) f_x(\bar{\pi}, w) + (\hat{c} - \bar{c}) f_c(\bar{\pi}, w) \Big] dt.
\end{aligned}$$

Definition 5.3.0.2 The Lagrange functional $L((x, c), \nu, \gamma, w, u, v) : \mathcal{X} \times \mathcal{C} \times \mathbb{R}^m_+ \times \mathbb{R}^{pq} \times \mathcal{W} \times \mathcal{U} \times \mathcal{V} \to \mathbb{R}^s$ associated with the multi-dimensional multi-objective optimization problem $(\text{MMOPU})_{\Omega}$ is defined as

$$\begin{aligned}
L((x, c), \nu, \gamma, w, u, v) = \int_{\Gamma} & \left\{ \max_{w \in \mathcal{W}} \Big[(x - \bar{x}) f_x(t, \bar{x}, \bar{c}, w) + (c - \bar{c}) f_c(t, \bar{x}, \bar{c}, w) \Big] \right. \\
& \left. + \nu^T G(t, x, c, u) e + \gamma^T (H(t, x, c, v) - x_\alpha) e \right\} dt,
\end{aligned}$$

where $e = (1, \ldots, 1) \in \mathbb{R}^s$.

Definition 5.3.0.3 A point $((\bar{x}, \bar{c}), \bar{\nu}, \bar{\gamma}, \bar{w}, \bar{u}, \bar{v}) \in \mathcal{D} \times \mathbb{R}^m_+ \times \mathbb{R}^{pq} \times \mathcal{W} \times \mathcal{U} \times \mathcal{V}$ is said to be a robust saddle-point in the multi-dimensional multi-objective optimization problem (MMOPU), iff

(i) $L((\bar{x}, \bar{c}), \nu, \gamma, w, u, v) \leqq L((\bar{x}, \bar{c}), \bar{\nu}, \bar{\gamma}, w, \bar{u}, \bar{v}), \ \forall \nu \in \mathbb{R}^m_+, \ \forall \gamma \in \mathbb{R}^{pq}, \ \forall u \in \mathcal{U}, \ \forall v \in \mathcal{V}$,
(ii) $L((x, c), \bar{\nu}, \bar{\gamma}, w, \bar{u}, \bar{v}) \geqq L((\bar{x}, \bar{c}), \bar{\nu}, \bar{\gamma}, w, \bar{u}, \bar{v}), \ \forall (x, c) \in \mathcal{X} \times \mathcal{C}$.

Now, we shall prove that the problems (MMOPU) and $(\text{MMOPU})_{\Omega}$ attain their solution set at the same point under appropriate convexity assumptions.

Theorem 5.3.0.1 *Let $(\bar{x}, \bar{c})$ be a normal robust efficient solution to the multi-dimensional multi-objective optimization problem (MMOPU). Further, assume that $\int_{\Gamma} \bar{\nu}^T G(\pi, \bar{u}) dt$, $\int_{\Gamma} \bar{\gamma}^T (H(\pi, \bar{v}) - x_\alpha) dt$ are convex at $(\bar{x}, \bar{c})$ on $\mathcal{D}$. Then $(\bar{x}, \bar{c})$ is* *also a weak robust efficient solution to the modified multi-dimensional multi-objective optimization problem (MMOPU)$_\Omega$.*

Proof By assumption, $(\bar{x}, \bar{c})$ is a normal robust efficient solution to the problem (MMOPU). Therefore, by Theorem 5.2.0.1, the robust necessary efficiency conditions are satisfied at $(\bar{x}, \bar{c})$. Hence, the conditions (5.2.1) and (5.2.2) yield

$$\int_\Gamma (\tilde{x} - \bar{x})\Big\{\bar{\mu}^T f_x(\bar{\pi}, \bar{w}) + \bar{\nu}^T G_x(\bar{\pi}, \bar{u}) + \bar{\gamma}^T H_x(\bar{\pi}, \bar{v}) + \bar{\gamma}_t\Big\}dt \\ + \int_\Gamma (\tilde{c} - \bar{c})\Big\{\bar{\mu}^T f_c(\bar{\pi}, \bar{w}) + \bar{\nu}^T G_c(\bar{\pi}, \bar{u}) + \bar{\gamma}^T H_c(\bar{\pi}, \bar{v})\Big\}dt = 0. \tag{5.3.1}$$

Let us proceed by contradiction, that is, we consider $(\bar{x}, \bar{c})$ is not a weak robust efficient solution to the modified multi-dimensional multi-objective optimization problem (MMOPU)$_\Omega$. Then, there exists $(\tilde{x}, \tilde{c}) \in \mathcal{D}$ such that

$$\int_\Gamma \max_{w \in \mathcal{W}} \Big[(\tilde{x} - \bar{x}) f_x(\bar{\pi}, w) + (\tilde{c} - \bar{c}) f_c(\bar{\pi}, w)\Big]dt \\ < \int_\Gamma \max_{w \in \mathcal{W}} \Big[(\bar{x} - \bar{x}) f_x(\bar{\pi}, w) + (\bar{c} - \bar{c}) f_c(\bar{\pi}, w)\Big]dt.$$

Since $\max_{w \in \mathcal{W}} f(\pi, w) = f(\pi, \bar{w})$, we get

$$\int_\Gamma \Big[(\tilde{x} - \bar{x}) f_x(\bar{\pi}, \bar{w}) + (\tilde{c} - \bar{c}) f_c(\bar{\pi}, \bar{w})\Big]dt < \int_\Gamma \Big[(\bar{x} - \bar{x}) f_x(\bar{\pi}, \bar{w}) + (\bar{c} - \bar{c}) f_c(\bar{\pi}, \bar{w})\Big]dt,$$

equivalent with

$$\int_\Gamma \Big[(\tilde{x} - \bar{x}) f_x(\bar{\pi}, \bar{w}) + (\tilde{c} - \bar{c}) f_c(\bar{\pi}, \bar{w})\Big]dt < 0. \tag{5.3.2}$$

The robust feasibility of $(\tilde{x}, \tilde{c})$ in the problem (MMOPU) and the robust necessary efficiency condition (5.2.3) yields

$$\int_\Gamma \bar{\nu}^T G(\tilde{\pi}, \bar{u}) dt - \int_\Gamma \bar{\nu}^T G(\bar{\pi}, \bar{u}) dt \leqq 0. \tag{5.3.3}$$

By hypothesis, since $\int_\Gamma \bar{\nu}^T G(\pi, \bar{u}) dt$ is convex at $(\bar{x}, \bar{c})$ on $\mathcal{D}$, we have

$$\int_\Gamma \Big\{\bar{\nu}^T G(\tilde{\pi}, \bar{u}) - \bar{\nu}^T G(\bar{\pi}, \bar{u})\Big\}dt \geqq \int_\Gamma (\tilde{x} - \bar{x}) \bar{\nu}^T G_x(\bar{\pi}, \bar{u}) dt + \int_\Gamma (\tilde{c} - \bar{c}) \bar{\nu}^T G_c(\bar{\pi}, \bar{u}) dt,$$

which together with the inequality (5.3.3) gives

$$\int_{\Gamma}(\tilde{x}-\bar{x})\bar{\upsilon}^T G_x(\bar{\pi},\bar{u})dt+\int_{\Gamma}(\tilde{c}-\bar{c})\bar{\upsilon}^T G_c(\bar{\pi},\bar{u})dt \leqq 0. \tag{5.3.4}$$

Again, by hypothesis $\int_{\Gamma}\bar{\gamma}^T(H(\pi,\bar{v})-x_{\alpha})dt$ is convex at $(\bar{x},\bar{c})$ on $\mathcal{D}$, we have

$$\int_{\Gamma}\Big\{\bar{\gamma}^T\Big(H(\tilde{\pi},\bar{v})-x_{\alpha}\Big)-\bar{\gamma}^T\Big(H(\bar{\pi},\bar{v})-x_{\alpha}\Big)\Big\}dt$$
$$\geqq \int_{\Gamma}(\tilde{x}-\bar{x})\Big\{\bar{\gamma}^T H_x(\bar{\pi},\bar{v})+\gamma_t\Big\}dt+\int_{\Gamma}(\tilde{c}-\bar{c})\bar{\gamma}^T H_c(\bar{\pi},\bar{v})dt.$$

By the robust feasibility of $(\tilde{x},\tilde{c})$ and $(\bar{x},\bar{c})$ in the problem (MMOPU), the above inequality reduces to

$$\int_{\Gamma}(\tilde{x}-\bar{x})\Big\{\bar{\gamma}^T H_x(\bar{\pi},\bar{v})+\gamma_t\Big\}dt+\int_{\Gamma}(\tilde{c}-\bar{c})\bar{\gamma}^T H_c(\bar{\pi},v)dt \leqq 0. \tag{5.3.5}$$

On combining the inequalities (5.3.2), (5.3.4) and (5.3.5), we get

$$\int_{\Gamma}(\tilde{x}-\bar{x})\Big\{\bar{\mu}^T f_x(\bar{\pi},\bar{w})+\bar{\upsilon}^T G_x(\bar{\pi},\bar{u})+\bar{\gamma}^T H_x(\bar{\pi},\bar{v})+\bar{\gamma}_t\Big\}dt$$
$$+\int_{\Gamma}(\tilde{c}-\bar{c})\Big\{\bar{\mu}^T f_c(\bar{\pi},\bar{w})+\bar{\upsilon}^T G_c(\bar{\pi},\bar{u})+\bar{\gamma}^T H_c(\bar{\pi},\bar{v})\Big\}dt<0,$$

which contradicts the relation (5.3.1). Thus, $(\bar{x},\bar{c})$ is weak robust efficient solution to the problem $(\text{MMOPU})_{\Omega}$ and this completes the proof. □

Now, we construct an example to illustrate the result established in Theorem 5.3.0.1.

Example 5.3.0.1 Let $p=2, q=r=1, k=2, \mathcal{W}_1=[-1,1], \mathcal{W}_2=[0,1], \mathcal{U}=[0,\frac{1}{2}], \mathcal{V}_1^1=\mathcal{V}_2^1=[\frac{1}{2},1]$ and Γ_{t_0,t_1} is fixed by the diagonally opposite points $t_0=(t_0^1,t_0^2)$ and $t_1=(t_1^1,t_1^2)$ in $\mathbb{R}^2$.

We consider the following multi-dimensional multi-objective optimization problem:

$$\text{(MMOPU1)}\quad \min_{(x(\cdot),c(\cdot))}\int_{\Gamma_{0,1}}(x^2+c^2w_1,2w_2c+xe^x)\,dt^1dt^2$$
$$\text{subject to}\quad ux(x-1)\leq 0, \tag{5.3.6}$$
$$x_{t^1}=\frac{3}{2}-2c+v_1^1, \tag{5.3.7}$$
$$x_{t^2}=\frac{3}{2}-2c+v_2^1, \tag{5.3.8}$$
$$x(0,0)=0,\ x(1,1)=2, \tag{5.3.9}$$

where $t=(t^1,t^2)\in\Gamma_{0,1}$.

The robust counterpart of the multi-dimensional multi-objective optimization problem (MMOPU1) is defined as follows:

$$
\begin{aligned}
\text{(RMMOPU1)} \quad & \min_{(x(\cdot),c(\cdot))} \int_{\Gamma_{0,1}} \left(\max_{w_1 \in \mathcal{W}_1} (x^2 + c^2 w_1),\ \max_{w_2 \in \mathcal{W}_2} (2w_2 c + x e^x) \right) dt^1 dt^2 \\
\text{subject to} \quad & u x(x-1) \leq 0,\ \forall u \in \mathcal{U}, \\
& x_{t^1} = \frac{3}{2} - 2c + v_1^1,\ \forall v_1^1 \in \mathcal{V}_1^1, \\
& x_{t^2} = \frac{3}{2} - 2c + v_2^1,\ \forall v_2^1 \in \mathcal{V}_2^1, \\
& x(0,0) = 0,\ x(1,1) = 2,
\end{aligned}
$$

where $t = (t^1, t^2) \in \Gamma_{0,1}$. Clearly $\mathcal{D} = \{(x, c) \in \mathcal{X} \times \mathcal{C} : 0 \leq x \leq 1 \wedge [\frac{3}{2} - 2c + v_1^1 = x_{t^1}] \wedge [\frac{3}{2} - 2c + v_2^1 = x_{t^2}], x(0,0) = 0, x(1,1) = 2\}$ is the robust feasible solution set to (MMOPU1).

Now, we find a weak robust efficient solution to the problem (MMOPU1), i.e. find the control function $\bar{c} : \Gamma_{0,1} \to \mathbb{R}$ (that determines the state function $\bar{x} : \Gamma_{0,1} \to \mathbb{R}$) satisfying the dynamical systems (5.3.7) and (5.3.8) with respect to boundary conditions (5.3.9). Additionally, we assume that the state function is affine.

Let $(\bar{x}, \bar{c})$ be a robust feasible solution to the problem (MMOPU1), the Lagrange multipliers $\bar{\mu}_1, \bar{\mu}_2, \bar{\nu}, \bar{\gamma}_1^1, \bar{\gamma}_2^1$, and the uncertain parameters $\bar{u}, \bar{v}_1^1, \bar{v}_2^1$. Let $\max_{w \in \mathcal{W}} f(\pi, w) = f(\pi, \bar{w})$. Therefore, the robust necessary efficiency conditions are given as follows:

$$2\bar{\mu}_1 \bar{x} + \bar{\mu}_2 e^{\bar{x}}(1 + \bar{x}) + \bar{\nu}\bar{u}(2\bar{x} - 1) + (\bar{\gamma}_1^1)_{t^1} + (\bar{\gamma}_2^1)_{t^2} = 0, \tag{5.3.10}$$

$$2\bar{\mu}_1 \bar{c}\bar{w}_1 + 2\bar{\mu}_2 \bar{w}_2 - 2\bar{\gamma}_1^1 - 2\bar{\gamma}_2^1 = 0, \tag{5.3.11}$$

$$\bar{\nu}\bar{u}\bar{x}(\bar{x} - 1) = 0. \tag{5.3.12}$$

Clearly, the relation (5.3.12) is satisfied if either $\bar{\nu} = 0$ or $\bar{u}\bar{x}(\bar{x} - 1) = 0$, for all $t = (t^1, t^2) \in \text{int}\Gamma_{0,1}$ (interior of hyperparallelepiped Γ). Let $\bar{\nu} \neq 0$, then $\bar{u}\bar{x}(\bar{x} - 1) = 0$ implies that $\bar{x} = 0, 1$. The boundary conditions (5.3.9) are failing at $\bar{x} = 0, 1$ for all $\bar{u} \in \mathcal{U}$. Consequently, $\bar{\nu} = 0$.

Case 1. Assume that the Lagrange multipliers $\bar{\gamma}_\alpha^\tau, \tau = 1, \alpha = 1, 2$ are constant functions, say $\bar{\gamma}_1^1 = h_1$ and $\bar{\gamma}_2^1 = h_2$, where $h_1, h_2 \in \mathbb{R}_+$. Therefore, Eq. (5.3.11) gives

$$\bar{c} = \frac{h_1 + h_2 - \bar{\mu}_2 \bar{w}_2}{\bar{\mu}_1 \bar{w}_1},\ \forall t \in \Gamma_{0,1},$$

which shows that $\bar{c}$ is a constant function on $\Gamma_{0,1}$. Taking into account of equality constraints (5.3.7) and (5.3.8), it follows that

$$\bar{x} = \Big(\frac{3}{2} - 2\frac{h_1 + h_2 - \bar{\mu}_2\bar{w}_2}{\bar{\mu}_1\bar{w}_1} + \bar{v}_1^1\Big)t^1 + \Big(\frac{3}{2} - 2\frac{h_1 + h_2 - \bar{\mu}_2\bar{w}_2}{\bar{\mu}_1\bar{w}_1} + \bar{v}_2^1\Big)t^2 + h_3, \ h_3 \in \mathbb{R}_+.$$

By using the boundary condition (5.3.9) along with Lagrange multiplier $\bar{\mu}_1 = 1, \bar{\mu}_2 = 0$ and the uncertain parameters $\bar{w}_1 = \bar{w}_2 = 1, \bar{v}_1^1 = \bar{v}_2^1 = 1$, we get $h_3 = 0$ and $h_1 + h_2 = \frac{3}{4}$, and therefore,

$$\bar{x} = t^1 + t^2, \ \bar{c} = \frac{3}{4}, \ \forall t \in \Gamma_{0,1}.$$

Now, we shall show that $(\bar{x}, \bar{c}) = (t^1 + t^2, \frac{3}{4})$ at $t^1 = t^2 = 0$, i.e. $(\bar{x}, \bar{c}) = (0, \frac{3}{4})$ is a weak robust efficient solution to the problem (MMOPU1). In virtue of Definition 5.2.0.1, for all $(x, c) \in \mathcal{D}$, we have

$$\begin{aligned}
&\int_{\Gamma_{0,1}} \max_{w \in \mathcal{W}} f(\pi, w)dt - \int_{\Gamma_{0,1}} \max_{w \in \mathcal{W}} f(\bar{\pi}, w)dt \\
&\quad = \int_{\Gamma_{0,1}} f(\pi, \bar{w})dt - \int_{\Gamma_{0,1}} f(\bar{\pi}, \bar{w})dt \\
&\quad = \int_{\Gamma_{0,1}} \Big(x^2 + c^2\bar{w}_1, 2\bar{w}_2 c + xe^x\Big)dt^1dt^2 \\
&\qquad\quad - \int_{\Gamma_{0,1}} \Big(\bar{x}^2 + \bar{c}^2\bar{w}_1, 2\bar{w}_2\bar{c} + \bar{x}e^{\bar{x}}\Big)dt^1dt^2 \\
&\quad = \int_{\Gamma_{0,1}} \Big(x^2 + c^2, 2c + xe^x\Big)dt^1dt^2 - \int_{\Gamma_{0,1}} \Big(\frac{9}{16}, \frac{3}{2}\Big)dt^1dt^2 \\
&\quad = \int_{\Gamma_{0,1}} \Big\{x^2 + c^2 - \frac{9}{16}, 2c + xe^x - \frac{3}{2}\Big\}dt^1dt^2. \qquad (5.3.13)
\end{aligned}$$

On solving Eq. (5.3.7), we have $c = \frac{1}{2}\Big(\frac{3}{2} + v_1^1 - x_{t^1}\Big)$. Therefore, the above Eq. (5.3.13) can be written as

$$\begin{aligned}
&\int_{\Gamma_{0,1}} \max_{w \in \mathcal{W}} f(\pi, w)dt - \int_{\Gamma_{0,1}} \max_{w \in \mathcal{W}} f(\bar{\pi}, w)dt \\
&\quad = \int_{\Gamma_{0,1}} \Big\{x^2 + \frac{1}{4}\Big(\frac{3}{2} + v_1^1 - x_{t^1}\Big)^2 - \frac{9}{16}, \Big(\frac{3}{2} + v_1^1 - x_{t^1}\Big) + xe^x - \frac{3}{2}\Big\}dt^1dt^2. \\
&\qquad (5.3.14)
\end{aligned}$$

Now, it remains to find the state function x which satisfies the boundary conditions (5.3.9).

If we set $v_1^1 = v_2^1$, then it follows from the equality constraints (5.3.7) and (5.3.8) that

$$x_{t^1} = x_{t^2}, \ \forall t \in \Gamma_{0,1}.$$

Applying the necessary Euler-Lagrange PDE (see, for instance, [18]) for Eq. (5.3.14), it follows

$$x = \phi(t^1 + t^2) + \psi, \ \phi, \psi \in \mathbb{R}, \ \forall t \in \Gamma_{0,1}.$$

The above solution together with the boundary conditions (5.3.9) imply that $\phi = 1$ and $\psi = 0$. Therefore, $x = t^1 + t^2$. Hence, Eq. (5.3.14) yields

$$\begin{aligned}
&\int_{\Gamma_{0,1}} \max_{w \in \mathcal{W}} f(\pi, w)dt - \int_{\Gamma_{0,1}} \max_{w \in \mathcal{W}} f(\bar{\pi}, w)dt \\
&\quad = \int_{\Gamma_{0,1}} \left\{ (t^1 + t^2)^2 + \frac{1}{4}\left(\frac{1}{2} + v_1^1\right)^2 - \frac{9}{16}, \left(\frac{1}{2} + v_1^1\right) + (t^1 + t^2)e^{(t^1+t^2)} - \frac{3}{2} \right\} dt^1 dt^2 \\
&\quad = \left(\left(\frac{1}{4} + \frac{1}{2}v_1^1\right)^2 + \frac{29}{48}, \ v_1^1 + 3.94 \right) \geqq 0, \ \forall v_1^1 \in \mathcal{V}_1^1.
\end{aligned}$$

Similarly, on solving Eq. (5.3.8), we have $c = \frac{1}{2}\left(\frac{3}{2} + v_2^1 - x_{t^2}\right)$ and the inequality (5.3.13) concludes that $\int_{\Gamma_{0,1}} \max_{w \in \mathcal{W}} f(\pi, w)dt - \int_{\Gamma_{0,1}} \max_{w \in \mathcal{W}} f(\bar{\pi}, w)dt \geq 0$. In both the cases, it shows that $(0, \frac{3}{4})$ is a weak robust efficient solution to the problem (MMOPU) at which the necessary efficiency conditions (5.3.10)–(5.3.12) are satisfied with the Lagrange multipliers $\bar{\mu}_1 = 1, \bar{\mu}_2 = 0, \bar{\nu} = 0$ and $\bar{\gamma}_1^1 + \bar{\gamma}_2^1 = \frac{3}{4}$ and the uncertain parameters $\bar{w}_1 = \bar{w}_2 = 1, \bar{u} = \frac{1}{2}, \bar{v}_1^1 = \bar{v}_2^1 = 1$.

At $(\bar{x}, \bar{c}) = (0, \frac{3}{4})$, we formulate the modified multi-dimensional multi-objective optimization problem $(\text{MMOPU1})_\Omega$ associated to the multi-dimensional multi-objective optimization problem (MMOPU1) as follows:

$$\begin{aligned}
(\text{MMOPU1})_\Omega \quad & \min_{(x(\cdot),c(\cdot))} \int_{\Gamma_{0,1}} \left\{ \frac{3}{2}w_1\left(c - \frac{3}{4}\right), x + 2w_2\left(c - \frac{3}{4}\right) \right\} dt^1 dt^2 \\
\text{subject to} \quad & ux(x-1) \leq 0, \\
& x_{t^1} = \frac{3}{2} - 2c + v_1^1, \\
& x_{t^2} = \frac{3}{2} - 2c + v_2^1, \\
& x(0,0) = 0, \ x(1,1) = 2,
\end{aligned}$$

where $t = (t^1, t^2) \in \Gamma_{0,1}$.
Therefore, the robust counterpart for the problem $(\text{MMOPU1})_\Omega$ is defined as follows:

$$(\text{RMMOPU1})_\Omega \quad \min_{(x(\cdot),c(\cdot))} \int_{\Gamma_{0,1}} \left\{ \max_{w_1 \in \mathcal{W}_1} \left(\frac{3}{2}w_1\Big(c - \frac{3}{4}\Big)\right), \max_{w_2 \in \mathcal{W}_2} \left(x + 2w_2\Big(c - \frac{3}{4}\Big)\right) \right\} dt^1 dt^2$$

$$\text{subject to} \quad ux(x-1) \le 0, \ \forall u \in \mathcal{U},$$

$$x_{t^1} = \frac{3}{2} - 2c + v_1^1, \ \forall v_1^1 \in \mathcal{V}_1^1,$$

$$x_{t^2} = \frac{3}{2} - 2c + v_2^1, \ \forall v_2^1 \in \mathcal{V}_2^1,$$

$$x(0,0) = 0, \ x(1,1) = 2,$$

where $t = (t^1, t^2) \in \Gamma_{0,1}$. Note that the robust feasible solution set to the problem $(\text{MMOPU1})_\Omega$ is the same as the robust feasible solution set associated with the problem (MMOPU1).
In virtue of Definition 5.3.0.1, for all $(x, c) \in \mathcal{D}$, we have

$$\begin{aligned}
&\int_\Gamma \max_{w \in \mathcal{W}} \Big\{(x - \bar{x})f_x(\bar{\pi}, w) + (c - \bar{c})f_c(\bar{\pi}, w)\Big\} dt^1 dt^2 \\
&\qquad - \int_\Gamma \max_{w \in \mathcal{W}} \Big\{(\bar{x} - \bar{x})f_x(\bar{\pi}, w) + (\bar{c} - \bar{c})f_c(\bar{\pi}, w)\Big\} dt^1 dt^2 \\
&= \int_\Gamma \Big\{(x - \bar{x})f_x(\bar{\pi}, \bar{w}) + (c - \bar{c})f_c(\bar{\pi}, \bar{w})\Big\} dt^1 dt^2 \\
&\qquad - \int_\Gamma \Big[(\bar{x} - \bar{x})f_x(\bar{\pi}, \bar{w}) + (\bar{c} - \bar{c})f_c(\bar{\pi}, \bar{w})\Big] dt^1 dt^2 \\
&= \int_\Gamma \Big\{(x - \bar{x})f_x(\bar{\pi}, \bar{w}) + (c - \bar{c})f_c(\bar{\pi}, \bar{w})\Big\} dt^1 dt^2 \\
&= \int_\Gamma \Big\{\frac{3}{2}\bar{w}_1\Big(c - \frac{3}{4}\Big), x + 2\bar{w}_2\Big(c - \frac{3}{4}\Big)\Big\} dt^1 dt^2 \\
&= \int_\Gamma \Big\{\frac{3}{2}\Big(c - \frac{3}{4}\Big), x + 2\Big(c - \frac{3}{4}\Big)\Big\} dt^1 dt^2 \\
&= \Big(\frac{3}{4}\Big(v_1^1 - \frac{1}{2}\Big), v_1^1 + \frac{1}{2}\Big) \ge 0, \ \forall v_1^1 \in [\frac{1}{2}, 1], \\
&\Big(or \ = \Big(\frac{3}{4}\Big(v_2^1 - \frac{1}{2}\Big), v_2^1 + \frac{1}{2}\Big) \ge 0, \ \forall v_2^1 \in [\frac{1}{2}, 1]\Big).
\end{aligned}$$

This shows that $(0, \frac{3}{4})$ is a weak robust efficient solution to the modified multi-dimensional optimization problem $(\text{MMOPU1})_\Omega$. Hence, since the functionals mentioned in Theorem 5.3.0.1 are convex at $(0, \frac{3}{4})$, this case is completed.

Case 2. Assume that the Lagrange multipliers $\bar{\gamma}_\alpha^\tau$, $\tau = 1, \alpha = 1, 2$ are functions of t, say $\bar{\gamma}_1^1 = t^2, \bar{\gamma}_2^1 = t^1$. Thus, by Eq. (5.3.11), we have

$$\bar{c} = \frac{t^2 + t^1 - \bar{\mu}_2 \bar{w}_2}{2\bar{\mu}_1 \bar{w}_1}, \ \forall t \in \Gamma_{0,1}. \tag{5.3.15}$$

On solving Eqs. (5.3.7) and (5.3.8) with the help of Eq. (5.3.15), we have

$$\bar{x} = \frac{5}{2}(t^1 + t^2) - (t^1)^2 - (t^2)^2 - 4t^1t^2 + h_4,\ h_4 \in \mathbb{R}_+,$$

and, by considering the boundary conditions (5.3.9), we get

$$\bar{x} = \frac{5}{2}(t^1 + t^2) - (t^1)^2 - (t^2)^2 - 4t^1t^2,$$

which shows that the state function is not an affine function. Consequently, we get a contradiction with the hypothesis.

Theorem 5.3.0.2 *Let $(\bar{x}, \bar{c})$ be a weak robust efficient solution to the modified multi-dimensional multi-objective optimization problem $(MMOPU)_{\Omega}$. Further, assume that $\max_{w \in \mathcal{W}} f(\pi, w) = f(\pi, \bar{w})$ and the objective functional $\int_{\Gamma} f(\pi, \bar{w})dt$ is convex at $(\bar{x}, \bar{c})$ on $\mathcal{D}$. Then $(\bar{x}, \bar{c})$ is a weak robust efficient solution to the multi-dimensional multi-objective optimization problem (MMOPU).*

Proof Contrary to the result, we assume that $(\bar{x}, \bar{c})$ is not a weak robust efficient solution to multi-dimensional multi-objective optimization problem (MMOPU). Then, there exists another point $(\tilde{x}, \tilde{c}) \in \mathcal{D}$ such that

$$\int_{\Gamma} \max_{w \in \mathcal{W}} f(\tilde{\pi}, w)dt < \int_{\Gamma} \max_{w \in \mathcal{W}} f(\bar{\pi}, w)dt.$$

Since $\max_{w \in \mathcal{W}} f(\pi, w) = f(\pi, \bar{w})$, we get

$$\int_{\Gamma} f(\tilde{\pi}, \bar{w})dt < \int_{\Gamma} f(\bar{\pi}, \bar{w})dt. \tag{5.3.16}$$

By assumption, since $\int_{\Gamma} f(\pi, \bar{w})dt$ is convex functional at $(\bar{x}, \bar{c})$ on $\mathcal{D}$, we have

$$\int_{\Gamma} \Big\{ f(\tilde{\pi}, \bar{w}) - f(\bar{\pi}, \bar{w}) \Big\} dt \geqq \int_{\Gamma} (\tilde{x} - \bar{x}) f_x(\bar{\pi}, \bar{w})dt + \int_{\Gamma} (\tilde{c} - \bar{c}) f_c(\bar{\pi}, \bar{w})dt,$$

which along with the inequality (5.3.16) yields

$$\int_{\Gamma} (\tilde{x} - \bar{x}) f_x(\bar{\pi}, \bar{w})dt + \int_{\Gamma} (\tilde{c} - \bar{c}) f_c(\bar{\pi}, \bar{w})dt < 0,$$

equivalently with

$$\int_{\Gamma} \Big\{ (\tilde{x} - \bar{x}) f_x(\bar{\pi}, \bar{w}) + (\tilde{c} - \bar{c}) f_c(\bar{\pi}, \bar{w}) \Big\} dt < \int_{\Gamma} \Big\{ (\bar{x} - \bar{x}) f_x(\bar{\pi}, \bar{w}) + (\bar{c} - \bar{c}) f_c(\bar{\pi}, \bar{w}) \Big\} dt.$$

The above inequality can be written as

$$\int_{\Gamma} \max_{w \in \mathcal{W}} \Big\{ (\tilde{x} - \bar{x}) f_x(\bar{\pi}, w) + (\tilde{c} - \bar{c}) f_c(\bar{\pi}, w) \Big\} dt$$
$$< \int_{\Gamma} \max_{w \in \mathcal{W}} \Big\{ (\bar{x} - \bar{x}) f_x(\bar{\pi}, w) + (\bar{c} - \bar{c}) f_c(\bar{\pi}, w) \Big\} dt,$$

which contradicts our assumption that $(\bar{x}, \bar{c})$ is a weak robust efficient solution to the modified multi-dimensional multi-objective optimization problem $(\text{MMOPU})_\Omega$. Hence, this completes the proof. □

Remark 5.3.0.2 It has been observed that to derive the equivalence between the original problem (MMOPU) and its associated modified problem $(\text{MMOPU})_\Omega$, we imposed the convexity assumption only on the constraint functionals. Whereas, to establish the converse of Theorem 5.3.0.1, it is sufficient to assume the convexity assumption of the objective functional.

Theorem 5.3.0.3 *Let $(\bar{x}, \bar{c})$ be a robust feasible solution to the multi-dimensional multi-objective optimization problem (MMOPU). Assume that $\max_{w \in \mathcal{W}} f(\pi, w) = f(\pi, \bar{w})$ and the objective functional $\int_\Gamma f(\pi, \bar{w}) dt$ is convex at $(\bar{x}, \bar{c})$ on $\mathcal{D}$. If $((\bar{x}, \bar{c}), \bar{\nu}, \bar{\gamma}, \bar{w}, \bar{u}, \bar{v})$ is a robust saddle-point to the modified multi-dimensional multi-objective optimization problem $(MMOPU)_\Omega$, then $(\bar{x}, \bar{c})$ is a weak robust efficient solution to the multi-dimensional multi-objective optimization problem (MMOPU).*

Proof Contrary to the result and assume that $(\bar{x}, \bar{c})$ is not a weak robust efficient solution to the multi-dimensional multi-objective optimization problem (MMOPU). Proceeding on the line of Theorem 5.3.0.2, we get

$$\int_{\Gamma} \max_{w \in \mathcal{W}} \Big\{ (\tilde{x} - \bar{x}) f_x(\bar{\pi}, w) + (\tilde{c} - \bar{c}) f_c(\bar{\pi}, w) \Big\} dt$$
$$< \int_{\Gamma} \max_{w \in \mathcal{W}} \Big\{ (\bar{x} - \bar{x}) f_x(\bar{\pi}, w) + (\bar{c} - \bar{c}) f_c(\bar{\pi}, w) \Big\} dt, \text{ for some } (\tilde{x}, \tilde{c}) \in \mathcal{D}. \tag{5.3.17}$$

From the feasibility of $(\tilde{x}, \tilde{c})$ to the problem (MMOPU) and $\bar{v} \in \mathbb{R}^m_+$, we get

$$\int_{\Gamma} \bar{\nu}^T G(\tilde{\pi}, \bar{u}) dt \leqq 0. \tag{5.3.18}$$

On the other hand, by assumption, $((\bar{x}, \bar{c}), \bar{\nu}, \bar{\gamma}, \bar{w}, \bar{u}, \bar{v})$ is a robust saddle-point to the modified multi-dimensional multi-objective optimization problem $(\text{MMOPU})_\Omega$. Then, from Definition 5.3.0.3 (i), we have

$$L((\bar{x}, \bar{c}), \nu, \gamma, w, u, v) \leqq L((\bar{x}, \bar{c}), \bar{\nu}, \bar{\gamma}, w, \bar{u}, \bar{v}), \ \forall \nu \in \mathbb{R}^m_+, \ \forall \gamma \in \mathbb{R}^{pq}, \forall u \in \mathcal{U}, \ \forall v \in \mathcal{V},$$

which in virtue of the definition of Lagrange functional can be written as

$$\int_{\Gamma}\left\{\max_{w\in\mathcal{W}}\left[(\tilde{x}-\bar{x})f_x(\bar{\pi},w)+(\tilde{c}-\bar{c})f_c(\bar{\pi},w)\right]+\nu^T G(\tilde{\pi},u)e+\gamma^T\left(H(\tilde{\pi},v)-x_\alpha\right)e\right\}dt$$
$$\leqq\int_{\Gamma}\left\{\max_{w\in\mathcal{W}}\left[(\tilde{x}-\bar{x})f_x(\bar{\pi},w)+(\tilde{c}-\bar{c})f_c(\bar{\pi},w)\right]+\bar{\nu}^T G(\bar{\pi},\bar{u})e+\bar{\gamma}^T\left(H(\tilde{\pi},\bar{v})-x_\alpha\right)e\right\}dt.$$

Since $\max_{w\in\mathcal{W}} f(\pi,w)=f(\pi,\bar{w})$, it follows that

$$\int_{\Gamma}\left\{(\tilde{x}-\bar{x})f_x(\bar{\pi},\bar{w})+(\tilde{c}-\bar{c})f_c(\bar{\pi},\bar{w})+\nu^T G(\tilde{\pi},u)e+\gamma^T\left(H(\tilde{\pi},v)-x_\alpha\right)e\right\}dt$$
$$\leqq\int_{\Gamma}\left\{(\tilde{x}-\bar{x})f_x(\bar{\pi},\bar{w})+(\tilde{c}-\bar{c})f_c(\bar{\pi},\bar{w})+\bar{\nu}^T G(\bar{\pi},\bar{u})e+\bar{\gamma}^T\left(H(\tilde{\pi},\bar{v})-x_\alpha\right)e\right\}dt.$$

If we set $\nu=0$ and $\gamma=0$, we obtain

$$\int_{\Gamma}\bar{\nu}^T G(\bar{\pi},\bar{u})dt\geqq 0. \tag{5.3.19}$$

From (5.3.18) and (5.3.19), it follows that

$$\int_{\Gamma}\bar{\nu}^T G(\tilde{\pi},\bar{u})dt\leqq\int_{\Gamma}\bar{\nu}^T G(\bar{\pi},\bar{u})dt,$$

which along with the inequality (5.3.17) gives

$$\int_{\Gamma}\left\{\max_{w\in\mathcal{W}}\left[(\tilde{x}-\bar{x})f_x(\bar{\pi},\bar{w})+(\tilde{c}-\bar{c})f_c(\bar{\pi},\bar{w})\right]+\bar{\nu}^T G(\tilde{\pi},\bar{u})e\right\}dt$$
$$<\int_{\Gamma}\left\{\max_{w\in\mathcal{W}}\left[(\bar{x}-\bar{x})f_x(\bar{\pi},\bar{w})+(\bar{c}-\bar{c})f_c(\bar{\pi},\bar{w})\right]+\bar{\nu}^T G(\bar{\pi},\bar{u})e\right\}dt,$$

equivalently with

$$\int_{\Gamma}\left\{\max_{w\in\mathcal{W}}\left[(\tilde{x}-\bar{x})f_x(\bar{\pi},w)+(\tilde{c}-\bar{c})f_c(\bar{\pi},w)\right]+\bar{\nu}^T G(\tilde{\pi},\bar{u})e+\bar{\gamma}^T\left(H(\tilde{\pi},\bar{v})-x_\alpha\right)e\right\}dt$$
$$<\int_{\Gamma}\left\{\max_{w\in\mathcal{W}}\left[(\bar{x}-\bar{x})f_x(\bar{\pi},w)+(\bar{c}-\bar{c})f_c(\bar{\pi},w)\right]+\bar{\nu}^T G(\bar{\pi},\bar{u})e+\bar{\gamma}^T\left(H(\bar{\pi},\bar{v})-x_\alpha\right)e\right\}dt.$$

In virtue of the definition of Lagrange functional, it can be written as

$$L((\tilde{x},\tilde{c}),\bar{\nu},\bar{\gamma},\bar{w},\bar{u},\bar{v})\leqq L((\bar{x},\bar{c}),\bar{\nu},\bar{\gamma},\bar{w},\bar{u},\bar{v}),\ \forall(\tilde{x},\tilde{c})\in\mathscr{X}\times\mathscr{C},$$

which contradicts Definition 5.3.0.3. Hence, this completes the proof. □

5.4 The Exact l_1 Penalty Function Method

In this section, we use the exact l_1 penalty function method to solve the modified multi-dimensional multi-objective optimization problem $(\text{MMOPU})_\Omega$. We also establish the equivalence between the robust solution sets to the modified problem $(\text{MMOPU})_\Omega$ and its associated penalized problem $(\text{MMOPU})_{\Omega\rho}$ via robust saddle-point theory.

The penalized multi-dimensional multi-objective optimization problem $(\text{MMOPU})_{\Omega\rho}$ associated with the modified problem $(\text{MMOPU})_\Omega$ is defined as follows:

$$(\text{MMOPU})_{\Omega\rho} \quad \min_{(x(\cdot),c(\cdot))} \mathscr{P}(x,c,w,u,v,\rho) = \int_{\Gamma_{t_0,t_1}} \Big\{(x(t)-\bar{x}(t))f_x(t,\bar{x}(t),\bar{c}(t),w)$$
$$+ (c(t)-\bar{c}(t))f_c(t,\bar{x}(t),\bar{c}(t),w) + \rho(t)\Big[\sum_{l=1}^{m} G_l^+(t,x(t),c(t),u)$$
$$+ \sum_{\tau=1}^{q}\sum_{\alpha=1}^{p}\Big|H_\alpha^\tau(t,x(t),c(t),v) - \frac{\partial x^\tau}{\partial t^\alpha}(t)\Big|\Big]e\Big\}dt,$$

where $e=(1,\ldots,1)\in\mathbb{R}^s$, $w\in\mathcal{W}$, $u\in\mathcal{U}$ and $v\in\mathcal{V}$ are the uncertain parameters, $\rho(t)$ is a non-zero non-negative penalty parameter and the penalty function $G_l^+(t,x(t),c(t),u)$ is defined as

$$G_l^+(t,x(t),c(t),u) = \begin{cases} 0, & \text{if } G_l(t,x(t),c(t),u) \leqq 0, \\ G_l(t,x(t),c(t),u), & \text{if } G_l(t,x(t),c(t),u) > 0. \end{cases} \tag{5.4.1}$$

The associated robust counterpart to the problem $(\text{MMOPU})_{\Omega\rho}$ is given by

$$(\text{RMMOPU})_{\Omega\rho} \quad \min_{(x(\cdot),c(\cdot))}\max_{w\in\mathcal{W}} \mathscr{P}(x,c,w,u,v,\rho) = \int_{\Gamma_{t_0,t_1}} \Big\{\max_{w\in\mathcal{W}}\Big((x(t)-\bar{x}(t))f_x(t,\bar{x}(t),\bar{c}(t),w)$$
$$+ (c(t)-\bar{c}(t))f_c(t,\bar{x}(t),\bar{c}(t),w)\Big) + \rho(t)\Big[\sum_{l=1}^{m} G_l^+(t,x(t),c(t),u)$$
$$+ \sum_{\tau=1}^{q}\sum_{\alpha=1}^{p}\Big|H_\alpha^\tau(t,x(t),c(t),v) - \frac{\partial x^\tau}{\partial t^\alpha}(t)\Big|\Big]e\Big\}dt,$$

for all $u\in\mathcal{U}$ and $v\in\mathcal{V}$.

Definition 5.4.0.1 A point $(\bar{x},\bar{c})$ is said to be a weak robust efficient solution to $(\text{MMOPU})_{\Omega\rho}$, if there does not exist another point $(x,c)\in\mathscr{X}\times\mathscr{C}$ such that

$$\max_{w\in\mathcal{W}} \mathscr{P}(x,c,w,\bar{u},\bar{v},\rho) < \max_{w\in\mathcal{W}} \mathscr{P}(\bar{x},\bar{c},w,\bar{u},\bar{v},\rho),$$

where $\bar{u}\in\mathcal{U}$ and $\bar{v}\in\mathcal{V}$.

Now, we shall establish the equivalence between the robust solution set to the problem $(\text{MMOPU})_{\Omega}$ and its associated penalized problem $(\text{MMOPU})_{\Omega\rho}$ via saddle-point criteria.

Theorem 5.4.0.1 *Let* $((\bar{x}, \bar{c}), \bar{v}, \bar{\gamma}, \bar{w}, \bar{u}, \bar{v})$ *be a robust saddle-point to the modified multi-dimensional multi-objective optimization problem* $(\textit{MMOPU})_{\Omega}$ *and* $\max_{w\in\mathcal{W}} f(\pi, w) = f(\pi, \bar{w})$. *If* $\rho \geq \max\{\bar{v}, |\bar{\gamma}|\})$, *then* $(\bar{x}, \bar{c})$ *is a weak robust efficient solution to the penalized multi-dimensional multi-objective optimization problem* $(\textit{MMOPU})_{\Omega\rho}$.

Proof Contrary to the result and assume that $(\bar{x}, \bar{c})$ is not a weak robust efficient solution to the penalized problem $(\text{MMOPU})_{\Omega\rho}$, and therefore, there exists $(\hat{x}, \hat{c}) \in \mathscr{X} \times \mathscr{C}$ such that

$$\max_{w\in\mathcal{W}} \mathscr{P}(\hat{x}, \hat{c}, w, \bar{u}, \bar{v}, \rho) < \max_{w\in\mathcal{W}} \mathscr{P}(\bar{x}, \bar{c}, w, \bar{u}, \bar{v}, \rho).$$

In virtue of the definition of the penalized multi-dimensional multi-objective optimization problem $(\text{MMOPU})_{\Omega\rho}$, we get

$$\int_{\Gamma}\Big\{\max_{w\in\mathcal{W}}\Big[(\hat{x}-\bar{x})f_x(\bar{\pi}, w) + (\hat{c}-\bar{c})f_c(\bar{\pi}, w)\Big] + \rho\Big[\sum_{l=1}^{m} G_l^{+}(\hat{\pi}, \bar{u})$$
$$+ \sum_{\tau=1}^{q}\sum_{\alpha=1}^{p}\Big|H_{\alpha}^{\tau}(\hat{\pi}, \bar{v}) - \frac{\partial x^{\tau}}{\partial t^{\alpha}}\Big|\Big]\Big\}dt < \int_{\Gamma}\Big\{\max_{w\in\mathcal{W}}\Big[(\bar{x}-\bar{x})f_x(\bar{\pi}, w)$$
$$+ (\bar{c}-\bar{c})f_c(\bar{\pi}, w)\Big] + \rho\Big[\sum_{l=1}^{m} G_l^{+}(\bar{\pi}, \bar{u}) + \sum_{\tau=1}^{q}\sum_{\alpha=1}^{p}\Big|H_{\alpha}^{\tau}(\bar{\pi}, \bar{v}) - \frac{\partial x^{\tau}}{\partial t^{\alpha}}\Big|\Big]\Big\}dt.$$

Since $\max_{w\in\mathcal{W}} f(\pi, w) = f(\pi, \bar{w})$, we have

$$\int_{\Gamma}\Big\{\Big[(\hat{x}-\bar{x})f_x(\bar{\pi}, \bar{w}) + (\hat{c}-\bar{c})f_c(\bar{\pi}, \bar{w})\Big] + \rho\Big[\sum_{l=1}^{m} G_l^{+}(\hat{\pi}, \bar{u})$$
$$+ \sum_{\tau=1}^{q}\sum_{\alpha=1}^{p}\Big|H_{\alpha}^{\tau}(\hat{\pi}, \bar{v}) - \frac{\partial x^{\tau}}{\partial t^{\alpha}}\Big|\Big]\Big\}dt < \int_{\Gamma}\Big\{\Big[(\bar{x}-\bar{x})f_x(\bar{\pi}, \bar{w})$$
$$+ (\bar{c}-\bar{c})f_c(\bar{\pi}, \bar{w})\Big] + \rho\Big[\sum_{l=1}^{m} G_l^{+}(\bar{\pi}, \bar{u}) + \sum_{\tau=1}^{q}\sum_{\alpha=1}^{p}\Big|H_{\alpha}^{\tau}(\bar{\pi}, \bar{v}) - \frac{\partial x^{\tau}}{\partial t^{\alpha}}\Big|\Big]\Big\}dt.$$

The above inequality together with the robust feasibility of $(\bar{x}, \bar{c})$ in the multi-dimensional multi-objective optimization problem $(\text{MMOPU})_{\Omega}$ and the relation (5.4.1) yields

$$\int_{\Gamma}\Big\{(\hat{x}-\bar{x})f_x(\bar{\pi},\bar{w})+(\hat{c}-\bar{c})f_c(\bar{\pi},\bar{w})+\rho\Big[\sum_{l=1}^{m}G_l^{+}(\hat{\pi},\bar{u})$$
$$+\sum_{\tau=1}^{q}\sum_{\alpha=1}^{p}\left|H_\alpha^\tau(\hat{\pi},\bar{v})-\frac{\partial x^\tau}{\partial t^\alpha}\right|\Big]\Big\}dt<\int_{\Gamma}\Big\{(\bar{x}-\bar{x})f_x(\bar{\pi},\bar{w})+(\bar{c}-\bar{c})f_c(\bar{\pi},\bar{w})\Big\}dt.$$

Since $\rho \geq \max\{\bar{\nu}, |\bar{\gamma}|\}$, we get

$$\int_{\Gamma}\Big\{(\hat{x}-\bar{x})f_x(\bar{\pi},\bar{w})+(\hat{c}-\bar{c})f_c(\bar{\pi},\bar{w})+\sum_{l=1}^{m}\bar{\nu}_l G_l^{+}(\hat{\pi},\bar{u})$$
$$+\sum_{\tau=1}^{q}\sum_{\alpha=1}^{p}\left|\bar{\gamma}_\alpha^\tau\Big(H_\alpha^\tau(\hat{\pi},\bar{v})-\frac{\partial x^\tau}{\partial t^\alpha}\Big)\right|\Big\}dt<\int_{\Gamma}\Big\{(\bar{x}-\bar{x})f_x(\bar{\pi},\bar{w})+(\bar{c}-\bar{c})f_c(\bar{\pi},\bar{w})\Big\}dt.$$

Again, using the relation (5.4.1) and the definition of the modulus functional, we have

$$\int_{\Gamma}\Big\{(\hat{x}-\bar{x})f_x(\bar{\pi},\bar{w})+(\hat{c}-\bar{c})f_c(\bar{\pi},\bar{w})+\sum_{l=1}^{m}\bar{\nu}_l G_l(\hat{\pi},\bar{u})+\sum_{\tau=1}^{q}\sum_{\alpha=1}^{p}\bar{\gamma}_\alpha^\tau\Big(H_\alpha^\tau(\hat{\pi},\bar{v})-\frac{\partial x^\tau}{\partial t^\alpha}\Big)\Big\}dt$$
$$<\int_{\Gamma}\Big\{(\bar{x}-\bar{x})f_x(\bar{\pi},\bar{w})+(\bar{c}-\bar{c})f_c(\bar{\pi},\bar{w})\Big\}dt,$$

equivalently,

$$\int_{\Gamma}\Big\{(\hat{x}-\bar{x})f_x(\bar{\pi},\bar{w})+(\hat{c}-\bar{c})f_c(\bar{\pi},\bar{w})+\bar{\nu}^T G(\hat{\pi},\bar{u})+\bar{\gamma}^T\Big(H(\hat{\pi},\bar{v})-x_\alpha\Big)\Big\}dt$$
$$<\int_{\Gamma}\Big\{(\bar{x}-\bar{x})f_x(\bar{\pi},\bar{w})+(\bar{c}-\bar{c})f_c(\bar{\pi},\bar{w})\Big\}dt. \qquad (5.4.2)$$

On the other hand, since $((\bar{x},\bar{c}),\bar{\nu},\bar{\gamma},\bar{w},\bar{u},\bar{v})$ is a robust saddle-point to the modified multi-dimensional multi-objective optimization problem (MMOPU)$_\Omega$, then, from Definition 5.3.0.3 (i), we have

$$L((\bar{x},\bar{c}),\nu,\gamma,w,u,v)\leqq L((\bar{x},\bar{c}),\bar{\nu},\bar{\gamma},w,\bar{u},\bar{v}),\ \forall\nu\in\mathbb{R}_{+}^{m},\ \forall\gamma\in\mathbb{R}^{pq},\ \forall u\in\mathcal{U},\ \forall v\in\mathcal{V},$$

which in virtue of the definition of Lagrange functional can be written as

$$\int_{\Gamma}\Big\{\max_{w\in\mathcal{W}}\Big[(\bar{x}-\bar{x})f_x(\bar{\pi},w)+(\bar{c}-\bar{c})f_c(\bar{\pi},w)\Big]+\nu^T G(\bar{\pi},u)e+\gamma^T\Big(H(\bar{\pi},v)-x_\alpha\Big)e\Big\}dt$$
$$\leqq\int_{\Gamma}\Big\{\max_{w\in\mathcal{W}}\Big[(\bar{x}-\bar{x})f_x(\bar{\pi},w)+(\bar{c}-\bar{c})f_c(\bar{\pi},w)\Big]+\bar{\nu}^T G(\bar{\pi},\bar{u})e+\bar{\gamma}^T\Big(H(\bar{\pi},\bar{v})-x_\alpha\Big)e\Big\}dt.$$

Since $\max_{w\in\mathcal{W}} f(\pi,w)=f(\pi,\bar{w})$, it follows that

$$\int_{\Gamma}\Big\{(\bar{x}-\bar{x})f_x(\bar{\pi},\bar{w})+(\bar{c}-\bar{c})f_c(\bar{\pi},\bar{w})+v_l^T G(\bar{\pi},u)e+\gamma^T\Big(H(\bar{\pi},v)-x_\alpha\Big)e\Big\}dt$$
$$\leqq \int_{\Gamma}\Big\{(\bar{x}-\bar{x})f_x(\bar{\pi},\bar{w})+(\bar{c}-\bar{c})f_c(\bar{\pi},\bar{w})+\bar{v}_l^T G(\bar{\pi},\bar{u})e+\bar{\gamma}^T\Big(H(\bar{\pi},\bar{v})-x_\alpha\Big)e\Big\}dt.$$

We set $\nu = 0$ and $\gamma = 0$, and we obtain

$$\int_{\Gamma}\bar{v}^T G(\bar{\pi},\bar{u})dt \geqq 0. \tag{5.4.3}$$

Also, from the robust feasibility of $(\bar{x},\bar{c})$ to the problem $(\text{MMOPU})_\Omega$, we have

$$\int_{\Gamma}\bar{v}^T G(\bar{\pi},\bar{u})dt \leqq 0. \tag{5.4.4}$$

On combining the inequalities (5.4.3) and (5.4.4), we get $\int_{\Gamma}\bar{v}^T G(\bar{\pi},\bar{u})dt = 0$. Further, the robust feasibility of $(\bar{x},\bar{c})$ to the problem $(\text{MMOPU})_\Omega$ also admits that $\int_{\Gamma}\bar{\gamma}^T(H(\bar{\pi},\bar{u})-x_\alpha)dt = 0$. Therefore, the inequality (5.4.2) can be written as

$$\int_{\Gamma}\Big\{(\hat{x}-\bar{x})f_x(\bar{\pi},\bar{w})+(\hat{c}-\bar{c})f_c(\bar{\pi},\bar{w})+\bar{v}^T G(\hat{\pi},\bar{u})+\bar{\gamma}^T\Big(H(\hat{\pi},\bar{v})-x_\alpha\Big)\Big\}dt$$
$$< \int_{\Gamma}\Big\{(\bar{x}-\bar{x})f_x(\bar{\pi},\bar{w})+(\bar{c}-\bar{c})f_c(\bar{\pi},\bar{w})\Big\}+\bar{v}^T G(\bar{\pi},\bar{u})+\bar{\gamma}^T\Big(H(\bar{\pi},\bar{v})-x_\alpha\Big)\Big\}dt.$$

From the virtue of the definition of the Lagrange functional, we get

$$L((\hat{x},\hat{c}),\bar{v},\bar{\gamma},w,\bar{u},\bar{v}) < L((\bar{x},\bar{c}),\bar{v},\bar{\gamma},w,\bar{u},\bar{v}),\ \forall(\hat{x},\hat{c})\in\mathscr{X}\times\mathscr{C},$$

which contradicts Definition 5.3.0.3 (ii). Hence, this completes the proof. □

Now, we give an example to illustrate the above theorem.

Example 5.4.0.1 Let $p=2, q=r=1, k=2, \mathcal{W}_1=[0,1], \mathcal{W}_2=[-1,1], \mathcal{U}=[1,2], \mathcal{V}_1^1=\mathcal{V}_2^1=[\frac{1}{4},1]$ and Γ_{t_0,t_1} is fixed by the diagonally opposite points $t_0=(t_0^1,t_0^2)$ and $t_1=(t_1^1,t_1^2)$ in $\mathbb{R}^2$.

We consider the following multi-dimensional multi-objective optimization problem:

$$(\text{MMOPU2}) \quad \min_{(x(\cdot),c(\cdot))}\int_{\Gamma_{0,\frac{1}{2}}}\Big(x\tan^{-1}x+c^2w_1,\, x+c+w_2\Big)\,dt^1dt^2$$

$$\text{subject to}\quad x^2-4u^2\leq 0, \tag{5.4.5}$$
$$x_{t^1}=2-cv_1^1, \tag{5.4.6}$$
$$x_{t^2}=2-cv_2^1, \tag{5.4.7}$$
$$x(0,0)=0,\ x\Big(\frac{1}{2},\frac{1}{2}\Big)=1, \tag{5.4.8}$$

where $t=(t^1,t^2)\in\Gamma_{0,\frac{1}{2}}$.

The robust counterpart to the problem (MMOPU2) is formulated as follows:

$$\text{(RMMOPU2)} \quad \min_{(x(\cdot),c(\cdot))} \int_{\Gamma_{0,\frac{1}{2}}} \Big(\max_{w_1 \in \mathcal{W}_1} (x \tan^{-1} x + c^2 w_1), \max_{w_2 \in \mathcal{W}_2} (x + c + w_2) \Big) \, dt^1 dt^2$$

$$\begin{aligned}
\text{subject to} \quad & x^2 - 4u^2 \leq 0, \ \forall u \in \mathcal{U}, \\
& x_{t^1} = 2 - c v_1^1, \ \forall v_1^1 \in \mathcal{V}_1^1, \\
& x_{t^2} = 2 - c v_2^1, \ \forall v_2^1 \in \mathcal{V}_2^1, \\
& x(0,0) = 0, \ x\Big(\frac{1}{2}, \frac{1}{2}\Big) = 1,
\end{aligned}$$

where $t = (t^1, t^2) \in \Gamma_{0,\frac{1}{2}}$. Clearly, $\mathcal{D} = \{(x, c) \in \mathcal{X} \times \mathcal{C} : -2 \leq x \leq 2 \wedge x_{t^1} - 2 + v_1^1 c = 0 \wedge x_{t^2} - 2 + v_2^1 c = 0, \ x(0,0) = 0, x\big(\frac{1}{2}, \frac{1}{2}\big) = 1\}$ is a robust feasible solution set to (MMOPU2).

Now, we find a weak robust efficient solution to the problem (MMOPU2), i.e. find the control function $\bar{c} : \Gamma_{0,\frac{1}{2}} \to \mathbb{R}$ (that determines the state function $\bar{x} : \Gamma_{0,\frac{1}{2}} \to \mathbb{R}$) satisfying the dynamical systems (5.4.6) and (5.4.7) with respect to boundary conditions (5.4.8). Additionally, we assume that the state function is affine.

Let $(\bar{x}, \bar{c})$ be a robust feasible solution to the problem (MMOPU2), the Lagrange multipliers $\bar{\mu}_1, \bar{\mu}_2, \bar{\nu}, \bar{\gamma}_1^1, \bar{\gamma}_2^1$, and the uncertain parameters $\bar{u}, \bar{v}_1^1, \bar{v}_2^1$. Let $\max_{w \in \mathcal{W}} f(\pi, w) = f(\pi, \bar{w})$. Therefore, the robust necessary efficiency conditions are given as follows:

$$\bar{\mu}_1 \Big(\frac{\bar{x}}{1 + \bar{x}^2} + \tan^1(\bar{x}) \Big) + \bar{\mu}_2 + 2\bar{\nu}\bar{x} + (\bar{\gamma}_1^1)_{t^1} + (\bar{\gamma}_2^1)_{t^2} = 0, \tag{5.4.9}$$

$$2\bar{\mu}_2 \bar{c} \bar{w}_1 + \bar{\mu}_2 - \bar{\gamma}_1^1 \bar{v}_1^1 - \bar{\gamma}_2^1 \bar{v}_2^1 = 0, \tag{5.4.10}$$

$$\bar{\nu}(\bar{x}^2 - 4\bar{u}^2) = 0. \tag{5.4.11}$$

Clearly, the relation (5.4.11) is satisfied if either $\bar{\nu} = 0$ or $\bar{x}^2 - 4\bar{u}^2 = 0$, for all $t = (t^1, t^2) \in \text{int}\Gamma_{0,\frac{1}{2}}$ (interior of hyperparallelepiped Γ). Let $\bar{\nu} \neq 0$, then $\bar{x}^2 - 4\bar{u}^2 = 0$ implies that $\bar{x} = \pm 2\bar{u}$. The boundary conditions (5.4.8) are failing at $\bar{x} = \pm 4$ with $\bar{u} = 2$. Consequently, $\bar{\nu} = 0$.

Case 1. Assume that the Lagrange multipliers $\bar{\gamma}_\alpha^\tau, \tau = 1, \alpha = 1, 2$ are constant functions, say $\bar{\gamma}_1^1 = h_1$ and $\bar{\gamma}_2^1 = h_2$, where $h_1, h_2 \in \mathbb{R}_+$. Therefore, Eq. (5.4.10) gives

$$\bar{c} = \frac{\bar{v}_1^1 h_1 + \bar{v}_2^1 h_2 - \bar{\mu}_2}{2\bar{\mu}_1 \bar{w}_1}, \ \forall t \in \Gamma_{0,\frac{1}{2}},$$

which shows that $\bar{c}$ is a constant function on $\Gamma_{0,\frac{1}{2}}$. Taking into account of equality constraints (5.4.6) and (5.4.7), it follows that

$$\bar{x}=\Big(2-\bar{v}_1^1\frac{\bar{v}_1^1h_1+\bar{v}_2^1h_2-\bar{\mu}_2}{2\bar{\mu}_1\bar{w}_1}\Big)t^1+\Big(2-\bar{v}_2^1\frac{\bar{v}_1^1h_1+\bar{v}_2^1h_2-\bar{\mu}_2}{2\bar{\mu}_1\bar{w}_1}\Big)t^2+h_3,\ h_3\in\mathbb{R}_+.$$

By using the boundary condition (5.4.8) along with Lagrange multiplier $\bar{\mu}_1=1, \bar{\mu}_2=0$ and the uncertain parameters $\bar{w}_2=1, \bar{v}_1^1=\bar{v}_2^1=1$, we get $h_3=0$ and $h_1+h_2=2$, and therefore,

$$\bar{x}=t^1+t^2,\ \bar{c}=1,\ \forall t\in\Gamma_{0,\frac{1}{2}}.$$

Now, we shall show that $(\bar{x},\bar{c})=(t^1+t^2,1)$ at $t^1=t^2=0$, i.e. $(\bar{x},\bar{c})=(0,1)$ is a weak robust efficient solution to the problem (MMOPU2). In virtue of Definition 5.2.0.1, for all $(x,c)\in\mathcal{D}$, we have

$$\begin{aligned}
&\int_{\Gamma_{0,\frac{1}{2}}}\max_{w\in\mathcal{W}}f(\pi,w)dt-\int_{\Gamma_{0,\frac{1}{2}}}\max_{w\in\mathcal{W}}f(\bar{\pi},w)dt\\
&=\int_{\Gamma_{0,\frac{1}{2}}}f(\pi,\bar{w})dt-\int_{\Gamma_{0,\frac{1}{2}}}f(\bar{\pi},\bar{w})dt\\
&=\int_{\Gamma_{0,\frac{1}{2}}}\Big(x\tan^{-1}x+c^2\bar{w}_1,x+c+\bar{w}_2\Big)dt^1dt^2\\
&\quad-\int_{\Gamma_{0,\frac{1}{2}}}\Big(\bar{x}\tan^{-1}\bar{x}+\bar{c}^2\bar{w}_1,\bar{x}+\bar{c}+\bar{w}_2\Big)dt^1dt^2\\
&=\int_{\Gamma_{0,\frac{1}{2}}}\Big(x\tan^{-1}x+c^2,x+c+1\Big)dt^1dt^2-\int_{\Gamma_{0,\frac{1}{2}}}\Big(1,2\Big)dt^1dt^2\\
&=\int_{\Gamma_{0,\frac{1}{2}}}\Big\{x\tan^{-1}x+c^2-1,x+c-1\Big\}dt^1dt^2.
\end{aligned}\tag{5.4.12}$$

On solving Eq. (5.4.6), we have $c=\frac{1}{v_1^1}(2-x_{t^1})$. Therefore, the above Eq. (5.4.12) can be written as

$$\begin{aligned}
&\int_{\Gamma_{0,\frac{1}{2}}}\max_{w\in\mathcal{W}}f(\pi,w)dt-\int_{\Gamma_{0,\frac{1}{2}}}\max_{w\in\mathcal{W}}f(\bar{\pi},w)dt\\
&=\int_{\Gamma_{0,\frac{1}{2}}}\Big\{x\tan^{-1}x+\Big(\frac{1}{v_1^1}(2-x_{t^1})\Big)^2-1,x+\frac{1}{v_1^1}(2-x_{t^1})-1\Big\}dt^1dt^2.
\end{aligned}\tag{5.4.13}$$

Now, it remains to find the state function x which satisfies the boundary conditions (5.4.8).

If we set $v_1^1=v_2^2$, then it follows from the equality constraints (5.4.6) and (5.4.7) that

$$x_{t^1} = x_{t^2}, \ \forall t \in \Gamma_{0,\frac{1}{2}}.$$

Applying the necessary Euler-Lagrange PDE (see, for instance, [18]) for Eq. (5.4.13), it follows

$$x = \phi(t^1 + t^2) + \psi, \ \phi, \psi \in \mathbb{R}, \ \forall t \in \Gamma_{0,\frac{1}{2}}.$$

The above solution together with the boundary conditions (5.4.8) imply that $\phi = 1$ and $\psi = 0$. Therefore, $x = t^1 + t^2$. Hence, Eq. (5.4.13) yields

$$\begin{aligned}
&\int_{\Gamma_{0,\frac{1}{2}}} \max_{w \in \mathcal{W}} f(\pi, w)dt - \int_{\Gamma_{0,\frac{1}{2}}} \max_{w \in \mathcal{W}} f(\bar{\pi}, w)dt \\
&\quad = \int_{\Gamma_{0,\frac{1}{2}}} \left\{(t^1 + t^2)\tan^{-1}(t^1 + t^2) + \left(\frac{1}{v_1^1}\right)^2 - 1, (t^1 + t^2) + \frac{1}{v_1^1} - 1\right\} dt^1 dt^2 \\
&\quad = \left(\frac{1}{4(v_1^1)^2} - 0.185, \ \frac{1}{4v_1^1} - \frac{1}{8}\right) \geq 0, \ \forall v_1^1 \in \mathcal{V}_1^1.
\end{aligned}$$

Similarly, on solving Eq. (5.4.7), we have $c = \frac{1}{v_2^1}(2 - x_{t^2})$ and the inequality (5.4.12) concludes that $\int_{\Gamma_{0,\frac{1}{2}}} \max_{w \in \mathcal{W}} f(\pi, w)dt - \int_{\Gamma_{0,\frac{1}{2}}} \max_{w \in \mathcal{W}} f(\bar{\pi}, w)dt \geq 0$. In both the cases, it shows that (0, 1) is a weak robust efficient solution to the problem (MMOPU2) at which the necessary efficiency conditions (5.4.9)–(5.4.11) are satisfied with the Lagrange multipliers $\bar{\mu}_1 = 1$, $\bar{\mu}_2 = 0$, $\bar{\nu} = 0$ and $\bar{\gamma}_1^1 + \bar{\gamma}_2^1 = 2$ and the uncertain parameters $\bar{w}_1 = \bar{w}_2 = 1$, $\bar{u} = 2$, $\bar{v}_1^1 = \bar{v}_2^1 = 1$. Further, it can be seen that the Lagrange functional is also convex at $(\bar{x}, \bar{c}) = (0, 1)$ on $\mathscr{X} \times \mathscr{C}$.

At $(\bar{x}, \bar{c}) = (0, 1)$, we formulate the modified multi-dimensional multi-objective optimization problem $(\text{MMOPU2})_\Omega$ associated to the multi-dimensional multi-objective optimization problem (MMOPU2) as follows:

$$\begin{aligned}
(\text{MMOPU2})_\Omega \quad &\min_{(x(\cdot),c(\cdot))} \int_{\Gamma_{0,\frac{1}{2}}} \Big(2(c-1)w_1, x + c - 1\Big) dt^1 dt^2 \\
\text{subject to} \quad & x^2 - 4u^2 \leq 0, \\
& x_{t^1} = 2 - cv_1^1, \\
& x_{t^2} = 2 - cv_2^1, \\
& x(0,0) = 0, \ x\left(\frac{1}{2}, \frac{1}{2}\right) = 1,
\end{aligned}$$

where $t = (t^1, t^2) \in \Gamma_{0,\frac{1}{2}}$.

The robust counterpart to the problem $(\text{MMOPU2})_\Omega$ is formulated as follows:

$$
\begin{aligned}
(\text{RMMOPU2})_\Omega \quad & \min_{(x(\cdot),c(\cdot))} \int_{\Gamma_{0,\frac{1}{2}}} \left(\max_{w_1 \in \mathcal{W}_1} (2(c-1)w_1),\ \max_{w_2 \in \mathcal{W}_2} (x+c-1) \right) dt^1 dt^2 \\
\text{subject to} \quad & x^2 - 4u^2 \leq 0,\ \forall u \in \mathcal{U}, \\
& x_{t^1} = 2 - cv_1^1,\ \forall v_1^1 \in \mathcal{V}_1^1, \\
& x_{t^2} = 2 - cv_2^1,\ \forall v_2^1 \in \mathcal{V}_2^1, \\
& x(0,0) = 0,\ x\left(\frac{1}{2}, \frac{1}{2}\right) = 1,
\end{aligned}
$$

where $t = (t^1, t^2) \in \Gamma_{0,\frac{1}{2}}$. Clearly, the robust feasible solution set to (MMOPU2)Ω is the same as the robust feasible solution set to (MMOPU2).
The uncertain Lagrange functional to the problem $(\text{MMOPU2})_\Omega$ is given as follows:

$$
\begin{aligned}
L((x,c),\nu,\gamma,w,u,v) = \int_{\Gamma_{0,\frac{1}{2}}} \Big\{ & \max_{w_1 \in \mathcal{W}_1} (2(c-1)w_1) + \nu(x^2 - 4u^2) + \gamma_1^1 \left(x_{t^1} - 2 + cv_1^1 \right) \\
& + \gamma_2^1 \left(x_{t^2} - 2 + cv_2^1 \right),\ \max_{w_2 \in \mathcal{W}_2} (x+c-1) + \nu(x^2 - 4u^2) \\
& + \gamma_1^1 \left(x_{t^1} - 2 + cv_1^1 \right) + \gamma_2^1 \left(x_{t^2} - 2 + cv_2^1 \right) \Big\} dt^1 dt^2.
\end{aligned}
$$

Now, the following inequalities:

$$
\begin{aligned}
& L((\bar{x},\bar{c}),\nu,\gamma,w,\bar{u},\bar{v}) - L((\bar{x},\bar{c}),\bar{\nu},\bar{\gamma},w,\bar{u},\bar{v}) \leqq (0,0), \\
& \qquad \forall \nu \in \mathbb{R},\ \gamma = (\gamma_1^1, \gamma_2^1) \in \mathbb{R}^2,\ u \in \mathcal{U},\ v_1^1 \in \mathcal{V}_1^1,\ v_2^1 \in \mathcal{V}_2^1, \\
\text{and } & L((x,c),\bar{\nu},\bar{\gamma},w,\bar{u},\bar{v}) - L((\bar{x},\bar{c}),\bar{\nu},\bar{\gamma},w,\bar{u},\bar{v}) \geqq (0,0),\ \forall (x,c) \in \mathscr{X} \times \mathscr{C},
\end{aligned}
$$

are satisfied at $((\bar{x},\bar{c}),\bar{\nu},\bar{\gamma},w,\bar{u},\bar{v}) = ((0,1),0,(1,0),(1,1),2,(1,1))$. Thus, $((0,1),0,(1,0),(1,1),2,(1,1))$ is a robust saddle-point of $(\text{MMOPU2})_\Omega$.

Now, we formulate the penalized multi-dimensional multi-objective optimization problem $(\text{MMOPU2})_{\Omega\rho}$ associated to the modified multi-dimensional multi-objective optimization problem $(\text{MMOPU2})_\Omega$ as follows:

$$
\begin{aligned}
(\text{MMOPU2})_{\Omega\rho} \quad \min_{(x(\cdot),c(\cdot))} \int_{\Gamma_{0,\frac{1}{2}}} \Big(& 2(c-1)w_1 + \rho \Big\{ \{\max\{0, x^2 - 4u^2\} + |x_{t^1} - 2 + cv_1^1| \\
& + |x_{t^2} - 2 + cv_2^1| \Big\},\ x + c - 1 + \rho \Big\{ \max\{0, x^2 - 4u^2\} \\
& + |x_{t^1} - 2 + cv_1^1| + |x_{t^2} - 2 + cv_2^1| \Big\} \Big)\, dt^1 dt^2,
\end{aligned}
$$

where $t = (t^1, t^2) \in \Gamma_{0,\frac{1}{2}}$.

The robust counterpart to the problem $(\text{MMOPU2})_{\Omega\rho}$ is formulated as follows:

$$(\text{RMMOPU2})_{\Omega\rho} \quad \min_{(x(\cdot),c(\cdot))} \int_{\Gamma_{0,\frac{1}{2}}} \Big(\max_{w_1 \in \mathcal{W}_1} (2(c-1)w_1) + \rho\Big\{\{\max\{0, x^2 - 4u\} + |x_{t^1} - 2 + cv_1^1| + |x_{t^2} - 2 + cv_2^1|\Big\}, \max_{w_2 \in \mathcal{W}_2} (x + c - 1) + \rho\Big\{ \max\{0, x^2 - 4u\} + |x_{t^1} - 2 + cv_1^1| + |x_{t^2} - 2 + cv_2^1|\Big\}\Big) dt^1 dt^2,$$

where $t = (t^1, t^2) \in \Gamma_{0,\frac{1}{2}}, u \in \mathcal{U}, v \in \mathcal{V}$. It can be observed that the inequality $\max_{w \in \mathcal{W}} \mathscr{P}(\pi, w, \bar{u}, \bar{v}, \rho) - \max_{w \in \mathcal{W}} \mathscr{P}(\bar{\pi}, w, \bar{u}, \bar{v}, \rho) \geq 0$ at $(\bar{x}, \bar{c}) = (0, 1)$ for all $(x, c) \in \mathscr{X} \times \mathscr{C}$. This shows that $(0, 1)$ is a weak robust efficient solution to the penalized multi-dimensional optimization problem $(\text{MMOPU2})_\rho$ with $\rho \geq 1$. Hence, Theorem 5.4.0.2 is verified.

Case 2. Assume that the Lagrange multipliers $\bar{\gamma}_\alpha^\tau$, $\tau = 1, \alpha = 1, 2$ are functions of t, say $\bar{\gamma}_1^1 = t^2, \bar{\gamma}_2^1 = t^1$. Thus, by Eq. (5.4.10), we have

$$\bar{c} = \frac{\bar{v}_1^1 t^2 + \bar{v}_2^1 t^1}{2\bar{\mu}_1 \bar{w}_1}, \quad \forall t \in \Gamma_{0,\frac{1}{2}}. \tag{5.4.14}$$

On solving Eqs. (5.4.6) and (5.4.7) with the help of Eq. (5.4.14), we have

$$\bar{x} = 2(t^1 + t^2) - \frac{1}{4}((t^1) + (t^2)^2) - t^1 t^2 + h_4, \ h_4 \in \mathbb{R}_+,$$

and, by considering the boundary conditions (5.4.8), we get

$$\bar{x} = 2(t^1 + t^2) - \frac{1}{4}((t^1) + (t^2)^2) - t^1 t^2,$$

which shows that the state function is not an affine function. Consequently, we get a contradiction with the hypothesis.

Proposition 5.4.0.1 *Let $(\bar{x}, \bar{c}) \in \mathscr{X} \times \mathscr{C}$ be a weak robust efficient solution to the problem $(MMOPU)_{\Omega\bar{\rho}}$ and $\max_{w \in \mathcal{W}} f(\pi, w) = f(\pi, \bar{w})$. Then, there does not exist another point $(x, c) \in \mathcal{D}$ such that*

$$\int_\Gamma \Big\{(x - \bar{x}) f_x(\bar{\pi}, \bar{w}) + (c - \bar{c}) f_c(\bar{\pi}, \bar{w})\Big\} dt < \int_\Gamma \Big\{(\bar{x} - \bar{x}) f_x(\bar{\pi}, \bar{w}) + (\bar{c} - \bar{c}) f_c(\bar{\pi}, \bar{w})\Big\} dt.$$

Proof Since $(\bar{x}, \bar{c})$ is a weak robust efficient solution to the problem $(\text{MMOPU})_{\Omega\bar{\rho}}$, then there does not exist another point $(x, c) \in \mathscr{X} \times \mathscr{C}$ such that

$$\max_{w \in \mathcal{W}} \mathscr{P}(\pi, w, \bar{u}, \bar{v}, \bar{\rho}) < \max_{w \in \mathcal{W}} \mathscr{P}(\bar{\pi}, w, \bar{u}, \bar{v}, \bar{\rho}).$$

In virtue of the definition of the penalized multi-dimensional multi-objective optimization problem $(\text{MMOPU})_{\Omega\bar{\rho}}$, we get

$$\int_{\Gamma}\Big\{\max_{w\in\mathcal{W}}\Big[(x-\bar{x})f_x(\bar{\pi},w)+(c-\bar{c})f_c(\bar{\pi},w)\Big]+\bar{\rho}\Big[\sum_{l=1}^{m}G_l^{+}(\pi,\bar{u})$$
$$+\sum_{\tau=1}^{q}\sum_{\alpha=1}^{p}\Big|H_{\alpha}^{\tau}(\pi,\bar{v})-\frac{\partial x^{\tau}}{\partial t^{\alpha}}\Big|\Big]\Big\}dt<\int_{\Gamma}\Big\{\max_{w\in\mathcal{W}}\Big[(\bar{x}-\bar{x})f_x(\bar{\pi},w)$$
$$+(\bar{c}-\bar{c})f_c(\bar{\pi},w)\Big]+\bar{\rho}\Big[\sum_{l=1}^{m}G_l^{+}(\bar{\pi},\bar{u})+\sum_{\tau=1}^{q}\sum_{\alpha=1}^{p}\Big|H_{\alpha}^{\tau}(\bar{\pi},\bar{v})-\frac{\partial x^{\tau}}{\partial t^{\alpha}}\Big|\Big]\Big\}dt.$$

By assumption, $\max_{w\in\mathcal{W}}f(\pi,w)=f(\pi,\bar{w})$, we have

$$\int_{\Gamma}\Big\{(x-\bar{x})f_x(\bar{\pi},\bar{w})+(c-\bar{c})f_c(\bar{\pi},\bar{w})+\bar{\rho}\Big[\sum_{l=1}^{m}G_l^{+}(\pi,\bar{u})$$
$$+\sum_{\tau=1}^{q}\sum_{\alpha=1}^{p}\Big|H_{\alpha}^{\tau}(\pi,\bar{v})-\frac{\partial x^{\tau}}{\partial t^{\alpha}}\Big|\Big]\Big\}dt<\int_{\Gamma}\Big\{(\bar{x}-\bar{x})f_x(\bar{\pi},\bar{w})$$
$$+(\bar{c}-\bar{c})f_c(\bar{\pi},\bar{w})+\bar{\rho}\Big[\sum_{l=1}^{m}G_l^{+}(\bar{\pi},\bar{u})+\sum_{\tau=1}^{q}\sum_{\alpha=1}^{p}\Big|H_{\alpha}^{\tau}(\bar{\pi},\bar{v})-\frac{\partial x^{\tau}}{\partial t^{\alpha}}\Big|\Big]\Big\}dt. \tag{5.4.15}$$

Since $\mathcal{D}\subseteq\mathcal{X}\times\mathcal{C}$, therefore, it follows that there does not exist another point $(x,c)\in\mathcal{D}$ such that

$$\int_{\Gamma}\Big\{(x-\bar{x})f_x(\bar{\pi},\bar{w})+(c-\bar{c})f_c(\bar{\pi},\bar{w})+\bar{\rho}\Big[\sum_{l=1}^{m}G_l^{+}(\pi,\bar{u})$$
$$+\sum_{\tau=1}^{q}\sum_{\alpha=1}^{p}\Big|H_{\alpha}^{\tau}(\pi,\bar{v})-\frac{\partial x^{\tau}}{\partial t^{\alpha}}\Big|\Big]\Big\}dt<\int_{\Gamma}\Big\{(\bar{x}-\bar{x})f_x(\bar{\pi},\bar{w})$$
$$+(\bar{c}-\bar{c})f_c(\bar{\pi},\bar{w})+\bar{\rho}\Big[\sum_{l=1}^{m}G_l^{+}(\bar{\pi},\bar{u})+\sum_{\tau=1}^{q}\sum_{\alpha=1}^{p}\Big|H_{\alpha}^{\tau}(\bar{\pi},\bar{v})-\frac{\partial x^{\tau}}{\partial t^{\alpha}}\Big|\Big]\Big\}dt,$$

equivalent with

$$\int_\Gamma \Big\{(x-\bar{x})(f_k)_x(\bar{\pi},\bar{w}) + (c-\bar{c})(f_k)_c(\bar{\pi},\bar{w}) + \bar{\rho}\Big[\sum_{l=1}^{m} G_l^+(\pi,\bar{u}) + \sum_{\tau=1}^{q}\sum_{\alpha=1}^{p}\Big|H_\alpha^\tau(\pi,\bar{v}) - \frac{\partial x^\tau}{\partial t^\alpha}\Big|\Big]\Big\}dt \geqq \int_\Gamma \Big\{(\bar{x}-\bar{x})(f_k)_x(\bar{\pi},\bar{w}) + (\bar{c}-\bar{c})(f_k)_c(\bar{\pi},\bar{w}) + \bar{\rho}\Big[\sum_{l=1}^{m} G_l^+(\bar{\pi},\bar{u}) + \sum_{\tau=1}^{q}\sum_{\alpha=1}^{p}\Big|H_\alpha^\tau(\bar{\pi},\bar{v}) - \frac{\partial x^\tau}{\partial t^\alpha}\Big|\Big]\Big\}dt, \ \forall (x,c)\in\mathcal{D}.$$

By virtue of the relation (5.4.1), we get

$$\int_\Gamma \Big\{(x-\bar{x})(f_k)_x(\bar{\pi},\bar{w}) + (c-\bar{c})(f_k)_c(\bar{\pi},\bar{w})\Big\}dt \geqq \int_\Gamma \Big\{(\bar{x}-\bar{x})(f_k)_x(\bar{\pi},\bar{w}) + (\bar{c}-\bar{c})(f_k)_c(\bar{\pi},\bar{w})\Big\}dt,$$

for all $(x,c)\in\mathcal{D}$.
Hence, we conclude that there does not exist another point $(x,c)\in\mathcal{D}$ such that

$$\int_\Gamma \Big\{(x-\bar{x})f_x(\bar{\pi},\bar{w}) + (c-\bar{c})f_c(\bar{\pi},\bar{w})\Big\}dt < \int_\Gamma \Big\{(\bar{x}-\bar{x})f_x(\bar{\pi},\bar{w}) + (\bar{c}-\bar{c})f_c(\bar{\pi},\bar{w})\Big\}dt.$$

Hence, this completes the proof. □

Now, we shall prove the converse result of Theorem 5.4.0.1.

Theorem 5.4.0.2 *Let $(\bar{x},\bar{c})$ be a weak robust efficient solution to the penalized multi-dimensional multi-objective optimization problem $(MMOPU)_{\Omega\bar{\rho}}$. Assume that $\mathcal{D}$ is a compact subset of $\mathcal{X}\times\mathcal{C}$ and the inequality $\max_{w\in W}\mathcal{P}(\pi,w,\bar{u},\bar{v},\rho)dt \geqq \max_{w\in W}\mathcal{P}(\bar{\pi},\bar{w},\bar{u},\bar{v},\rho)dt$ holds for any $\rho\geq\bar{\rho}$ and for all $(x,c)\in\mathcal{D}$. Further, assume that $\max_{w\in W} f(\pi,w) = f(\pi,\bar{w})$ and $\int_\Gamma \bar{\upsilon}^T G(\pi,u)dt$, $\int_\Gamma \bar{\gamma}^T\big(H(\pi,v) - x_\alpha\big)dt$ are convex at $(\bar{x},\bar{c})$ on $\mathcal{X}\times\mathcal{C}$. Then $((\bar{x},\bar{c}),\bar{\upsilon},\bar{\gamma},\bar{w},\bar{u},\bar{v})$ is a robust saddle-point to the modified multi-dimensional multi-objective optimization problem $text(MMOPU)_\Omega$.*

Moreover, if we assume that $\int_\Gamma f(\pi,w)dt$ is convex at $(\bar{x},\bar{c})$ on $\mathcal{X}\times\mathcal{C}$, then $(\bar{x},\bar{c})$ is a weak robust efficient solution to the multi-dimensional multi-objective optimization problem (MMOPU).

Proof Since $(\bar{x},\bar{c})$ is a weak robust efficient solution to the penalized multi-dimensional multi-objective optimization problem $(\text{MMOPU})_{\Omega\bar{\rho}}$, and $\max_{w\in W} f(\pi,w) = f(\pi,\bar{w})$, therefore, by Proposition 5.4.0.1, there does not exist another point $(x,c)\in\mathcal{D}$ such that

$$\int_\Gamma \Big\{(x-\bar{x})f_x(\bar{\pi},\bar{w}) + (c-\bar{c})f_c(\bar{\pi},\bar{w})\Big\}dt < \int_\Gamma \Big\{(\bar{x}-\bar{x})f_x(\bar{\pi},\bar{w}) + (\bar{c}-\bar{c})f_c(\bar{\pi},\bar{w})\Big\}dt. \tag{5.4.16}$$

In order to prove that $(\bar{x}, \bar{c})$ is a weak robust efficient solution to the problem $(\text{MMOPU})_{\Omega}$, firstly, we show that it is a robust feasible solution to the problem $(\text{MMOPU})_{\Omega}$. We proceed by contradiction and assume that $(\bar{x}, \bar{c})$ is not a robust feasible solution to the problem $(\text{MMOPU})_{\Omega}$. Therefore, by virtue of the relation (5.4.1), we have

$$\int_{\Gamma} \Big\{ \sum_{l=1}^{m} G_l^{+}(\bar{\pi}, \bar{u}) + \sum_{\tau=1}^{q} \sum_{\alpha=1}^{p} \Big| H_{\alpha}^{\tau}(\bar{\pi}, \bar{v}) - \frac{\partial x^{\tau}}{\partial t^{\alpha}} \Big| \Big\} dt > 0. \tag{5.4.17}$$

Let (x, c) be any robust feasible solution to the problem $(\text{MMOPU})_{\Omega}$. By assumptions, the following inequality:

$$\max_{w \in \mathcal{W}} \mathscr{P}(\pi, w, \bar{u}, \bar{v}, \rho) dt \geqq \max_{w \in \mathcal{W}} \mathscr{P}(\bar{\pi}, w, \bar{u}, \bar{v}, \rho) dt$$

holds for all $\rho \geq \bar{\rho}$, and for all $(x, c) \in \mathcal{D}$.
In virtue of the definition of the penalized multi-dimensional multi-objective optimization problem $(\text{MMOPU})_{\Omega\rho}$, we get

$$\begin{aligned}
&\int_{\Gamma} \Big\{ \max_{w \in \mathcal{W}} \Big[(x - \bar{x}) f_x(\bar{\pi}, w) + (c - \bar{c}) f_c(\bar{\pi}, w) \Big] + \rho \Big[\sum_{l=1}^{m} G_l^{+}(\pi, \bar{u}) \\
&\quad + \sum_{\tau=1}^{q} \sum_{\alpha=1}^{p} \Big| H_{\alpha}^{\tau}(\pi, \bar{v}) - \frac{\partial x^{\tau}}{\partial t^{\alpha}} \Big| \Big] \Big\} dt \geq \int_{\Gamma} \Big\{ \max_{w \in \mathcal{W}} \Big[(\bar{x} - \bar{x}) f_x(\bar{\pi}, w) \\
&\quad + (\bar{c} - \bar{c}) f_c(\bar{\pi}, w) \Big] + \rho \Big[\sum_{l=1}^{m} G_l^{+}(\bar{\pi}, \bar{u}) + \sum_{\tau=1}^{q} \sum_{\alpha=1}^{p} \Big| H_{\alpha}^{\tau}(\bar{\pi}, \bar{v}) - \frac{\partial x^{\tau}}{\partial t^{\alpha}} \Big| \Big] \Big\} dt.
\end{aligned}$$

Since $\max_{w \in \mathcal{W}} f(\pi, w) = f(\pi, \bar{w})$, we have

$$\begin{aligned}
&\int_{\Gamma} \Big\{ (x - \bar{x}) f_x(\bar{\pi}, \bar{w}) + (c - \bar{c}) f_c(\bar{\pi}, \bar{w}) + \rho \Big[\sum_{l=1}^{m} G_l^{+}(\pi, \bar{u}) \\
&\quad + \sum_{\tau=1}^{q} \sum_{\alpha=1}^{p} \Big| H_{\alpha}^{\tau}(\pi, \bar{v}) - \frac{\partial x^{\tau}}{\partial t^{\alpha}} \Big| \Big] \Big\} dt \geq \int_{\Gamma} \Big\{ (\bar{x} - \bar{x}) f_x(\bar{\pi}, \bar{w}) \\
&\quad + (\bar{c} - \bar{c}) f_c(\bar{\pi}, \bar{w}) + \rho \Big[\sum_{l=1}^{m} G_l^{+}(\bar{\pi}, \bar{u}) + \sum_{\tau=1}^{q} \sum_{\alpha=1}^{p} \Big| H_{\alpha}^{\tau}(\bar{\pi}, \bar{v}) - \frac{\partial x^{\tau}}{\partial t^{\alpha}} \Big| \Big] \Big\} dt.
\end{aligned} \tag{5.4.18}$$

On the other hand, we set

$$\rho > \max\left\{\frac{\begin{array}{c}\int_\Gamma\{(x-\bar{x})f_x(\bar{\pi},\bar{w})+(c-\bar{c})f_c(\bar{\pi},\bar{w})\}dt\\ -\int_\Gamma\{(\bar{x}-\bar{x})f_x(\bar{\pi},\bar{w})+(\bar{c}-\bar{c})f_c(\bar{\pi},\bar{w})\}dt\end{array}}{\int_\Gamma\left\{\sum_{l=1}^m G_l^+(\bar{\pi},\bar{u})+\sum_{\tau=1}^q\sum_{\alpha=1}^p\left|H_\alpha^\tau(\bar{\pi},\bar{v})-\frac{\partial x^\tau}{\partial t^\alpha}\right|\right\}edt}, \bar{\rho}; (x,c)\in\mathcal{D}\right\}.$$

In the view of the inequalities (5.4.16) and (5.4.17), it follows that ρ is a finite positive real number, and therefore,

$$\int_\Gamma\Big\{(\bar{x}-\bar{x})f_x(\bar{\pi},\bar{w})+(\bar{c}-\bar{c})f_c(\bar{\pi},\bar{w})+\rho\Big[\sum_{l=1}^m G_l^+(\bar{\pi},\bar{u})+\sum_{\tau=1}^q\sum_{\alpha=1}^p\Big|H_\alpha^\tau(\bar{\pi},\bar{v})-\frac{\partial x^\tau}{\partial t^\alpha}\Big|\Big]e\Big\}dt$$
$$>\int_\Gamma(x-\bar{x})f_x(\bar{\pi},\bar{w})+(c-\bar{c})f_c(\bar{\pi},\bar{w})dt. \tag{5.4.19}$$

Further, the robust feasibility of (x,c) to the problem $(\text{MMOPU})_\Omega$ and by virtue of the relation (5.4.1), it follows from the above inequality

$$\int_\Gamma\Big\{(\bar{x}-\bar{x})f_x(\bar{\pi},\bar{w})+(\bar{c}-\bar{c})f_c(\bar{\pi},\bar{w})+\rho\Big[\sum_{l=1}^m G_l^+(\bar{\pi},\bar{u})+\sum_{\tau=1}^q\sum_{\alpha=1}^p\Big|H_\alpha^\tau(\bar{\pi},\bar{v})-\frac{\partial x^\tau}{\partial t^\alpha}\Big|\Big]e\Big\}dt$$
$$>\int_\Gamma\Big\{(x-\bar{x})f_x(\bar{\pi},\bar{w})+(c-\bar{c})f_c(\bar{\pi},\bar{w})+\rho\Big[\sum_{l=1}^m G_l^+(\pi,\bar{u})+\sum_{\tau=1}^q\sum_{\alpha=1}^p\Big|H_\alpha^\tau(\pi,\bar{v})-\frac{\partial x^\tau}{\partial t^\alpha}\Big|\Big]e\Big\}dt,$$

which contradicts the inequality (5.4.18). Therefore, $(\bar{x},\bar{c})$ is a robust feasible solution to the problem $(\text{MMOPU})_\Omega$. Hence, the inequality (5.4.16) directly concludes that $(\bar{x},\bar{c})$ is a weak robust efficient solution to the problem $(\text{MMOPU})_\Omega$.

Now, we shall prove that $((\bar{x},\bar{c}),\bar{\nu},\bar{\gamma},\bar{w},\bar{u},\bar{v})$ is a robust saddle-point to the problem $(\text{MMOPU})_\Omega$. Since $(\bar{x},\bar{c})$ is a weak robust efficient solution to the problem $(\text{MMOPU})_\Omega$, there exist Lagrange multipliers $\bar{\mu}\in\mathbb{R}^s$, $\bar{\nu}\in\mathbb{R}_+^m$ and $\bar{\gamma}\in\mathbb{R}^{pq}$ and the uncertain parameters $\bar{u}\in\mathcal{U}$, $\bar{v}\in\mathcal{V}$. Let $\max_{w\in\mathcal{W}}f(\bar{\pi},w)=f(\bar{\pi},\bar{w})$. Therefore, the robust necessary efficiency conditions (5.2.1)–(5.2.4) are satisfied at $(\bar{x},\bar{c})$ on $\mathscr{X}\times\mathscr{C}$.

From the robust feasibility of $(\bar{x},\bar{c})$ to the problem $(\text{MMOPU})_\Omega$ together with the necessary efficiency condition (5.2.3), we have

$$\int_\Gamma \nu^T G(\bar{\pi},u)dt\le\int_\Gamma\bar{\nu}^T G(\bar{\pi},\bar{u})dt,\ \forall\nu\in\mathbb{R}_+^m, u\in\mathcal{U},$$

equivalently,

$$\int_{\Gamma}\Big\{(\bar{x}-\bar{x})f_x(\bar{\pi},\bar{w})+(\bar{c}-\bar{c})f_c(\bar{\pi},\bar{w})+\nu^T G(\bar{\pi},u)e+\gamma^T(H(\bar{\pi},v)-x_\alpha)e\Big\}dt$$
$$\leqq \int_{\Gamma}\Big\{(\bar{x}-\bar{x})f_x(\bar{\pi},\bar{w})+(\bar{c}-\bar{c})f_c(\bar{\pi},\bar{w})+\bar{\nu}^T G(\bar{\pi},\bar{u})e+\bar{\gamma}^T(H(\bar{\pi},\bar{v})-x_\alpha)e\Big\}dt,$$

for all $\nu \in \mathbb{R}^m_+$, $\gamma \in \mathbb{R}^{pq}$, $u \in \mathcal{U}$, $v \in \mathcal{V}$.
From the virtue of the definition of Lagrange functional, we get

$$L((\bar{x},\bar{c}),\nu,\gamma,w,u,v) \leqq L((\bar{x},\bar{c}),\bar{\nu},\bar{\gamma},w,\bar{u},\bar{v}),\ \forall \nu \in \mathbb{R}^m_+,\ \forall \gamma \in \mathbb{R}^{pq},\ \forall u \in \mathcal{U},\ \forall v \in \mathcal{V}. \tag{5.4.20}$$

Now, we prove the second condition of robust saddle-point to the problem $(\text{MMCOP})_{\Omega\rho}$. By assumption, $\int_{\Gamma}\bar{\nu}^T G(\pi,u)dt$ and $\int_{\Gamma}\int_{\Gamma}\bar{\gamma}^T(H(\pi,v)-x_\alpha)dt$ are convex at $(\bar{x},\bar{c})$ on $\mathscr{X}\times\mathscr{C}$, we have

$$\int_{\Gamma}\Big\{\bar{\nu}^T G(\pi,\bar{u})-\bar{\nu}^T G(\bar{\pi},\bar{u})\Big\}dt \geqq \int_{\Gamma}(x-\bar{x})\bar{\nu}^T G_x(\bar{\pi},\bar{u})dt+\int_{\Gamma}(c-\bar{c})\bar{\nu}^T G_c(\bar{\pi},\bar{u})dt,$$

and

$$\int_{\Gamma}\Big\{\bar{\gamma}^T(H(\pi,\bar{v})-x_\alpha)-\bar{\gamma}^T(H(\bar{\pi},\bar{v})-x_\alpha)\Big\}dt$$
$$\geqq \int_{\Gamma}(x-\bar{x})\Big\{\bar{\gamma}^T H_x(\bar{\pi},\bar{v})+\gamma_t\Big\}dt+\int_{\Gamma}(c-\bar{c})\bar{\gamma}^T H_c(\bar{\pi},\bar{v})dt.$$

On combining the above two inequalities with the robust necessary efficiency conditions (5.2.1) and (5.2.2), we get

$$\int_{\Gamma}\Big\{(x-\bar{x})\bar{\mu}^T f_x(\bar{\pi},\bar{w})+(c-\bar{c})\bar{\mu}^T f_c(\bar{\pi},\bar{w})+\bar{\nu}^T G(\pi,\bar{u})+\bar{\gamma}^T(H_\alpha(\pi,\bar{v})-x_\alpha)dt$$
$$\geqq \int_{\Gamma}\Big\{\bar{\nu}^T G(\bar{\pi},\bar{u})+\bar{\gamma}^T(H(\bar{\pi},\bar{v})-x_\alpha)\Big\}dt, \tag{5.4.21}$$

equivalent with

$$\int_{\Gamma}\Big\{(x-\bar{x})f_x(\bar{\pi},\bar{w})+(c-\bar{c})f_c(\bar{\pi},\bar{w})+\bar{\nu}^T G(\pi,\bar{u})e+\bar{\gamma}^T(H_\alpha(\pi,\bar{v})-x_\alpha)e\Big\}dt$$
$$\geqq \int_{\Gamma}\Big\{(\bar{x}-\bar{x})f_x(\bar{\pi},\bar{w})+(\bar{c}-\bar{c})f_c(\bar{\pi},\bar{w})+\bar{\nu}^T G(\bar{\pi},\bar{u})e+\bar{\gamma}^T(H_\alpha(\bar{\pi},\bar{v})-x_\alpha)e\Big\}dt,$$

for all $(x,c)\in\mathscr{X}\times\mathscr{C}$.
Again, from the virtue of the definition of Lagrange functional, we get

$$L((x,c),\bar{\nu},\bar{\gamma},w,\bar{u},\bar{v}) \geqq L((\bar{x},\bar{c}),\bar{\nu},\bar{\gamma},w,\bar{u},\bar{v}),\ \forall (x,c)\in\mathscr{X}\times\mathscr{C}. \tag{5.4.22}$$

The inequalities (5.4.20) and (5.4.22) together conclude that $((\bar{x}, \bar{c}), \bar{\nu}, \bar{\gamma}, \bar{w}, \bar{u}, \bar{v})$ is a robust saddle-point to the modified multi-dimensional multi-objective optimization problem $(\text{MMOPU})_\Omega$.

Now, it remains to prove that $(\bar{x}, \bar{c})$ is a weak robust efficient solution to the original problem (MMOPU). Since $((\bar{x}, \bar{c}), \bar{\nu}, \bar{\gamma}, \bar{w}, \bar{u}, \bar{v})$ is a robust saddle-point to the modified multi-dimensional multi-objective optimization problem $(\text{MMOPU})_\Omega$ and $\int_\Gamma f(\pi, w)dt$ is convex at $(\bar{x}, \bar{c})$ on $\mathscr{X} \times \mathscr{C}$, therefore, in the view of Theorem 5.3.0.3, the point $(\bar{x}, \bar{c})$ is a weak robust efficient solution to the multi-dimensional multi-objective optimization problem (MMOPU). Hence, the proof is completed. □

References

1. M.A. Hanson, Bounds for functionally convex optimal control problems. J. Math. Anal. Appl. **8**, 84–89 (1964)
2. V.A. de Oliveira, G.N. Silva, New optimality conditions for nonsmooth control problems. J. Global Optim. **57**, 1465–1484 (2013)
3. V.A. de Oliveira, G.N. Silva, M.A. Rojas-Medar, KT-invexity in optimal control problems. Nonlinear Anal. Theory Methods Appl. **71**, 4790–4797 (2009)
4. K. Khazafi, N. Rueda, P. Enflo, Sufficiency and duality for multiobjective control problems under generalized (B;ρ)-type I functions. J. Global Optim. **46**, 111–132 (2010)
5. Şt. Mititelu, Optimality and duality for invex multi-dimensional control problems with mixed constraints, J. Adv. Math. Stud. **2**, 25–34 (2009)
6. Şt. Mititelu, V. Preda and M. Postolache, Duality of multitime vector integral programming with quasiinvexity, J. Adv. Math. Stud. **4** 59–72 (2011)
7. S. Treanţă, Efficiency in generalized V-KT-pseudoinvex control problems. Internat. J. Control **93**, 611–618 (2020)
8. S. Treanţă, M. Arana-Jiménez, KT-pseudoinvex multidimensional control problem. Optim. Control Appl. Meth. **39**, 1291–1300 (2018)
9. Şt. Mititelu, S. Treanţă, Efficiency conditions in vector control problems governed by multiple integrals. J. Appl. Math. Comput. **57**, 647–665 (2018)
10. T. Antczak, A new approach to multiobjective programming with a modified objective function. J. Global Optim. **27**, 485–495 (2003)
11. T. Antczak, A Modified objective function method in mathematical programming with second order invexity. Numer. Funct. Anal. Optim. **28**, 1–12 (2007)
12. T. Antczak, A. Jayswal, S. Jha, The modified objective function method for univex multiobjective variational problems. Bull. Iranian Math. Soc. **45**, 267–282 (2019)
13. A. Jayswal, T. Antczak, S. Jha, On equivalence between a variational problem and its modified variational problem with the η objective function under invexity. Int. Trans. Op. Res. **26**, 2053–2070 (2019)
14. A. Jayswal, T. Antczak, S. Jha, Modified objective function approach for multitime variational problems. Turkish J. Math. **42**, 1111–1129 (2018)
15. S. Treanţă, Saddle-point optimality criteria in modified variational control problems with PDE constraints. Optimal Control. Appl. Meth. **41**, 1160–1175 (2020)
16. S. Treanţă, On Modified Interval-Valued Variational Control Problems with First-Order PDE Constraints. Symmetry-Basel **12**, 472 (2020)
17. S. Treanţă, On a modified optimal control problem with first-order PDE constraints and the associated saddle-point optimality criterion. Eur. J. Control **51**, 1–9 (2020)
18. S. Treanţă, Higher-order Hamilton dynamics and Hamilton-Jacobi divergence PDE. Comput. Math. Appl. **75**, 547–560 (2018)

Chapter 6
Semi-infinite Multi-dimensional Variational Control Problem with Data Uncertainty

6.1 Introduction

Semi-infinite optimization problem is one of the most important optimization problem that optimizes the objective function with respect to possibly an infinite number of constraints. This fact motivates the researchers to study the semi-infinite optimization problems and to develop new results under several hypotheses. López and G. Still [1] explored the semi-infinite problem and established the compromise between an introduction and a survey, proving the theoretical basis, numerical methods, applications and historical background. Ye and Wu [2] examined the generalized semi-infinite programming problems and derived the first-order optimality conditions. Ştefănescu and Ştefănescu [3] studied the semi-infinite minmax programming problem and obtained the optimality conditions. Further, the authors also formulated the dual problem and derived the duality results under invexity assumptions. Jayswal *et al.* [4] considered the semi-infinite programming problem, modulated its Wolfe and Mond-Weir-type dual programs and derived the duality results under generalized (H_p, r)-invexity.

6.2 Problem Description

Throughout the chapter, we are considering the following notations and working hypotheses as:

• $\mathbb{R}^p$, $\mathbb{R}^q$, $\mathbb{R}^r$ and $\mathbb{R}^n$ are Euclidean spaces of dimensions p, q, r and n, respectively.

• $\Gamma_{t_0,t_1} \subset \mathbb{R}^p$ is a hyperparallelepiped fixed by the diagonally opposite points $t_0 = (t_0^\alpha)$ and $t_1 = (t_1^\alpha)$, $\alpha = \overline{1, p}$ and the point $t = ((t^\alpha), \alpha = \overline{1, p}) \in \Gamma_{t_0,t_1} \subset \mathbb{R}^p$.

• $\mathscr{X}$ is the space of state functions (piecewise smooth) $x = (x^\tau) : \Gamma_{t_0,t_1} \subset \mathbb{R}^p \to \mathbb{R}^q$ and $\frac{\partial x}{\partial t^\alpha} = x_\alpha$ denotes the first-order partial derivative of x with respect to t^α.

• $\mathscr{C}$ is the space of control functions (piecewise continuous) $c = (c^j) : \Gamma_{t_0,t_1} \subset \mathbb{R}^p \to \mathbb{R}^r$.

A. Jayswal et al., *Multi-dimensional Control Problems*, Industrial and Applied Mathematics, https://doi.org/10.1007/978-981-19-6561-6_6

• $dt = dt^1 \dots dt^p$ is the volume element on $\mathbb{R}^p \supset \Gamma_{t_0,t_1}$.
• The index set Λ is a arbitrary sets not necessarily finite.
• T denotes the transpose of a vector.
• We use the following convention for inequalities and equalities: for $x, y \in \mathbb{R}^n$, we have

$$(i)\ x < y \Leftrightarrow x_i < y_i, \forall i = \overline{1, n},$$
$$(ii)\ x = y \Leftrightarrow x_i = y_i,\ \forall i = \overline{1, n},$$
$$(iii)\ x \leqq y \Leftrightarrow x_i \leq y_i,\ \forall i = \overline{1, n},$$
$$(iv)\ x \leq y \Leftrightarrow x_i \leq y_i,\ \forall i = \overline{1, n} \text{ and } x_i < y_i \text{ for some } i.$$

The semi-infinite multi-dimensional multi-objective optimization problem with data uncertainty in the objective and constraint functionals is defined as follows:

$$\text{(SMMOPU)} \quad \min_{(x(\cdot),c(\cdot))} \left\{ \int_{\Gamma_{t_0,t_1}} f(t, x(t), c(t), w)dt \right.$$
$$\left. = \left(\int_{\Gamma_{t_0,t_1}} f_1(t, x(t), c(t), w_1)dt, \dots, \int_{\Gamma_{t_0,t_1}} f_s(t, x(t), c(t), w_s)dt \right) \right\}$$
$$\text{subject to} \quad G(t, x(t), c(t), u) \leqq 0,$$
$$x_\alpha = H(t, x(t), c(t), v),$$
$$x(t_0) = x_0,\ x(t_1) = x_1,$$

where $f = (f_k) : \Gamma_{t_0,t_1} \times \mathscr{X} \times \mathscr{C} \times \mathcal{W}_k \to \mathbb{R}^s, k = \overline{1, s},\ G = (G_l) : \Gamma_{t_0,t_1} \times \mathscr{X} \times \mathscr{C} \times \mathcal{U}_l \to \mathbb{R}^\Lambda,\ l \in \Lambda,\ H = (H_\alpha^\tau) : \Gamma_{t_0,t_1} \times \mathscr{X} \times \mathscr{C} \times \mathcal{V}_\alpha^\tau \to \mathbb{R}^{pq}, \alpha = \overline{1, p}, \tau = \overline{1, q}$ are C^∞-class functionals. $w = (w_k)$, $u = (u_l)$ and $v = (v_\alpha^\tau)$ are the uncertain parameters for some convex compact subsets $\mathcal{W} = (\mathcal{W}_k) \subset \mathbb{R}^s$, $\mathcal{U} = (\mathcal{U}_l) \subset \mathbb{R}^\Lambda, l \in \Lambda$ and $\mathcal{V} = (\mathcal{V}_\alpha^\tau) \subset \mathbb{R}^{pq}$, respectively. The functions H_α satisfy the closeness conditions (complete integrability conditions) $D_\beta H_\alpha = D_\alpha H_\beta, \alpha, \beta = \overline{1, p}, \alpha \neq \beta$, where the total derivative is denoted by D_β.

The associated robust counterpart of the semi-infinite multi-dimensional multi-objective optimization problem (SMMOPU) is defined as:

$$\text{(RSMMOPU)} \quad \min_{(x(\cdot),c(\cdot))} \left\{ \int_{\Gamma_{t_0,t_1}} \max_{w \in \mathcal{W}} f(t, x(t), c(t), w)dt \right.$$
$$\left. = \left(\int_{\Gamma_{t_0,t_1}} \max_{w_1 \in \mathcal{W}_1} f_1(t, x(t), c(t), w_1)dt, \dots, \int_{\Gamma_{t_0,t_1}} \max_{w_s \in \mathcal{W}_s} f_s(t, x(t), c(t), w_s)dt \right) \right\}$$
$$\text{subject to} \quad G(t, x(t), c(t), u) \leqq 0, \forall u \in \mathcal{U},$$
$$x_\alpha = H(t, x(t), c(t), v), \forall v \in \mathcal{V},$$
$$x(t_0) = x_0,\ x(t_1) = x_1,$$

where f, G and H are defined as above in the semi-infinite multi-dimensional multi-objective optimization problem (SMMOPU).

We denote $\mathcal{D} = \{(x, c) \in \mathscr{X} \times \mathscr{C} : \quad G(t, x(t), c(t), u) \leqq 0, x_\alpha = H(t, x(t), c(t), v), x(t_0) = x_0, x(t_1) = x_1\}$ the set of all feasible solutions in (RSMMOPU), and we say that it is the robust feasible solution set to the problem (SMMOPU).

From now on, to simplify our presentation, we introduce some notations as follows: $x = x(t), \hat{x} = \hat{x}(t), \bar{x} = \bar{x}(t), c = c(t), \hat{c} = \hat{c}(t), \bar{c} = \bar{c}(t), \pi = (t, x(t), c(t))$, $\bar{\pi} = (t, \bar{x}(t), \bar{c}(t)), \hat{\pi} = (t, \hat{x}(t), \hat{c}(t)), \rho = \rho(t), \bar{\rho} = \bar{\rho}(t)$ and $\Gamma = \Gamma_{t_0, t_1}$.

The partial derivatives of Lagrange multiplier γ and the partial derivatives associated with f are defined as

$$\gamma_t = \begin{bmatrix} \frac{\partial \gamma_1^\tau}{\partial t^1} & \cdots & \frac{\partial \gamma_1^\tau}{\partial t^p} \\ \vdots & \ddots & \vdots \\ \frac{\partial \gamma_p^\tau}{\partial t^1} & \cdots & \frac{\partial \gamma_p^\tau}{\partial t^p} \end{bmatrix}, \quad f_x = \begin{bmatrix} \frac{\partial f_1}{\partial x^1} & \cdots & \frac{\partial f_1}{\partial x^q} \\ \vdots & \ddots & \vdots \\ \frac{\partial f_s}{\partial x^1} & \cdots & \frac{\partial f_s}{\partial x^q} \end{bmatrix}, \quad f_c = \begin{bmatrix} \frac{\partial f_1}{\partial c^1} & \cdots & \frac{\partial f_1}{\partial c^r} \\ \vdots & \ddots & \vdots \\ \frac{\partial f_s}{\partial c^1} & \cdots & \frac{\partial f_s}{\partial c^r} \end{bmatrix}.$$

Similarly, we have G_x and G_c using matrices with m rows and H_x, H_c, x_α using matrices with p rows.

In general, for a vector optimization problem, it is not so easy to find a point which optimizes more than one objective simultaneously. Therefore, the notion of an efficient solution is used to determine an optimality in vector optimization problems.

Definition 6.2.0.1 A point $(\bar{x}, \bar{c}) \in \mathcal{D}$ is said to be weak robust efficient solution to the semi-infinite multi-dimensional multi-objective optimization problem (SMMOPU), if there does not exist another point $(x, c) \in \mathcal{D}$ such that

$$\int_\Gamma \max_{w \in \mathcal{W}} f(\pi, w) dt < \int_\Gamma \max_{w \in \mathcal{W}} f(\bar{\pi}, w) dt.$$

Definition 6.2.0.1 A functional $F(x, c, \bar{w}) = \int_\Gamma f(t, x(t), x_\sigma(t), c(t), \bar{w}) dt$ is said to be convex at $(\bar{x}, \bar{c})$, if the following inequality

$$\begin{aligned} F(x, c, \bar{w}) - F(\bar{x}, \bar{c}, \bar{w}) \geqq & \int_\Gamma (x(t) - \bar{x}(t)) f_x(t, \bar{x}(t), \bar{x}_\sigma(t), \bar{c}(t), \bar{w}) dt \\ & + \int_\Gamma (x_\sigma(t) - \bar{x}_\sigma(t)) f_{x_\sigma}(t, \bar{x}(t), \bar{x}_\sigma(t), \bar{c}(t), \bar{w}) dt \\ & + \int_\Gamma (c(t) - \bar{c}(t)) f_c(t, \bar{x}(t), \bar{x}_\sigma(t), \bar{c}(t), \bar{w}) dt \end{aligned}$$

holds for all $(x(t), c)$ on $\mathscr{X} \times \mathscr{C}$.

Following the same lines formulated in Mititelu [5], we establish the following result. Thus, the proof is omitted. The proof can be read (it is free) at the following link: https://www.journal.fairpartners.ro/volume-22009-no-1_3.html.

Theorem 6.2.0.1 *Let* $(\bar{x}, \bar{c}) \in \mathcal{D}$ *be a weak robust efficient solution to the problem* (SMMOPU). *Further assume that* $\max_{w \in \mathcal{W}} f(\pi, w) = f(\pi, \bar{w})$. *If the constraint conditions (for the existence of the multipliers) hold, then there exist the scalar vector*

$\bar{\mu} \in \mathbb{R}^s$, *the piecewise smooth Lagrange multipliers* $\bar{\nu} = (\bar{\nu}_l(t)) \in \mathbb{R}^{\Lambda}_{+}, l \in \mathbb{R}^{\Lambda}, \bar{\gamma} = (\bar{\gamma}^{\tau}_{\alpha}(t)) \in \mathbb{R}^{pq}, \tau = \overline{1,q}, \alpha = \overline{1,p}$, *and the uncertain parameters* $\bar{u} \in \mathcal{U}, \bar{v} \in \mathcal{V}$ *such that* $(\bar{x}, \bar{c})$ *satisfies the following conditions:*

$$\bar{\mu}^T f_x(\bar{\pi}, \bar{w}) + \bar{\nu}^T G_x(\bar{\pi}, \bar{u}) + \bar{\gamma}^T H_x(\bar{\pi}, \bar{v}) + \bar{\gamma}_t = 0, \tag{6.2.1}$$

$$\bar{\mu}^T f_c(\bar{\pi}, \bar{w}) + \bar{\nu}^T G_c(\bar{\pi}, \bar{v}) + \bar{\gamma}^T H_c(\bar{\pi}, \bar{v}) = 0, \tag{6.2.2}$$

$$\bar{\nu}^T G(\bar{\pi}, \bar{u}) = 0, \tag{6.2.3}$$

$$\bar{\mu} \geq 0, \bar{\nu} \geqq 0, \tag{6.2.4}$$

hold for all $t \in \Gamma$, *except at discontinuities.*

Remark 6.2.0.1 The conditions (6.2.1)–(6.2.4) are known as robust necessary efficiency conditions for the semi-infinite multi-dimensional multi-objective optimization problem (SMMOPU).

6.3 Robust Sufficient Efficiency Conditions

In this section, we will discuss the robust sufficient efficiency conditions for the semi-infinite multi-dimensional multi-objective optimization problem with data uncertainty (SMMOPU) which states that any robust feasible solution to the problem (SMMOPU) will also be a weak robust efficient solution to the considered problem (SMMOPU) if the involved functionals are convex.

Theorem 6.3.0.1 *Let* $(\bar{x}, \bar{c})$ *be a robust feasible solution to the semi-infinite multi-dimensional multi-objective optimization problem* (SMMOPU) *and the robust necessary efficiency conditions* (6.2.1)–(6.2.4) *be satisfied at* $(\bar{x}, \bar{c})$ *and* $\max_{w \in \mathcal{W}} f(\pi, w) = f(\pi, \bar{w})$. *Further, assume that* $\int_{\Gamma} \bar{\mu}^T f(\pi, \bar{w})dt$, $\int_{\Gamma} \bar{\nu}^T G(\pi, \bar{u})dt$, *and* $\int_{\Gamma} \bar{\gamma}^T (H(\pi, \bar{v}) - x_{\alpha})dt$ *are convex at* $(\bar{x}, \bar{c})$. *Then,* $(\bar{x}, \bar{c})$ *is a weak robust efficient solution to the semi-infinite multi-dimensional multi-objective optimization problem* (SMMOPU).

Proof Contrary to the result, assume that $(\bar{x}, \bar{c})$ is not a weak robust efficient solution to the semi-infinite multi-dimensional multi-objective optimization problem (SMMOPU). Then, there exists $(\hat{x}, \hat{c}) \in \mathcal{D}$ such that

$$\int_{\Gamma} \max_{w \in \mathcal{W}} f(\hat{\pi}, w)dt < \int_{\Gamma} \max_{w \in \mathcal{W}} f(\bar{\pi}, w)dt.$$

Because $\max_{w \in \mathcal{W}} f(\pi, w) = f(\pi, \bar{w})$, we get

$$\int_{\Gamma} f(\hat{\pi}, \bar{w})dt < \int_{\Gamma} f(\bar{\pi}, \bar{w})dt. \tag{6.3.1}$$

Since $(\bar{x}, \bar{c})$ satisfies the conditions (6.2.1)–(6.2.4), on multiplying Eqs. (6.2.1) and (6.2.2) by $(\hat{x} - \bar{x})$ and $(\hat{c} - \bar{c})$, respectively, and then integrating, we get

$$\int_{\Gamma}(\hat{x}-\bar{x})\Big\{\bar{\mu}^T f_x(\bar{\pi},\bar{w})+\bar{\nu}^T G_x(\bar{\pi},\bar{u})+\bar{\gamma}^T H_x(\bar{\pi},\bar{v})+\bar{\gamma}_t\Big\}dt$$
$$+\int_{\Gamma}(\hat{c}-\bar{c})\Big\{\bar{\mu}^T f_c(\bar{\pi},\bar{w})+\bar{\nu}^T G_c(\bar{\pi},\bar{u})+\bar{\gamma}^T H_c(\bar{\pi},\bar{v})\Big\}dt=0. \tag{6.3.2}$$

On the other hand, from the assumption $\int_{\Gamma}\bar{\mu}^T f(\pi,\bar{w})dt$ is convex at $(\bar{x},\bar{c})$, we have

$$\int_{\Gamma}\Big\{\bar{\mu}^T f(\hat{\pi},\bar{w})-\bar{\mu}^T f(\bar{\pi},\bar{w})\Big\}dt \geqq \int_{\Gamma}(\hat{x}-\bar{x})\bar{\mu}^T f_x(\bar{\pi},\bar{w})dt+\int_{\Gamma}(\hat{c}-\bar{c})\bar{\mu}^T f_c(\bar{\pi},\bar{w})dt,$$

which along with the inequality (6.3.1), gives

$$\int_{\Gamma}(\hat{x}-\bar{x})\bar{\mu}^T f_x(\bar{\pi},\bar{w})dt+\int_{\Gamma}(\hat{c}-\bar{c})\bar{\mu}^T f_c(\bar{\pi},\bar{w})dt<0. \tag{6.3.3}$$

Again, by the assumption $\int_{\Gamma}\bar{\nu}^T G(\pi,\bar{u})dt$ is convex at $(\bar{x},\bar{c})$, we obtain

$$\int_{\Gamma}\Big\{\bar{\nu}^T G(\hat{\pi},\bar{u})-\bar{\nu}^T G(\bar{\pi},\bar{u})\Big\}dt \geqq \int_{\Gamma}(\hat{x}-\bar{x})\bar{\nu}^T G_x(\bar{\pi},\bar{u})dt+\int_{\Gamma}(\hat{c}-\bar{c})\bar{\nu}^T G_c(\bar{\pi},\bar{u})dt.$$

The above inequality together with the robust feasibility of $(\hat{x},\hat{c})$ in the semi-infinite multi-dimensional multi-objective optimization problem (SMMOPU) and the condition (6.2.3) yields

$$\int_{\Gamma}(\hat{x}-\bar{x})\bar{\nu}^T G_x(\bar{\pi},\bar{u})dt+\int_{\Gamma}(\hat{c}-\bar{c})\bar{\nu}^T G_c(\bar{\pi},\bar{u})dt \leqq 0. \tag{6.3.4}$$

Further, from the assumption $\int_{\Gamma}\bar{\gamma}^T(H(\pi,\bar{v})-x_{\alpha})dt$ is convex at $(\bar{x},\bar{c})$ and the robust feasibility of $(\hat{x},\hat{c})$ in the semi-infinite multi-dimensional multi-objective optimization problem (SMMOPU), we obtain

$$\int_{\Gamma}(\hat{x}-\bar{x})\Big\{\bar{\gamma}^T H_x(\bar{\pi},\bar{v})+\bar{\gamma}_t\Big\}dt+\int_{\Gamma}(\hat{c}-\bar{c})\bar{\gamma}^T H_c(\bar{\pi},\bar{v})dt \leqq 0. \tag{6.3.5}$$

On adding the inequalities (6.3.3), (6.3.4) and (6.3.5), we have

$$\int_{\Gamma}(\hat{x}-\bar{x})\Big\{\bar{\mu}^T f_x(\bar{\pi},\bar{w})+\bar{\nu}^T G_x(\bar{\pi},\bar{u})+\bar{\gamma}^T H_x(\bar{\pi},\bar{v})+\bar{\gamma}_t\Big\}dt$$
$$+\int_{\Gamma}(\hat{c}-\bar{c})\Big\{\bar{\mu}^T f_c(\bar{\pi},\bar{w})+\bar{\nu}^T G_c(\bar{\pi},\bar{u})+\bar{\gamma}^T H_c(\bar{\pi},\bar{v})\Big\}dt<0,$$

which contradicts the relation (6.3.2). Hence, this completes the proof. □

The following example verifies the robust sufficient efficiency conditions for a weak robust efficient solution to the semi-infinite multi-dimensional multi-objective optimization problem (SMMOPU).

Example 6.3.0.1 Let $p = 2, q = r = 1, k = 2, \mathcal{W}_1 = \mathcal{W}_2 = [0, 1], \mathcal{U}_l = [1 - l, l + 2], l \in \Lambda = [0, 1], \mathcal{V}_1^1 = \mathcal{V}_2^1 = [0, 1]$ and Γ_{t_0,t_1} is fixed by the diagonally opposite points $t_0 = (t_0^1, t_0^2)$ and $t_1 = (t_1^1, t_1^2)$ in $\mathbb{R}^2$.

We consider the following semi-infinite multi-dimensional multi-objective optimization problem:

$$\text{(SMMOPU1)} \quad \min_{(x(\cdot),c(\cdot))} \int_{\Gamma_{0,\frac{1}{2}}} \left(\frac{1}{2}(x-1)^2 e^{x-1} + cw_1,\ c^2 + w_2\left(x - \frac{1}{2}\right)^2 e^{x-1}\right) dt^1 dt^2$$

$$\text{subject to} \quad -\frac{1}{2} l \log(2x - 1) u_l \leqq 0, \tag{6.3.6}$$

$$x_{t^1} = 1 - cv_1^1, \tag{6.3.7}$$

$$x_{t^2} = 1 - cv_2^1, \tag{6.3.8}$$

$$x(0,0) = 1,\ x\left(\frac{1}{2}, \frac{1}{2}\right) = \frac{3}{2}, \tag{6.3.9}$$

where $t = (t^1, t^2) \in \Gamma_{0,\frac{1}{2}}$.

The robust counterpart for the semi-infinite multi-dimensional multi-objective optimization problem (SMMOPU1) is defined as:

$$\text{(RSMMOPU1)} \quad \min_{(x(\cdot),c(\cdot))} \int_{\Gamma_{0,\frac{1}{2}}} \left(\max_{w_1 \in \mathcal{W}_1} \left\{\frac{1}{2}(x-1)^2 e^{x-1} + cw_1\right\}, \right.$$

$$\left. \max_{w_2 \in \mathcal{W}_2} \left\{c^2 + w_2\left(x - \frac{1}{2}\right)^2 e^{x-1}\right\}\right) dt^1 dt^2,$$

$$\text{subject to} \quad -\frac{1}{2} l \log(2x - 1) u_l \leq 0,\ \forall u_l \in \mathcal{U}_l,$$

$$x_{t^1} = 1 - cv_1^1,\ \forall v_1^1 \in \mathcal{V}_1^1,$$

$$x_{t^2} = 1 - cv_2^1,\ \forall v_2^1 \in \mathcal{V}_2^1,$$

$$x(0,0) = 0,\ x\left(\frac{1}{2}, \frac{1}{2}\right) = \frac{3}{2},$$

where $t = (t^1, t^2) \in \Gamma_{0,\frac{1}{2}}$. Clearly $\mathcal{D} = \{(x, c) \in \mathscr{X} \times \mathscr{C} : x \geq \frac{1}{2} \wedge x_{t^1} = 1 - cv_1^1 \wedge x_{t^2} = 1 - v_2^1, x(0,0) = 0, x\left(\frac{1}{2}, \frac{1}{2}\right) = \frac{3}{2}\}$ is a robust feasible solution set to (SMMOPU1).

Now, we are interested to find a weak robust efficient solution to the problem (SMMOPU1). It means that we have to find the control function $\bar{c} : \Gamma_{0,\frac{1}{2}} \to \mathbb{R}$ (that determines the state function $\bar{x} : \Gamma_{0,\frac{1}{2}} \to \mathbb{R}$) satisfying the dynamical system (6.3.7) and (6.3.8) with respect to the boundary conditions (6.3.9). Additionally, we assume that state function is affine.

Let $(\bar{x}, \bar{c})$ be a robust feasible solution to the problem (SMMOPU1) and the Lagrange multipliers $\bar{\mu}_1, \bar{\mu}_2, \bar{\nu}, \bar{\gamma}_1^1, \bar{\gamma}_2^1$, and the uncertain parameters $\bar{u}, \bar{v}_1^1, \bar{v}_2^1$. Let $\max_{w \in W} f(\pi, w) = f(\pi, \bar{w})$. Therefore, the robust necessary efficiency conditions are given as follows:

$$\bar{\mu}_1\Big((\bar{x}-1)e^{\bar{x}-1} + \frac{1}{2}(\bar{x}-1)^2 e^{\bar{x}-1}\Big) + \bar{\mu}_2\bar{w}_2\Big(2\Big(\bar{x}-\frac{1}{2}\Big)e^{\bar{x}-1} + \Big(\bar{x}-\frac{1}{2}\Big)^2 e^{\bar{x}-1}\Big)$$

$$+ \bar{\nu}_l\Big(\frac{-l\bar{u}_l}{2\bar{x}-1}\Big) + (\bar{\gamma}_1^1)_{t^1} + (\bar{\gamma}_2^1)_{t^2} = 0, \tag{6.3.10}$$

$$\bar{\mu}_1\bar{w}_1 + 2\bar{\mu}_2\bar{c} - \bar{v}_1^1\bar{\gamma}_1^1 - \bar{v}_2^1\bar{\gamma}_2^1 = 0, \tag{6.3.11}$$

$$-\bar{\nu}_l\frac{1}{2}l\log(2\bar{x}-1)\bar{u}_l = 0. \tag{6.3.12}$$

Let the Lagrange multipliers $\bar{\nu}_l = 0,\ \forall l \in [0, 1), \bar{\nu}_1 = \frac{1}{4}$ and the uncertain parameters $\bar{u}_l = [1-l, 2+l],\ \forall l \in [0, 1), \bar{u}_1 = \frac{5}{2}$.

Case 1. Assume that the Lagrange multipliers $\bar{\gamma}_\alpha^\tau, \tau = 1, \alpha = 1, 2$ are constant functions, say $\bar{\gamma}_1^1 = d_1$ and $\bar{\gamma}_2^1 = d_2$, where $d_1, d_2 \in \mathbb{R}_+$. Therefore, Eq. (6.3.11) gives

$$\bar{c} = \frac{d_1\bar{v}_1^1 + d_2\bar{v}_2^1 - \bar{w}_1\bar{\mu}_1}{2\bar{\mu}_2},\quad \forall t \in \Gamma_{0,\frac{1}{2}},$$

which shows that $\bar{c}$ is a constant function on $\Gamma_{0,\frac{1}{2}}$. Taking into account of equality constraints (6.3.7) and (6.3.8), it follows that

$$\bar{x} = \Big(1 - \bar{v}_1^1\frac{d_1\bar{v}_1^1 + d_2\bar{v}_2^1 - \bar{w}_1\bar{\mu}_1}{2\bar{\mu}_2}\Big)t^1 + \Big(1 - \bar{v}_2^1\frac{d_1\bar{\gamma}_1^1 + d_2\bar{\gamma}_2^1 - \bar{w}_1\bar{\mu}_1}{2\bar{\mu}_2}\Big)t^2 + d_3,\ d_3 \in \mathbb{R}_+.$$

By using the boundary condition (6.3.9) along with Lagrange multiplier $\bar{\mu}_1 = \bar{\mu}_2 = \frac{1}{2}$ and the uncertain parameters $\bar{w}_1 = \bar{w}_2 = 1, \bar{v}_1^1 = \bar{v}_2^1 = 1$, we get $d_3 = 0$ and $d_1 + d_2 = 1$, and therefore

$$\bar{x} = \frac{1}{2}(t^1 + t^2) + 1,\ \bar{c} = \frac{1}{2},\ \forall t \in \Gamma_{0,\frac{1}{2}}.$$

One can easily verify that the robust necessary efficiency conditions (6.3.10), (6.3.11) and (6.3.12) are satisfied at $(\bar{x}, \bar{c}) = (\frac{1}{2}(t^1 + t^2) + 1, \frac{1}{2})$ at $t^1 = t^2 = 0$, i.e., $(1, \frac{1}{2})$ with the Lagrange multipliers $\bar{\mu}_1 = \bar{\mu}_2 = \frac{1}{2}, \bar{\nu}_l = 0,\ \forall l \in [0, 1), \bar{\nu}_1 = \frac{1}{4}, \bar{\gamma}_1^1 + \bar{\gamma}_2^1 = 1$ and the uncertain parameters $\bar{w}_1 = \bar{w}_2 = 1, \bar{u}_l = [1-l, 2+l]$, $\forall l \in [0, 1), \bar{u}_1 = \frac{5}{2}, \bar{v}_1^1 = \bar{v}_2^1 = 1$. Further, it can be easily verified that the involved functionals are convex at $(\bar{x}, \bar{c}) = (1, \frac{1}{2})$. Hence, all the conditions of Theorem 6.3.0.1 are satisfied, which ensure that $(1, \frac{1}{2})$ is also a weak robust efficient solution to the problem (SMMOPU1).

Case 2. Assume that the Lagrange multipliers $\bar{\gamma}_\alpha^\tau$, $\tau = 1, \alpha = 1, 2$ are functions of t, say $\bar{\gamma}_1^1 = t^2$, $\bar{\gamma}_2^1 = t^1$. So, from Eq. (6.3.11), we have

$$\bar{c} = t_2 + t_1 - \frac{1}{2}, \ \forall t \in \Gamma_{0,\frac{1}{2}}. \tag{6.3.13}$$

On solving Eqs. (6.3.7) and (6.3.8) with the help of Eq. (6.3.13) together with the Lagrange multiplier $\bar{\mu}_1 = 1$, we have

$$\bar{x} = \frac{3}{2}(t^1 + t^2) - \frac{1}{2}\Big[(t^1)^2 + (t^2)^2\Big] - 2t^1t^2 + d_3, \ d_3 \in \mathbb{R}_+,$$

and, by considering the boundary conditions (6.3.9), we get

$$\bar{x} = \frac{3}{2}(t^1 + t^2) - \frac{1}{2}\Big[(t^1)^2 + (t^2)^2\Big] - 2t^1t^2 + 1.$$

This shows that the state function is not an affine function. Consequently, we get a contradiction with the hypothesis.

6.4 The Exact l_1 Penalty Function Method

In this section, we use the exact l_1 penalty function method to formulate the penalized semi-infinite multi-dimensional multi-objective optimization problem associated with the semi-infinite multi-dimensional multi-objective optimization problem with data uncertainty (SMMOPU).

The penalized semi-infinite multi-dimensional multi-objective optimization problem associated to the problem (SMMOPU) is defined as follows:

$$(\text{SMMOPU})_\rho \quad \min_{(x(\cdot),c(\cdot))} \Big\{ \mathscr{P}(t, x(t), c(t), w, u, v, \rho(t)) = \int_\Gamma \Big\{ f(t, x(t), c(t), w)$$
$$+ \rho(t)\Big[\sum_{l\in\Lambda} G_l^+(t, x(t), c(t), u) + \sum_{\tau=1}^{q}\sum_{\alpha=1}^{p} \Big| H_\alpha^\tau(t, x(t), c(t), v) - \frac{\partial x^\tau}{\partial t^\alpha} \Big| \Big] e \Big\} dt \Big\},$$

where $e = (1, \ldots, 1) \in \mathbb{R}^s$, $w \in \mathcal{W}$, $u \in \mathcal{U}$ and $v \in \mathcal{V}$ are the uncertain parameters, $\rho(t)$ is a non-zero non-negative penalty parameter and the penalty function $G_l^+(t, x(t), c(t), u)$ is defined as

$$G_l^+(t, x(t), c(t), u) = \begin{cases} 0, & \text{if } G_l(t, x(t), c(t), u) \leqq 0, \\ G_l(t, x(t), c(t), u), & \text{if } G_l(t, x(t), c(t), u) > 0. \end{cases} \tag{6.4.1}$$

The associated robust counterpart to the problem $(\text{SMMOPU})_\rho$ is given by

$$(\text{RSMMOPU})_\rho \quad \min_{(x(\cdot),c(\cdot))} \Big\{ \max_{w\in\mathcal{W}} \mathscr{P}(t,x(t),c(t),w,u,v,\rho(t)) = \int_\Gamma \Big\{ \max_{w\in\mathcal{W}} f(t,x(t),c(t),w)$$
$$+\rho(t)\Big[\sum_{l\in\Lambda} G_l^+(t,x(t),c(t),u) + \sum_{\tau=1}^{q}\sum_{\alpha=1}^{p}\Big|H_\alpha^\tau(t,x(t),c(t),v) - \frac{\partial x^\tau}{\partial t^\alpha}\Big|\Big]e\Big\}dt\Big\},$$

for all $u \in \mathcal{U}$ and $v \in \mathcal{V}$.

Definition 6.4.0.1 A point $(\bar{x},\bar{c}) \in \mathscr{X}\times\mathscr{C}$ is said to be a weak robust efficient solution to the problem $(\text{SMMOPU})_\rho$, if there does not exist another point $(x,c) \in \mathscr{X}\times\mathscr{C}$ such that

$$\max_{w\in\mathcal{W}} \mathscr{P}(\pi,w,\bar{u},\bar{v},\rho) < \max_{w\in\mathcal{W}} \mathscr{P}(\bar{\pi},w,\bar{u},\bar{v},\rho),$$

where $\bar{u} \in \mathcal{U}$ and $\bar{v} \in \mathcal{V}$.

Theorem 6.4.0.1 *Let $(\bar{x},\bar{c})$ be a robust feasible solution to the semi-infinite multi-dimensional multi-objective optimization problem* (SMMOPU) *and the robust necessary efficiency conditions* (6.2.1)–(6.2.4) *be satisfied at $(\bar{x},\bar{c})$ and $\max_{w\in\mathcal{W}} f(\pi,w) = f(\pi,\bar{w})$. Further, assume that $\int_\Gamma \bar{\mu}^T f(\pi,\bar{w})dt$, $\int_\Gamma \bar{v}^T G(\pi,\bar{u})dt$ and $\int_\Gamma \bar{\gamma}^T(H(\pi,\bar{v}) - x_\alpha)dt$ are convex at $(\bar{x},\bar{c})$ on $\mathscr{X}\times\mathscr{C}$. If $\rho \geq \max\{\bar{v},|\bar{\gamma}|\}$, then $(\bar{x},\bar{c})$ is a weak robust efficient solution to the penalized semi-infinite multi-dimensional multi-objective optimization problem* (SMMOPU)ρ.

Proof Contrary to the result and assume that $(\bar{x},\bar{c})$ is not a weak robust efficient solution to the penalized problem $(\text{SMMOPU})_\rho$. Then, there exists another point $(\hat{x},\hat{c}) \in \mathscr{X}\times\mathscr{C}$ such that

$$\max_{w\in\mathcal{W}} \mathscr{P}(\hat{\pi},w,\bar{u},\bar{v},\rho) < \max_{w\in\mathcal{W}} \mathscr{P}(\bar{\pi},w,\bar{u},\bar{v},\rho).$$

By using the definition of the penalized semi-infinite multi-dimensional multi-objective optimization problem $(\text{SMMOPU})_\rho$, we get

$$\int_\Gamma \Big\{ \max_{w\in\mathcal{W}} f(\hat{\pi},w) + \rho\Big[\sum_{l\in\Lambda} G_l^+(\hat{\pi},\bar{u}) + \sum_{\tau=1}^{q}\sum_{\alpha=1}^{p}\Big|H_\alpha^\tau(\hat{\pi},\bar{v}) - \frac{\partial x^\tau}{\partial t^\alpha}\Big|\Big]e\Big\}dt$$
$$< \int_\Gamma \Big\{ \max_{w\in\mathcal{W}} f(\bar{\pi},w) + \rho\Big[\sum_{l\in\Lambda} G_l^+(\bar{\pi},\bar{u}) + \sum_{\tau=1}^{q}\sum_{\alpha=1}^{p}\Big|H_\alpha^\tau(\bar{\pi},\bar{v}) - \frac{\partial x^\tau}{\partial t^\alpha}\Big|\Big]e\Big\}dt.$$

Since $\max_{p\in\mathcal{W}} f(\pi,w) = f(\pi,\bar{w})$, we have

$$\int_\Gamma \Big\{ f(\hat{\pi},\bar{w}) + \rho\Big[\sum_{l\in\Lambda} G_l^+(\hat{\pi},\bar{u}) + \sum_{\tau=1}^{q}\sum_{\alpha=1}^{p}\Big|H_\alpha^\tau(\hat{\pi},\bar{v}) - \frac{\partial x^\tau}{\partial t^\alpha}\Big|\Big]e\Big\}dt$$
$$< \int_\Gamma \Big\{ f(\bar{\pi},\bar{w}) + \rho\Big[\sum_{l\in\Lambda} G_l^+(\bar{\pi},\bar{u}) + \sum_{\tau=1}^{q}\sum_{\alpha=1}^{p}\Big|H_\alpha^\tau(\bar{\pi},\bar{v}) - \frac{\partial x^\tau}{\partial t^\alpha}\Big|\Big]e\Big\}dt.$$

The above inequality together with the robust feasibility of $(\bar{x}, \bar{c})$ to the problem (SMMOPU) and the relation (6.4.1) yields

$$\int_{\Gamma} \Big\{ f(\hat{\pi}, \bar{w}) + \rho \Big[\sum_{l \in \Lambda} G_l^{+}(\hat{\pi}, \bar{u}) + \sum_{\tau=1}^{q} \sum_{\alpha=1}^{p} \Big| H_{\alpha}^{\tau}(\hat{\pi}, \bar{v}) - \frac{\partial x^{\tau}}{\partial t^{\alpha}} \Big| \Big] e \Big\} dt < \int_{\Gamma} f(\bar{\pi}, \bar{w}) dt.$$

Multiplying the above inequality with the scalar vector $\bar{\mu}$ on both sides, we get

$$\int_{\Gamma} \Big\{ \bar{\mu}^T f(\hat{\pi}, \bar{w}) + \bar{\mu}^T \rho \Big[\sum_{l \in \Lambda} G_l^{+}(\hat{\pi}, \bar{u}) + \sum_{\tau=1}^{q} \sum_{\alpha=1}^{p} \Big| H_{\alpha}^{\tau}(\hat{\pi}, \bar{v}) - \frac{\partial x^{\tau}}{\partial t^{\alpha}} \Big| \Big] e \Big\} dt$$
$$< \int_{\Gamma} \bar{\mu}^T f(\bar{\pi}, \bar{w}) dt.$$

As $\bar{\mu}^T e = 1$, it follows that

$$\int_{\Gamma} \Big\{ \bar{\mu}^T f(\hat{\pi}, \bar{w}) + \rho \Big[\sum_{l \in \Lambda} G_l^{+}(\hat{\pi}, \bar{u}) + \sum_{\tau=1}^{q} \sum_{\alpha=1}^{p} \Big| H_{\alpha}^{\tau}(\hat{\pi}, \bar{v}) - \frac{\partial x^{\tau}}{\partial t^{\alpha}} \Big| \Big] \Big\} dt$$
$$< \int_{\Gamma} \bar{\mu}^T f(\bar{\pi}, \bar{w}) dt.$$

Since $\rho \geq \max\{\bar{\nu}, |\bar{\gamma}|\}$, the above inequality yields

$$\int_{\Gamma} \Big\{ \bar{\mu}^T f(\hat{\pi}, \bar{w}) + \sum_{l \in \Lambda} \bar{\nu}_l G_l^{+}(\hat{\pi}, \bar{u}) + \sum_{\tau=1}^{q} \sum_{\alpha=1}^{p} \bar{\gamma}_{\alpha}^{\tau} \Big| H_{\alpha}^{\tau}(\hat{\pi}, \bar{v}) - \frac{\partial x^{\tau}}{\partial t^{\alpha}} \Big| \Big\} dt$$
$$< \int_{\Gamma} \bar{\mu}^T f(\bar{\pi}, \bar{w}) dt. \tag{6.4.2}$$

Again, using the robust feasibility of $(\bar{x}, \bar{c})$ to the problem (SMMOPU) along with the condition (6.2.3), the inequality (6.4.2) can be written as

$$\int_{\Gamma} \Big\{ \bar{\mu}^T f(\hat{\pi}, \bar{w}) + \sum_{l \in \Lambda} \bar{\nu}_l G_l(\hat{\pi}, \bar{u}) + \sum_{\tau=1}^{q} \sum_{\alpha=1}^{p} \bar{\gamma}_{\alpha}^{\tau} \Big(H_{\alpha}^{\tau}(\hat{\pi}, \bar{v}) - \frac{\partial x^{\tau}}{\partial t^{\alpha}} \Big) \Big\} dt$$
$$< \int_{\Gamma} \Big\{ \bar{\mu}^T f(\bar{\pi}, \bar{w}) + \sum_{l \in \Lambda} \bar{\nu}_l G_l(\bar{\pi}, \bar{u}) + \sum_{\tau=1}^{q} \sum_{\alpha=1}^{p} \bar{\gamma}_{\alpha}^{\tau} \Big(H_{\alpha}^{\tau}(\bar{\pi}, \bar{v}) - \frac{\partial x^{\tau}}{\partial t^{\alpha}} \Big) \Big\} dt,$$

equivalently,

$$\int_{\Gamma}\left\{\bar{\mu}^T f(\hat{\pi},\bar{w})+\bar{v}^T G(\hat{\pi},\bar{u})+\bar{\gamma}^T(H(\hat{\pi},\bar{v})-x_\alpha)\right\}dt$$
$$<\int_{\Gamma}\left\{\bar{\mu}^T f(\bar{\pi},\bar{w})+\bar{v}^T G(\bar{\pi},\bar{u})+\bar{\gamma}^T(H(\bar{\pi},\bar{v})-x_\alpha)\right\}dt. \tag{6.4.3}$$

On the other hand, by assumption $\int_{\Gamma}\bar{\mu}^T f(\pi,\bar{w})dt$, $\int_{\Gamma}\bar{v}^T G(\pi,\bar{u})dt$ and $\int_{\Gamma}\bar{\gamma}^T(H(\pi,\bar{v})-x_\alpha)dt$ are convex at $(\bar{x},\bar{c})$ on $\mathscr{X}\times\mathscr{C}$, we have

$$\int_{\Gamma}\left\{\bar{\mu}^T f(\hat{\pi},\bar{w})-\bar{\mu}^T f(\bar{\pi},\bar{w})\right\}dt \geqq \int_{\Gamma}(\hat{x}-\bar{x})\bar{\mu}^T f_x(\bar{\pi},\bar{w})dt+\int_{\Gamma}(\hat{c}-\bar{c})\bar{\mu}^T f_c(\bar{\pi},\bar{w})dt, \tag{6.4.4}$$

$$\int_{\Gamma}\left\{\bar{v}^T G(\hat{\pi},\bar{u})-\bar{v}^T G(\bar{\pi},\bar{u})\right\}dt \geqq \int_{\Gamma}(\hat{x}-\bar{x})\bar{v}^T G_x(\bar{\pi},\bar{u})dt+\int_{\Gamma}(\hat{c}-\bar{c})\bar{v}^T G_c(\bar{\pi},\bar{u})dt, \tag{6.4.5}$$

and $$\int_{\Gamma}\left\{\bar{\gamma}^T(H(\hat{\pi},\bar{v})-x_\alpha)-\bar{\gamma}^T(H(\bar{\pi},\bar{v})-x_\alpha)\right\}dt$$
$$\geqq \int_{\Gamma}(\hat{x}-\bar{x})\left\{\bar{\gamma}^T H_x(\bar{\pi},\bar{v})+\bar{\gamma}_t\right\}dt+\int_{\Gamma}(\hat{c}-\bar{c})\bar{\gamma}^T H_c(\bar{\pi},\bar{v})dt. \tag{6.4.6}$$

On combining the inequalities (6.4.4)–(6.4.6) together with the robust necessary efficiency conditions (6.2.1)–(6.2.2), we have

$$\int_{\Gamma}\left\{\bar{\mu}^T f(\hat{\pi},\bar{w})+\bar{v}^T G(\hat{\pi},\bar{u})+\bar{\gamma}^T(H(\hat{\pi},\bar{v})-x_\alpha)\right\}dt$$
$$\geqq \int_{\Gamma}\left\{\bar{\mu}^T f(\bar{\pi},\bar{w})+\bar{v}^T G(\bar{\pi},\bar{u})+\bar{\gamma}^T(H(\bar{\pi},\bar{v})-x_\alpha)\right\}dt,$$

which contradicts the inequality (6.4.3). Hence, this completes the proof. □

Let us consider the Example 6.3.0.1 to show the relationship between the solution set of semi-infinite multi-dimensional multi-objective optimization problem (SMMOPU) and its associated penalized problem $(\text{SMMOPU})_\rho$.

Example 6.4.0.1 The penalized semi-infinite multi-dimensional multi-objective optimization problem in the face of data uncertainty $(\text{SMMOPU1})_\rho$ associated with the semi-infinite multi-dimensional multi-objective optimization problem (SMMOPU1) is defined as:

$$(\text{SMMOPU1})_\rho \quad \min_{(x(\cdot),c(\cdot))}\Bigg(\int_{\Gamma_{0,\frac{1}{2}}}\left\{\left(\frac{1}{2}(x-1)^2e^{x-1}+cw_1\right)+\rho\Big[\max\{0,-\frac{1}{2}l\log(2x-1)u_l\}\right.$$
$$\left.+\left|x_{t^1}-1+cv_1^1\right|+\left|x_{t^2}-1+cv_2^1\right|\Big]\right\}dt^1dt^2,$$
$$\int_{\Gamma_{0,\frac{1}{2}}}\left\{\left(c^2+w_2\left(x-\frac{1}{2}\right)^2e^{x-1}\right)+\rho\Big[\max\{0,-\frac{1}{2}l\log(2x-1)u_l\}\right.$$
$$\left.+\left|x_{t^1}-1+cv_1^1\right|+\left|x_{t^2}-1+cv_2^1\right|\Big]\right\}dt^1dt^2\Bigg).$$

The robust counterpart for $(\text{SMMOPU1})_\rho$ is formulated as:

$$(\text{RSMMOPU1})_\rho \quad \min_{(x(\cdot),c(\cdot))}\Bigg(\int_{\Gamma_{0,\frac{1}{2}}}\Big\{\max_{w_1\in\mathcal{W}_1}\Big(\frac{1}{2}(x-1)^2e^{x-1}+cw_1\Big)+\rho\Big[\max\{0,$$
$$-\frac{1}{2}l\log(2x-1)u_l\}+\Big|x_{t^1}-1+cv_1^1\Big|+\Big|x_{t^2}-1+cv_2^1\Big|\Big]\Big\}dt^1dt^2,$$
$$\int_{\Gamma_{0,\frac{1}{2}}}\Big\{\max_{w_2\in\mathcal{W}_2}\Big(c^2+w_2\Big(x-\frac{1}{2}\Big)^2e^{x-1}\Big)+\rho\Big[\max\{0,-\frac{1}{2}l\log(2x-1)u_l\}$$
$$+\Big|x_{t^1}-1+cv_1^1\Big|+\Big|x_{t^2}-1+cv_2^1\Big|\Big]\Big\}dt^1dt^2\Bigg).$$

In virtue of Definition 6.4.0.1, it can be observed that the inequality $\max_{w\in\mathcal{W}}\mathcal{P}(\pi, w, \bar{u}, \bar{v}, \rho) - \max_{w\in\mathcal{W}}\mathcal{P}(\bar{\pi}, w, \bar{u}, \bar{v}, \rho) \geq 0$ at $(\bar{x}, \bar{c}) = (1, \frac{1}{2})$. This shows that $(1, \frac{1}{2})$ is a weak robust efficient solution to $(\text{SMMOPU1})_\rho$ with $\rho \geq \frac{5}{2}$. Hence, Theorem 6.4.0.1 is verified.

Proposition 6.4.0.1 *Let $(\bar{x}, \bar{c}) \in \mathcal{X} \times \mathcal{C}$ be a weak robust efficient solution to the problem* $(\text{SMMOPU})_{\bar{\rho}}$ *and* $\max_{w\in\mathcal{W}} f(\pi, w) = f(\pi, \bar{w})$. *Then, there does not exist another point $(x, c) \in \mathcal{D}$ such that*

$$\int_\Gamma f(\pi, \bar{w})dt < \int_\Gamma f(\bar{\pi}, \bar{w})dt.$$

Proof Since $(\bar{x}, \bar{c})$ is a weak robust efficient solution to the problem $(\text{SMMOPU})_{\bar{\rho}}$, then there does not exist another point $(x, c) \in \mathcal{X} \times \mathcal{C}$ such that

$$\max_{w\in\mathcal{W}}\mathcal{P}(\pi, w, \bar{u}, \bar{v}, \bar{\rho}) < \max_{w\in\mathcal{W}}\mathcal{P}(\bar{\pi}, w, \bar{u}, \bar{v}, \bar{\rho}).$$

By using the definition of the penalized semi-infinite multi-dimensional multi-objective optimization problem $(\text{SMMOPU})_\rho$, we get

$$\int_\Gamma\Big\{\max_{w\in\mathcal{W}} f(\pi, w)+\bar{\rho}\Big[\sum_{l\in\Lambda}G_l^+(\pi, \bar{u})+\sum_{\tau=1}^{q}\sum_{\alpha=1}^{p}\Big|H_\alpha^\tau(\pi, \bar{v})-x_\alpha\Big|\Big]e\Big\}dt$$
$$<\int_\Gamma\Big\{\max_{w\in\mathcal{W}} f(\bar{\pi}, w)+\bar{\rho}\Big[\sum_{l\in\Lambda}G_l^+(\bar{\pi}, \bar{u})+\sum_{\tau=1}^{q}\sum_{\alpha=1}^{p}\Big|H_\alpha^\tau(\bar{\pi}, \bar{v})-x_\alpha\Big|\Big]e\Big\}dt.$$

Since $\max_{w\in\mathcal{W}} f(\pi, w) = f(\pi, \bar{w})$, we have

$$\int_\Gamma\Big\{f(\pi, \bar{w})+\bar{\rho}\Big[\sum_{l\in\Lambda}G_l^+(\pi, \bar{u})+\sum_{\tau=1}^{q}\sum_{\alpha=1}^{p}\Big|H_\alpha^\tau(\pi, \bar{v})-x_\alpha\Big|\Big]e\Big\}dt$$
$$<\int_\Gamma\Big\{f(\bar{\pi}, \bar{w})+\bar{\rho}\Big[\sum_{l\in\Lambda}G_l^+(\bar{\pi}, \bar{u})+\sum_{\tau=1}^{q}\sum_{\alpha=1}^{p}\Big|H_\alpha^\tau(\bar{\pi}, \bar{v})-x_\alpha\Big|\Big]e\Big\}dt.$$

Since $\mathcal{D} \subseteq \mathcal{X} \times \mathcal{C}$, therefore, it follows that there does not exist another point $(x, c) \in \mathcal{D}$ such that

$$\int_{\Gamma} \Big\{ f(\pi, \bar{w}) + \bar{\rho}\Big[\sum_{l \in \Lambda} G_l^{+}(\pi, \bar{u}) + \sum_{\tau=1}^{q} \sum_{\alpha=1}^{p} \Big| H_{\alpha}^{\tau}(\pi, \bar{v}) - x_{\alpha} \Big| \Big] e \Big\} dt$$
$$< \int_{\Gamma} \Big\{ f(\bar{\pi}, \bar{w}) + \bar{\rho}\Big[\sum_{l \in \Lambda} G_l^{+}(\bar{\pi}, \bar{u}) + \sum_{\tau=1}^{q} \sum_{\alpha=1}^{p} \Big| H_{\alpha}^{\tau}(\bar{\pi}, \bar{v}) - x_{\alpha} \Big| \Big] e \Big\} dt.$$

Equivalently, one can say that the following relation holds for all $(x, c) \in \mathcal{D}$

$$\int_{\Gamma} \Big\{ f_k(\pi, \bar{w}) + \bar{\rho}\Big[\sum_{l \in \Lambda} G_l^{+}(\pi, \bar{u}) + \sum_{\tau=1}^{q} \sum_{\alpha=1}^{p} \Big| H_{\alpha}^{\tau}(\pi, \bar{v}) - \frac{\partial x^{\tau}}{\partial t^{\alpha}} \Big| \Big] \Big\} dt$$
$$\geqq \int_{\Gamma} \Big\{ f_k(\bar{\pi}, \bar{w}) + \bar{\rho}\Big[\sum_{l \in \Lambda} G_l^{+}(\bar{\pi}, \bar{u}) + \sum_{\tau=1}^{q} \sum_{\alpha=1}^{p} \Big| H_{\alpha}^{\tau}(\bar{\pi}, \bar{v}) - \frac{\partial x^{\tau}}{\partial t^{\alpha}} \Big| \Big] \Big\} dt.$$

By virtue of the relation (6.4.1), we get

$$\int_{\Gamma} f_k(\pi, \bar{w}) dt \geqq \int_{\Gamma} f_k(\bar{\pi}, \bar{w}) dt.$$

Therefore, we conclude that there does not exist another point $(x, c) \in \mathcal{D}$ such that

$$\int_{\Gamma} f(\pi, \bar{w}) dt < \int_{\Gamma} f(\bar{\pi}, \bar{w}) dt.$$

This completes the proof. □

Now, we prove the converse result of Theorem 6.4.0.1 without imposing any convexity assumption on the involved functionals.

Theorem 6.4.0.2 *Let* $(\bar{x}, \bar{c})$ *be a weak robust efficient solution to the semi-infinite multi-dimensional multi-objective optimization problem* (SMMOPU)$_{\bar{\rho}}$. *Assume that* $\max_{w \in \mathcal{W}} \mathcal{P}(\pi, w, \bar{u}, \bar{v}, \rho) \geqq \max_{w \in \mathcal{W}} \mathcal{P}(\bar{\pi}, w, \bar{u}, \bar{v}, \rho)$ *holds for any* $\rho \geq \bar{\rho}$ *and for all* $(x, c) \in \mathcal{D}$. *Further, if* $\max_{w \in \mathcal{W}} f(\pi, w) = f(\pi, \bar{w})$ *and* $\rho \geq \max\{\bar{v}, |\bar{\gamma}|\}$, *then* $(\bar{x}, \bar{c})$ *is a weak robust efficient solution to the semi-infinite multi-dimensional multi-objective optimization problem* (SMMOPU).

Proof Since $(\bar{x}, \bar{c})$ is a weak robust efficient solution to the problem (SMMOPU)$_{\bar{\rho}}$ and $\max_{w \in \mathcal{W}} f(\pi, w) = f(\pi, \bar{w})$, thus, by Proposition 6.4.0.1, there does not exist another point $(x, c) \in \mathcal{D}$ such that

$$\int_{\Gamma} f(\pi, \bar{w}) dt < \int_{\Gamma} f(\bar{\pi}, \bar{w}) dt,$$

equivalently,

$$\int_{\Gamma} f_k(\pi, \bar{w})dt < \int_{\Gamma} f_k(\bar{\pi}, \bar{w})dt, \ \forall k = \overline{1, s}, \tag{6.4.7}$$

which specify that the objective functional $\int_{\Gamma} f_k(\pi, \bar{w})dt$ is bounded below in the compact subset $\mathcal{D}$.

Now, to show that $(\bar{x}, \bar{c})$ is a weak robust efficient solution to the problem (SMMOPU), firstly, we shall prove that $(\bar{x}, \bar{c}) \in \mathcal{D}$. We proceed by contradiction and assume that $(\bar{x}, \bar{c}) \notin \mathcal{D}$. Therefore, by virtue of the relation (6.4.1), we have

$$\int_{\Gamma} \Big\{ \sum_{l \in \Lambda} G_l^+(\bar{\pi}, \bar{u}) + \sum_{\tau=1}^{q} \sum_{\alpha=1}^{p} \Big| H_\alpha^\tau(\bar{\pi}, \bar{v}) - \frac{\partial x^\tau}{\partial t^\alpha} \Big| \Big\} dt > 0. \tag{6.4.8}$$

Let (x, c) be any robust feasible solution to the problem (SMMOPU). By assumption, the following inequality

$$\max_{w \in \mathcal{W}} \mathscr{P}(\pi, w, \bar{u}, \bar{v}, \rho)dt \geqq \max_{w \in \mathcal{W}} \mathscr{P}(\bar{\pi}, w, \bar{u}, \bar{v}, \rho)dt$$

holds for all $(x, c) \in \mathcal{D}$ and for any $\rho \geq \bar{\rho}$.

By using the definition of the penalized semi-infinite multi-dimensional multi-objective optimization problem (SMMOPU), we get

$$\int_{\Gamma} \Big\{ \max_{w \in \mathcal{W}} f(\pi, w) + \rho \Big[\sum_{l \in \Lambda} G_l^+(\pi, \bar{u}) + \sum_{\tau=1}^{q} \sum_{\alpha=1}^{p} \Big| H_\alpha^\tau(\pi, \bar{v}) - \frac{\partial x^\tau}{\partial t^\alpha} \Big| \Big] e \Big\} dt$$
$$\geqq \int_{\Gamma} \Big\{ \max_{w \in \mathcal{W}} f(\bar{\pi}, w) + \rho \Big[\sum_{l \in \Lambda} G_l^+(\bar{\pi}, \bar{u}) + \sum_{\tau=1}^{q} \sum_{\alpha=1}^{p} \Big| H_\alpha^\tau(\bar{\pi}, \bar{v}) - \frac{\partial x^\tau}{\partial t^\alpha} \Big| \Big] e \Big\} dt.$$

Since $\max_{w \in \mathcal{W}} f(\pi, w) = f(\pi, \bar{w})$, we get

$$\int_{\Gamma} \Big\{ f(\pi, \bar{w}) + \rho \Big[\sum_{l \in \Lambda} G_l^+(\pi, \bar{u}) + \sum_{\tau=1}^{q} \sum_{\alpha=1}^{p} \Big| H_\alpha^\tau(\pi, \bar{v}) - \frac{\partial x^\tau}{\partial t^\alpha} \Big| \Big] e \Big\} dt$$
$$\geqq \int_{\Gamma} \Big\{ f(\bar{\pi}, \bar{w}) + \rho \Big[\sum_{l \in \Lambda} G_l^+(\bar{\pi}, \bar{u}) + \sum_{\tau=1}^{q} \sum_{\alpha=1}^{p} \Big| H_\alpha^\tau(\bar{\pi}, \bar{v}) - \frac{\partial x^\tau}{\partial t^\alpha} \Big| \Big] e \Big\} dt. \tag{6.4.9}$$

On the other hand, if we set

$$\rho > \max \left\{ \frac{\int_{\Gamma} \{ f(\pi, \bar{w}) - f(\bar{\pi}, \bar{w}) \} dt}{\int_{\Gamma} \Big\{ \sum_{l \in \Lambda} G_l^+(\bar{\pi}, \bar{u}) + \sum_{\tau=1}^{q} \sum_{\alpha=1}^{p} \Big| H_\alpha^\tau(\bar{\pi}, \bar{v}) - \frac{\partial x^\tau}{\partial t^\alpha} \Big| \Big\} e \, dt}, \bar{\rho}; (x, c) \in \mathcal{D} \right\},$$

therefore, in view of the inequalities (6.4.7) and (6.4.8), it follows that ρ is a positive real number, and

$$\int_{\Gamma}\Big\{f(\bar{\pi},\bar{w})+\rho\Big[\sum_{l\in\Lambda}G_l^+(\bar{\pi},\bar{u})+\sum_{\tau=1}^{q}\sum_{\alpha=1}^{p}\Big|H_\alpha^\tau(\bar{\pi},\bar{v})-\frac{\partial x^\tau}{\partial t^\alpha}\Big|\Big]e\Big\}dt>\int_{\Gamma}f(\pi,\bar{w})dt.$$

Further, the robust feasibility of (x, c) to the problem (SMMOPU) and by virtue of the relation (6.4.1), it follows from the above inequality

$$\begin{aligned}\int_{\Gamma}\Big\{f(\bar{\pi},\bar{w})+\rho\Big[\sum_{l\in\Lambda}G_l^+(\bar{\pi},\bar{u})+\sum_{\tau=1}^{q}\sum_{\alpha=1}^{p}\Big|H_\alpha^\tau(\bar{\pi},\bar{v})-\frac{\partial x^\tau}{\partial t^\alpha}\Big|\Big]e\Big\}dt\\ >\int_{\Gamma}\Big\{f(\pi,\bar{w})+\rho\Big[\sum_{l\in\Lambda}G_l^+(\pi,\bar{u})+\sum_{\tau=1}^{q}\sum_{\alpha=1}^{p}\Big|H_\alpha^\tau(\pi,\bar{v})-\frac{\partial x^\tau}{\partial t^\alpha}\Big|\Big]e\Big\}dt,\end{aligned}$$

which contradicts the inequality (6.4.9). Therefore, $(\bar{x}, \bar{c}) \in \mathcal{D}$, i.e., $(\bar{x}, \bar{c})$ is a robust feasible solution to the problem (SMMOPU). Hence, the inequality (6.4.7) directly concludes that, $(\bar{x}, \bar{c})$ is a weak robust efficient solution to the problem (SMMOPU). □

6.5 Uncertain Lagrange Functional

In this section, we introduce the uncertain Lagrange functional for semi-infinite multi-dimensional multi-objective optimization problem (SMMOPU) and derive the relationship between a weak robust efficient solution to the problem (SMMOPU) and its associated penalized problem (SMMOPU)$_\rho$ under convexity assumption of the uncertain Lagrange functional.

Definition 6.5.0.1 The uncertain Lagrange functional $L((x, c), \nu, \gamma, w, u, v)$: $\mathscr{X} \times \mathscr{C} \times \mathbb{R}_+^{\Lambda} \times \mathbb{R}^{pq} \times \mathcal{W} \times \mathcal{U} \times \mathcal{V} \to \mathbb{R}$ to the problem (SMMOPU) is defined as:

$$L((x,c),\nu,\gamma,w,u,v)=\int_{\Gamma}\Big\{\max_{w\in\mathcal{W}}f(\pi,w)+\nu^TG(\pi,u)e+\gamma^T(H(\pi,v)-x_\alpha)e\Big\}\,dt.$$

Now, we shall show that the problem (SMMOPU) and its associated penalized problem (SMMOPU)$_\rho$ consist the same weak robust efficient solution set under the convexity assumption of the uncertain Lagrange functional.

Theorem 6.5.0.1 *Let $(\bar{x}, \bar{c})$ be a robust feasible solution to the semi-infinite multi-dimensional multi-objective optimization problem* (SMMOPU) *at which the robust necessary efficiency conditions* (6.2.1)–(6.2.4) *hold and* $\max_{w\in\mathcal{W}} f(\pi, w) = f(\pi, \bar{w})$. *Further, assume that, if the Lagrange functional $L((x, c), \nu, \gamma, w, u, v)$ is convex at $(\bar{x}, \bar{c})$ on $\mathscr{X} \times \mathscr{C}$ and $\rho \geq \max\{\bar{\nu}, |\bar{\gamma}|\}$, then $(\bar{x}, \bar{c})$ is also a weak robust efficient* *solution to the penalized semi-infinite multi-dimensional multi-objective optimization problem* $(\text{SMMOPU})_\rho$.

Proof We proceed by contradiction and assume that $(\bar{x}, \bar{c})$ is not a weak robust efficient solution to the problem $(\text{SMMOPU})_\rho$. Then, there exists another point $(\hat{x}, \hat{c}) \in \mathscr{X} \times \mathscr{C}$ such that

$$\max_{w \in \mathcal{W}} \mathscr{P}(\hat{\pi}, w, \bar{u}, \bar{v}, \rho) < \max_{w \in \mathcal{W}} \mathscr{P}(\bar{\pi}, w, \bar{u}, \bar{v}, \rho).$$

Proceeding on the lines of Theorem 6.4.0.1, we get

$$\int_\Gamma \Big\{ \sum_{k=1}^{s} \bar{\mu}_k f_k(\hat{\pi}, \bar{w}) + \sum_{l \in \Lambda} \bar{v}_l G_l(\hat{\pi}, \bar{u}) + \sum_{\tau=1}^{q} \sum_{\alpha=1}^{p} \bar{\gamma}_\alpha^\tau \Big(H_\alpha^\tau(\hat{\pi}, \bar{v}) - \frac{\partial x^\tau}{\partial t^\alpha} \Big) \Big\} dt$$
$$< \int_\Gamma \Big\{ \sum_{k=1}^{s} \bar{\mu}_k f_k(\bar{\pi}, \bar{w}) + \sum_{l \in \Lambda} \bar{v}_l G_l(\bar{\pi}, \bar{u}) + \sum_{\tau=1}^{q} \sum_{\alpha=1}^{p} \bar{\gamma}_\alpha^\tau \Big(H_\alpha^\tau(\bar{\pi}, \bar{v}) - \frac{\partial x^\tau}{\partial t^\alpha} \Big) \Big\} dt,$$

equivalently,

$$\int_\Gamma \Big\{ \bar{\mu}^T f(\hat{\pi}, \bar{w}) + \bar{v}^T G(\hat{\pi}, \bar{u}) + \bar{\gamma}^T (H(\hat{\pi}, \bar{v}) - x_\alpha) \Big\} dt$$
$$< \int_\Gamma \Big\{ \bar{\mu}^T f(\bar{\pi}, \bar{w}) + \bar{v}^T G(\bar{\pi}, \bar{u}) + \bar{\gamma}^T (H(\bar{\pi}, \bar{v}) - x_\alpha) \Big\} dt. \tag{6.5.1}$$

On the other hand, from the assumption that the uncertain Lagrange functional is convex at $(\bar{x}, \bar{c})$ on $\mathscr{X} \times \mathscr{C}$, we have

$$L((\hat{x}, \hat{c}), \bar{v}, \bar{\gamma}, w, \bar{u}, \bar{v}) - L((\bar{x}, \bar{c}), \bar{v}, \bar{\gamma}, w, \bar{u}, \bar{v})$$
$$\geqq (\hat{x} - \bar{x}) L_x((\bar{x}, \bar{c}), \bar{v}, \bar{\gamma}, w, \bar{u}, \bar{v}) + (\hat{c} - \bar{c}) L_c((\bar{x}, \bar{c}), \bar{v}, \bar{\gamma}, w, \bar{u}, \bar{v}).$$

In virtue of the definition of the Lagrange functional, we get

$$\int_\Gamma \Big\{ \max_{w \in \mathcal{W}} f(\hat{\pi}, w) + \bar{v}^T G(\hat{\pi}, \bar{u}) e + \bar{\gamma}^T (H(\hat{\pi}, \bar{v}) - x_\alpha) e \Big\} dt - \int_\Gamma \Big\{ \max_{w \in \mathcal{W}} f(\bar{\pi}, w)$$
$$+ \bar{v}^T G(\bar{\pi}, \bar{u}) e + \bar{\gamma}^T (H(\bar{\pi}, \bar{v}) - x_\alpha) e \Big\} dt \geqq \int_\Gamma (\hat{x} - \bar{x}) \Big\{ \max_{w \in \mathcal{W}} f_x(\bar{\pi}, w) + \bar{v}^T G_x(\bar{\pi}, \bar{u}) e$$
$$+ (\bar{\gamma}^T H_x(\bar{\pi}, \bar{v}) + \bar{\gamma}_t) e \Big\} dt + \int_\Gamma (\hat{c} - \bar{c}) \Big\{ \max_{w \in \mathcal{W}} f_c(\bar{\pi}, w) + \bar{v}^T G_c(\bar{\pi}, \bar{u}) e + \bar{\gamma}^T H_c(\bar{\pi}, \bar{v}) e \Big\} dt.$$

Because $\max_{w \in \mathcal{W}} f(\pi, w) = f(\pi, \bar{w})$, we have

$$\int_{\Gamma}\Big\{f(\hat{\pi},\bar{w})+\bar{v}^T G(\hat{\pi},\bar{u})e+\bar{\gamma}^T(H(\hat{\pi},\bar{v})-x_\alpha)e\Big\}dt-\int_{\Gamma}\Big\{f(\bar{\pi},\bar{w})$$
$$+\bar{v}^T G(\bar{\pi},\bar{u})e+\bar{\gamma}^T(H(\bar{\pi},\bar{v})-x_\alpha)e\Big\}dt \geqq \int_{\Gamma}(\hat{x}-\bar{x})\Big\{f_x(\bar{\pi},\bar{w})+\bar{v}^T G_x(\bar{\pi},\bar{u})e$$
$$+(\bar{\gamma}^T H_x(\bar{\pi},\bar{v})+\bar{\gamma}_t)e\Big\}dt+\int_{\Gamma}(\hat{c}-\bar{c})\Big\{f_c(\bar{\pi},\bar{w})+\bar{v}^T G_c(\bar{\pi},\bar{u})e+\bar{\gamma}^T H_c(\bar{\pi},\bar{v})e\Big\}dt.$$

Since $\bar{\mu}>0$ and $\bar{\mu}^T e=1$, multiplying the above inequality by $\bar{\mu}$ on the both side, we get

$$\int_{\Gamma}\Big\{\bar{\mu}^T f(\hat{\pi},\bar{w})+\bar{v}^T G(\hat{\pi},\bar{u})+\bar{\gamma}^T(H(\hat{\pi},\bar{v})-x_\alpha)\Big\}dt-\int_{\Gamma}\Big\{\bar{\mu}^T f(\bar{\pi},\bar{w})$$
$$+\bar{v}^T G(\bar{\pi},\bar{u})+\bar{\gamma}^T(H(\bar{\pi},\bar{v})-x_\alpha)\Big\}dt \geqq \int_{\Gamma}(\hat{x}-\bar{x})\Big\{\bar{\mu}^T f_x(\bar{\pi},\bar{w})+\bar{v}^T G_x(\bar{\pi},\bar{u})$$
$$+(\bar{\gamma}^T H_x(\bar{\pi},\bar{v})+\bar{\gamma}_t)\Big\}dt+\int_{\Gamma}(\hat{c}-\bar{c})\Big\{\bar{\mu}^T f_c(\bar{\pi},\bar{w})+\bar{v}^T G_c(\bar{\pi},\bar{u})+\bar{\gamma}^T H_c(\bar{\pi},\bar{v})\Big\}dt.$$

By using the robust necessary efficiency conditions (6.2.1) and (6.2.2), the above inequality can be written as

$$\int_{\Gamma}\Big\{\bar{\mu}^T f(\hat{\pi},\bar{w})+\bar{v}^T G(\hat{\pi},\bar{u})+\bar{\gamma}^T(H(\hat{\pi},\bar{v})-x_\alpha)\Big\}dt$$
$$\geqq \int_{\Gamma}\Big\{\bar{\mu}^T f(\bar{\pi},\bar{w})+\bar{v}^T G(\bar{\pi},\bar{u})+\bar{\gamma}^T(H(\bar{\pi},\bar{v})-x_\alpha)\Big\}dt,$$

which contradicts the inequality (6.5.1). Hence, this completes the proof. □

Now, we give the following example to validate Theorem 6.5.0.1.

Example 6.5.0.1 Let $p=2, q=r=1, k=2, \mathcal{W}_1=[0,1], \mathcal{W}_2=[1,2], \mathcal{U}_l=[2-l,l+2],\ \forall l\in\Lambda=[0,1], \mathcal{V}_1^1=\mathcal{V}_2^1=[0,\frac{1}{2}]$ and Γ_{t_0,t_1} is fixed by the diagonally opposite points $t_0=(t_0^1,t_0^2)$ and $t_1=(t_1^1,t_1^2)$ in $\mathbb{R}^2$.

We consider the following semi-infinite multi-dimensional multi-objective optimization problem:

$$\text{(SMMOPU2)}\quad \min_{(x(\cdot),c(\cdot))}\int_{\Gamma_{0,1}}(w_1c^2+x,\ c^2+w_2x)\,dt^1dt^2$$
$$\text{subject to}\quad l(e^{-x}-1)u_l\leqq 0, \tag{6.5.2}$$
$$x_{t^1}=1+v_1^1-c, \tag{6.5.3}$$
$$x_{t^2}=1+v_2^1-c, \tag{6.5.4}$$
$$x(0,0)=0,\ x(1,1)=2, \tag{6.5.5}$$

where $t=(t^1,t^2)\in\Gamma_{0,1}$.

Therefore, the robust counterpart for the problem (SMMOPU2) is defined as:

$$\text{(RSMMOPU2)} \quad \min_{(x(\cdot),c(\cdot))} \int_{\Gamma_{0,1}} \Big(\max_{w_1 \in \mathcal{W}_1} (w_1 c^2 + x),\ \max_{w_2 \in \mathcal{W}_2} (c^2 + w_2 x) \Big)\, dt^1 dt^2$$

$$\begin{aligned}\text{subject to} \quad & l(e^{-x} - 1)u_l \leqq 0,\ \forall u_l \in \mathcal{U}_l,\\ & x_{t^1} = 1 + v_1^1 - c,\ \forall v_1^1 \in \mathcal{V}_1^1,\\ & x_{t^2} = 1 + v_2^1 - c,\ \forall v_2^1 \in \mathcal{V}_2^1,\\ & x(0,0) = 0,\ x(1,1) = 2,\end{aligned}$$

where $t = (t^1, t^2) \in \Gamma_{0,1}$. Clearly $\mathcal{D} = \{(x, c) \in \mathscr{X} \times \mathscr{C} : x \geq 0 \wedge x_{t^1} = 1 + v_1^1 - c \wedge x_{t^2} = 1 + v_2^1 - c,\ x(0,0) = 0,\ x(1,1) = 2\}$ is a robust feasible solution set to (SMMOPU2).

Now, we find a weak robust efficient solution to the problem (SMMOPU2), i.e, find the control function $\bar{c} : \Gamma_{0,1} \to \mathbb{R}$ (that determines the state function $\bar{x} : \Gamma_{0,1} \to \mathbb{R}$) satisfying the dynamical system (6.5.3) and (6.5.4) with respect to boundary conditions (6.5.5). Additionally, we assume that state function is affine.

Let $(\bar{x}, \bar{c})$ be a robust feasible solution to the problem (SMMOPU2) and the Lagrange multipliers $\bar{\mu}_1, \bar{\mu}_2, \bar{\nu}, \bar{\gamma}_1^1, \bar{\gamma}_2^1$, and the uncertain parameters $\bar{u}, \bar{v}_1^1, \bar{v}_2^1$. Let $\max_{w \in \mathcal{W}} f(\pi, w) = f(\pi, \bar{w})$. Therefore, the robust necessary efficiency conditions are given as follows:

$$\bar{\mu}_1 + \bar{\mu}_2 \bar{w}_2 - \bar{\nu}_l l \bar{u}_l e^{-\bar{x}} + (\bar{\gamma}_1^1)_{t^1} + (\bar{\gamma}_2^1)_{t^2} = 0, \tag{6.5.6}$$

$$2\bar{\mu}_1 \bar{c} \bar{w}_1 + 2\bar{\mu}_2 \bar{c} - \bar{\gamma}_1^1 - \bar{\gamma}_2^1 = 0, \tag{6.5.7}$$

$$\bar{\nu}_l l(e^{-\bar{x}} - 1)\bar{u}_l = 0. \tag{6.5.8}$$

Let the Lagrange multipliers $\bar{\nu}_l = 0,\ \forall l \in [0, 1)$ and $\bar{\nu}_1 = \frac{1}{2}$ and the uncertain parameters $\bar{u}_l \in [2 - l, 2 + l],\ \forall l \in [0, 1)$ and $\bar{u}_1 = 3$.

Case 1. Assume that the Lagrange multipliers $\bar{\gamma}_\alpha^\tau,\ \tau = \overline{1, q}, \alpha = \overline{1, p}$ are constant functions, say $\bar{\gamma}_1^1 = h_1$ and $\bar{\gamma}_2^1 = h_2$, where $h_1, h_2 \in \mathbb{R}_+$. Therefore, Eq. (6.5.7) gives

$$\bar{c} = \frac{h_1 + h_2}{2(\bar{\mu}_1 \bar{w}_1 + \bar{\mu}_2)},\quad \forall t \in \Gamma_{0,1},$$

which shows that $\bar{c}$ is a constant function on $\Gamma_{0,1}$. Taking into account of equality constraints (6.5.3) and (6.5.4), it follows that

$$\bar{x} = \Big(1 + \bar{v}_1^1 - \frac{h_1 + h_2}{2(\bar{\mu}_1 \bar{w}_1 + \bar{\mu}_2)}\Big) t^1 + \Big(1 + \bar{v}_2^1 - \frac{h_1 + h_2}{2(\bar{\mu}_1 \bar{w}_1 + \bar{\mu}_2)}\Big) t^2 + h_3,\ h_3 \in \mathbb{R}_+.$$

By using the boundary condition (6.5.5) along with the Lagrange multiplier $\bar{\mu}_1 = \bar{\mu}_2 = \frac{1}{2}$ and the uncertain parameters $\bar{w}_1 = 1,\ \bar{v}_1^1 = \bar{v}_2^1 = \frac{1}{2}$, we get $h_3 = 0$ and $h_1 + h_2 = 1$, and therefore

$$\bar{x} = t^1 + t^2, \ \bar{c} = \frac{1}{2}, \ \forall t \in \Gamma_{0,1}.$$

Now, we shall show that $(\bar{x}, \bar{c}) = (t^1 + t^2, \frac{1}{2})$ at $t^1 = t^2 = 0$ i.e. $(\bar{x}, \bar{c}) = (0, \frac{1}{2})$ is a weak robust efficient solution to the problem (SMMOPU2). In virtue of Definition 6.2.0.1, for all $(x, c) \in \mathcal{D}$, we have

$$\begin{aligned}
&\int_{\Gamma_{0,1}} \max_{w \in \mathcal{W}} f(\pi, w)dt - \int_{\Gamma_{0,1}} \max_{w \in \mathcal{W}} f(\bar{\pi}, w)dt \\
&= \int_{\Gamma_{0,1}} f(\pi, \bar{w})dt - \int_{\Gamma_{0,1}} f(\bar{\pi}, \bar{w})dt \\
&= \int_{\Gamma_{0,1}} \Big(\bar{w}_1 c^2 + x, \ c^2 + \bar{w}_2 x\Big)dt^1dt^2 \\
&\quad - \int_{\Gamma_{0,1}} \Big(\bar{w}_1 \bar{c}^2 + \bar{x}, \ \bar{c}^2 + \bar{w}_2 \bar{x}\Big)dt^1dt^2 \\
&= \int_{\Gamma_{0,1}} \Big(c^2 + x, \ c^2 + 2x\Big)dt^1dt^2 - \int_{\Gamma_{0,1}} \Big(\frac{1}{4}, \frac{1}{4}\Big)dt^1dt^2 \\
&= \int_{\Gamma_{0,1}} \Big(c^2 + x - \frac{1}{4}, \ c^2 + 2x - \frac{1}{4}\Big)dt^1dt^2. \qquad (6.5.9)
\end{aligned}$$

On solving Eq. (6.5.3), we have $c = v_1^1 + 1 - x_{t^1}$. Therefore, the above Eq. (6.5.9) can be written as

$$\begin{aligned}
&\int_{\Gamma_{0,1}} \max_{w \in \mathcal{W}} f(\pi, w)dt - \int_{\Gamma_{0,1}} \max_{w \in \mathcal{W}} f(\bar{\pi}, w)dt \\
&= \int_{\Gamma_{0,1}} \Big((v_1^1 + 1 - x_{t^1})^2 + x - \frac{1}{4}, \ (v_1^1 + 1 - x_{t^1})^2 + 2x - \frac{1}{4}\Big)dt^1dt^2. \qquad (6.5.10)
\end{aligned}$$

Now, it remains to find the state function x which satisfies the boundary conditions (6.5.5). If we set $v_1^1 = v_2^1$, then, it follows from the equality constraints (6.5.3) and (6.5.4) that

$$x_{t^1} = x_{t^2}, \ \forall t \in \Gamma_{0,1}.$$

Applying the necessary Euler-Lagrange PDE (see, for instance, [6]) for Eq. (6.5.10), it follows

$$x = \phi(t^1 + t^2) + \psi, \ \phi, \psi \in \mathbb{R}, \ \forall t \in \Gamma_{0,1}.$$

The above solution together with the boundary conditions (6.5.5) imply that $\phi = 1$ and $\psi = 0$. Therefore, $x = t^1 + t^2$. Hence, Eq. (6.5.10) yields

$$\int_{\Gamma_{0,1}} \max_{w \in \mathcal{W}} f(\pi, w)dt - \int_{\Gamma_{0,1}} \max_{w \in \mathcal{W}} f(\bar{\pi}, w)dt$$
$$= \int_{\Gamma_{0,1}} \Big((v_1^1)^2 + (t^1 + t^1) - \frac{1}{4},\ (v_1^1)^2 + 2(t^1 + t^2) - \frac{1}{4}\Big)dt^1dt^2$$
$$= \Big((v_1^1)^2 + \frac{3}{4},\ (v_1^1)^2 + \frac{7}{4}\Big) \geq 0,\ \forall v_1^1 \in \mathcal{V}_1^1.$$

Similarly, by solving Eq. (6.5.4), we have $c = v_2^1 + 1 - x_{t^2}$ and the inequality (6.5.9) concludes that $\int_{\Gamma_{0,1}} \max_{w \in \mathcal{W}} f(\pi, w)dt - \int_{\Gamma_{0,1}} \max_{w \in \mathcal{W}} f(\bar{\pi}, w)dt \geq 0$. In both the cases, it shows that $(0, \frac{1}{2})$ is a weak robust efficient solution to the problem (SMMOPU2) at which the robust necessary efficiency conditions (6.5.6)–(6.5.8) are satisfied with the Lagrange multipliers $\bar{\mu}_1 = \bar{\mu}_2 = \frac{1}{2}, \bar{\nu}_l = 0,\ \forall l \in [0, 1), \bar{\nu}_1 = \frac{1}{2}$ and $\bar{\gamma}_1^1 + \bar{\gamma}_2^1 = 1$ and the uncertain parameters $\bar{w}_1 = 1, \bar{w}_2 = 2, \bar{\mu}_l = 0,\ \forall l \in [0, 1), \bar{\mu}_1 = 3, \bar{v}_1^1 = \bar{v}_2^1 = \frac{1}{2}$. Further, it can be seen that the Lagrange functional is also convex at $(\bar{x}, \bar{c}) = (0, \frac{1}{2})$ on $\mathscr{X} \times \mathscr{C}$.

Now, we formulate the penalized semi-infinite multi-dimensional multi-objective optimization problem $(\text{SMMOPU2})_\rho$ associated to the semi-infinite multi-dimensional multi-objective optimization problem (SMMOPU2) as follows:

$$(\text{SMMOPU2})_\rho \quad \min_{(x(\cdot),c(\cdot))} \Big(\int_{\Gamma_{0,1}} \Big\{(w_1c^2 + x) + \rho\Big[\max\{0, l(e^{-x} - 1)u_l\} + |x_{t^1} - 1 - v_1^1 + c| + |x_{t^2} - 1 - v_2^1 + c|\Big]\Big\}dt^1dt^2, \int_{\Gamma_{0,1}} \Big\{(c^2 + w_2x) + \rho\Big[\max\{0, l(e^{-x} - 1)u_l\} + |x_{t^1} - 1 - v_1^1 + c| + |x_{t^2} - 1 - v_2^1 + c|\Big]\Big\}dt^1dt^2\Big).$$

Therefore, the robust counterpart for $(\text{SMMOPU2})_\rho$ is defined as:

$$(\text{RSMMOPU2})_\rho \quad \min_{(x(\cdot),c(\cdot))} \Big(\int_{\Gamma_{0,1}} \Big\{\max_{w_1 \in \mathcal{W}_1} (w_1c^2 + x) + \rho\Big[\max\{0, l(e^{-x} - 1)u_l\} + |x_{t^1} - 1 - v_1^1 + c| + |x_{t^2} - 1 - v_2^1 + c|\Big]\Big\}dt^1dt^2, \int_{\Gamma_{0,1}} \Big\{\max_{w_2 \in \mathcal{W}_2} (c^2 + w_2x) + \rho\Big[\max\{0, l(e^{-x} - 1)u_l\} + |x_{t^1} - 1 - v_1^1 + c| + |x_{t^2} - 1 - v_2^1 + c|\Big]\Big\}dt^1dt^2\Big).$$

In virtue of Definition 6.4.0.1, it can be observed that the inequality $\mathscr{P}(\pi, w, \bar{u}, \bar{v}, \rho) - \mathscr{P}(\bar{\pi}, w, \bar{u}, \bar{v}, \rho) \geqq 0$ at $(\bar{x}, \bar{c}) = (0, \frac{1}{2})$ for all $(x, c) \in \mathscr{X} \times \mathscr{C}$. This shows that $(0, \frac{1}{2})$ is a weak robust efficient solution of the penalized semi-infinite multi-dimensional optimization problem $(\text{SMMOPU2})_\rho$ with $\rho \geq 3$. Hence, Theorem 6.5.0.1 is verified.

Case 2. Assume that the Lagrange multipliers $\bar{\gamma}_\alpha^\tau, \tau = 1, \alpha = 1, 2$ are functions of t, say $\bar{\gamma}_1^1 = t^2, \bar{\gamma}_2^1 = t^1$. Thus, by Eq. (6.5.7), we have

$$\bar{c} = \frac{1}{2}(t^2 + t^1), \ \forall t \in \Gamma_{0,1}. \quad (6.5.11)$$

On solving Eqs. (6.5.3) and (6.5.4) with the help of Eq. (6.5.11), we have

$$\bar{x} = \frac{3}{2}(t^1 + t^2) - \frac{1}{2}((t^1)^2 + (t^1)^2) - 2t^1t^2 + h_4, \ h_4 \in \mathbb{R}_+,$$

and, by considering the boundary conditions (6.5.5), we get

$$\bar{x} = \frac{3}{2}(t^1 + t^2) - \frac{1}{2}((t^1)^2 + (t^1)^2) - 2t^1t^2,$$

which shows that the state function is not affine function. Consequently, we get a contradiction with the hypothesis.

6.6 Robust Saddle-point Criteria

Definition 6.6.0.1 A point $((\bar{x}, \bar{c}), \bar{\nu}, \bar{\gamma}, \bar{w}, \bar{u}, \bar{v}) \in \mathcal{D} \times \mathbb{R}_+^{\Lambda} \times \mathbb{R}^{pq} \times \mathcal{W} \times \mathcal{U} \times \mathcal{V}$ is said to be a robust saddle-point to the multi-dimensional multi-objective optimization problem (SMMOPU), iff

(i) $L((\bar{x}, \bar{c}), \nu, \gamma, w, u, v) \leqq L((\bar{x}, \bar{c}), \bar{\nu}, \bar{\gamma}, w, \bar{u}, \bar{v}), \ \forall \nu \in \mathbb{R}_+^{\Lambda}, \ \forall \gamma \in \mathbb{R}^{pq}, \ \forall u \in \mathcal{U}, \ \forall v \in \mathcal{V}$,
(ii) $L((x, c), \bar{\nu}, \bar{\gamma}, w, \bar{u}, \bar{v}) \geqq L((\bar{x}, \bar{c}), \bar{\nu}, \bar{\gamma}, w, \bar{u}, \bar{v}), \ \forall (x, c) \in \mathscr{X} \times \mathscr{C}$.

In the following theorem, we shall show that the robust saddle-point of the multi-dimensional multi-objective optimization problem with data uncertainty (SMMOPU) and a weak robust efficient solution of its associated unconstrained multi-dimensional multi-objective optimization problem with data uncertainty $(\text{SMMOPU})_\rho$ are equivalent if the penalty parameter is assumed to be sufficiently large.

Theorem 6.6.0.1 *Let* $((\bar{x}, \bar{c}), \bar{\nu}, \bar{\gamma}, \bar{w}, \bar{u}, \bar{v})$ *be a robust saddle-point to the semi-infinite multi-dimensional multi-objective optimization problem* (SMMOPU) *and* $\max_{w \in \mathcal{W}} f(\pi, w) = f(\pi, \bar{w})$. *Then,* $(\bar{x}, \bar{c})$ *is a weak robust efficient solution to the penalized problem* $(\text{SMMOPU})_\rho$, *if the penalty parameter* ρ *is assumed to be sufficiently large, that is,* $\rho \geq \max\{\bar{\nu}, |\bar{\gamma}|\}$.

Proof We proceed by contradiction and consider $(\bar{x}, \bar{c})$ is not a weak robust efficient solution to the penalized problem $(\text{SMMOPU})_\rho$. Then, there exists $(\hat{x}, \hat{c}) \in \mathscr{X} \times \mathscr{C}$ such that

$$\max_{w \in \mathcal{W}} \mathscr{P}(\hat{\pi}, w, \bar{u}, \bar{v}, \rho) < \max_{w \in \mathcal{W}} \mathscr{P}(\bar{\pi}, w, \bar{u}, \bar{v}, \rho).$$

By using the definition of the penalized multi-dimensional multi-objective optimization problem $(\text{SMMOPU})_\rho$, we get

$$\int_{\Gamma}\Big\{\max_{w\in W} f(\hat{\pi},w)+\rho\Big[\sum_{l\in\Lambda}G_l^{+}(\hat{\pi},\bar{u})+\sum_{\tau=1}^{q}\sum_{\alpha=1}^{p}\Big|H_{\alpha}^{\tau}(\hat{\pi},\bar{v})-\frac{\partial x^{\tau}}{\partial t^{\alpha}}\Big|\Big]e\Big\}dt$$
$$<\int_{\Gamma}\Big\{\max_{w\in W} f(\bar{\pi},w)+\rho\Big[\sum_{l\in\Lambda}G_l^{+}(\bar{\pi},\bar{u})+\sum_{\tau=1}^{q}\sum_{\alpha=1}^{p}\Big|H_{\alpha}^{\tau}(\bar{\pi},\bar{v})-\frac{\partial x^{\tau}}{\partial t^{\alpha}}\Big|\Big]e\Big\}dt.$$

Since $\max_{w\in W} f(\pi,w)=f(\pi,\bar{w})$, we have

$$\int_{\Gamma}\Big\{f(\hat{\pi},\bar{w})+\rho\Big[\sum_{l\in\Lambda}G_l^{+}(\hat{\pi},\bar{u})+\sum_{\tau=1}^{q}\sum_{\alpha=1}^{p}\Big|H_{\alpha}^{\tau}(\hat{\pi},\bar{v})-\frac{\partial x^{\tau}}{\partial t^{\alpha}}\Big|\Big]e\Big\}dt$$
$$<\int_{\Gamma}\Big\{f(\bar{\pi},\bar{w})+\rho\Big[\sum_{l\in\Lambda}G_l^{+}(\bar{\pi},\bar{u})+\sum_{\tau=1}^{q}\sum_{\alpha=1}^{p}\Big|H_{\alpha}^{\tau}(\bar{\pi},\bar{v})-\frac{\partial x^{\tau}}{\partial t^{\alpha}}\Big|\Big]e\Big\}dt.$$

The above inequality together with the robust feasibility of $(\bar{x},\bar{c})$ in the multi-dimensional multi-objective optimization problem (SMMOPU) and the relation (6.4.1) yields

$$\int_{\Gamma}\Big\{f(\hat{\pi},\bar{w})+\rho\Big[\sum_{l\in\Lambda}G_l^{+}(\hat{\pi},\bar{u})+\sum_{\tau=1}^{q}\sum_{\alpha=1}^{p}\Big|H_{\alpha}^{\tau}(\hat{\pi},\bar{v})-\frac{\partial x^{\tau}}{\partial t^{\alpha}}\Big|\Big]e\Big\}dt<\int_{\Gamma}f(\bar{\pi},\bar{w})dt.$$

Since $\rho\geq\max\{\bar{\nu},|\bar{\gamma}|\}$, we have

$$\int_{\Gamma}\Big\{f(\hat{\pi},\bar{w})+\sum_{l\in\Lambda}\bar{\nu}_l G_l^{+}(\hat{\pi},\bar{u})e+\sum_{\tau=1}^{q}\sum_{\alpha=1}^{p}\Big|\bar{\gamma}_{\alpha}^{\tau}\Big(H_{\alpha}^{\tau}(\hat{\pi},\bar{v})-\frac{\partial x^{\tau}}{\partial t^{\alpha}}\Big)\Big|e\Big\}dt<\int_{\Gamma}f(\bar{\pi},\bar{w})dt.$$

Again using the relation (6.4.1), we get

$$\int_{\Gamma}\Big\{f(\hat{\pi},\bar{w})+\sum_{l\in\Lambda}\bar{\nu}_l G_l(\hat{\pi},\bar{u})e+\sum_{\tau=1}^{q}\sum_{\alpha=1}^{p}\bar{\gamma}_{\alpha}^{\tau}\Big(H_{\alpha}^{\tau}(\hat{\pi},\bar{v})-\frac{\partial x^{\tau}}{\partial t^{\alpha}}\Big)e\Big\}dt<\int_{\Gamma}f(\bar{\pi},\bar{w})dt,$$

equivalently,

$$\int_{\Gamma}\Big\{f(\hat{\pi},\bar{w})+\bar{\nu}^{T}G(\hat{\pi},\bar{u})e+\bar{\gamma}^{T}(H(\hat{\pi},\bar{v})-x_{\alpha})e\Big\}dt<\int_{\Gamma}f(\bar{\pi},\bar{w})dt. \quad (6.6.1)$$

On the other hand, since $((\bar{x},\bar{c}),\bar{\nu},\bar{\gamma},\bar{w},\bar{u},\bar{v})$ is a robust saddle-point to the semi-infinite multi-dimensional multi-objective optimization problem (SMMOPU), by Definition 6.6.0.1 (i), we have

$$L((\bar{x},\bar{c}),\nu,\gamma,w,u,v)\leqq L((\bar{x},\bar{c}),\bar{\nu},\bar{\gamma},w,\bar{u},\bar{v}),\ \forall\nu\in\mathbb{R}_{+}^{\Lambda},\ \forall\gamma\in\mathbb{R}^{pq},\ \forall u\in\mathcal{U},\ \forall v\in\mathcal{V}.$$

By considering the definition of the Lagrange functional, we get

$$\int_{\Gamma}\left\{\max_{w\in\mathcal{W}} f(\bar{\pi},w)+\nu^{T}G(\bar{\pi},u)e+\gamma^{T}(H(\bar{\pi},v)-x_{\alpha})e\right\}dt$$
$$\leqq\int_{\Gamma}\left\{\max_{w\in\mathcal{W}} f(\bar{\pi},w)+\bar{\nu}^{T}G(\bar{\pi},\bar{u})e+\bar{\gamma}^{T}(H(\bar{\pi},\bar{v})-x_{\alpha})e\right\}dt.$$

Since $\max_{w\in\mathcal{W}} f(\pi,w)=f(\pi,\bar{w})$, we have

$$\int_{\Gamma}\left\{f(\bar{\pi},\bar{w})+\nu^{T}G(\bar{\pi},u)e+\gamma^{T}(H(\bar{\pi},v)-x_{\alpha})e\right\}dt$$
$$\leqq\int_{\Gamma}\left\{f(\bar{\pi},\bar{w})+\bar{\nu}^{T}G(\bar{\pi},\bar{u})e+\bar{\gamma}^{T}(H(\bar{\pi},\bar{v})-x_{\alpha})e\right\}dt.$$

If we set $\nu=0$, $\gamma=0$, we obtain

$$\int_{\Gamma}\bar{\nu}^{T}G(\bar{\pi},\bar{u})\,dt\geqq 0.$$

Also, the robust feasibility of $(\bar{x},\bar{c})$ in the problem (SMMOPU) yields

$$\int_{\Gamma}\bar{\nu}^{T}G(\bar{\pi},\bar{u})\,dt\leqq 0.$$

On combining the above two inequalities, we get $\int_{\Gamma}\bar{\nu}^{T}G(\bar{\pi},\bar{u})\;dt=0$. Further, the robust feasibility of $(\bar{x},\bar{c})$ to the problem (SMMOPU) also admits that $\int_{\Gamma}\Big(H(\bar{\pi},\bar{v})-x_{\alpha}\Big)dt=0$. Therefore, the inequality (6.6.1) can be rewritten as

$$\int_{\Gamma}\left\{f(\hat{\pi},\bar{w})+\bar{\nu}^{T}G(\hat{\pi},\bar{u})e+\bar{\gamma}^{T}(H(\hat{\pi},\bar{v})-x_{\alpha})e\right\}dt$$
$$<\int_{\Gamma}\left\{f(\bar{\pi},\bar{w})+\bar{\nu}^{T}G(\bar{\pi},\bar{u})e+\bar{\gamma}^{T}(H(\bar{\pi},\bar{v})-x_{\alpha})e\right\}dt,$$

equivalently

$$\int_{\Gamma}\left\{\max_{w\in w} f(\hat{\pi},w)+\bar{\nu}^{T}G(\hat{\pi},\bar{u})e+\bar{\gamma}^{T}(H(\hat{\pi},\bar{v})-x_{\alpha})e\right\}dt$$
$$<\int_{\Gamma}\left\{\max_{w\in w} f(\bar{\pi},w)+\bar{\nu}^{T}G(\bar{\pi},\bar{u})e+\bar{\gamma}^{T}(H(\bar{\pi},\bar{v})-x_{\alpha})e\right\}dt.$$

By using the definition of the Lagrange functional, we get

$$L((\hat{x},\hat{c}),\bar{\nu},\bar{\gamma},w,\bar{u},\bar{v})<L((\bar{x},\bar{c}),\bar{\nu},\bar{\gamma},w,\bar{u},\bar{v}),\ \forall(\hat{x},\hat{c})\in\mathscr{X}\times\mathscr{C},$$

which contradicts Definition 6.6.0.1 (ii). Hence, the proof is complete. □

Theorem 6.6.0.2 *Let $(\bar{x}, \bar{c})$ be a weak robust efficient solution to the penalized multi-dimensional multi-objective optimization problem* (SMMOPU)$_{\bar{\rho}}$. *Assume that $\mathcal{D}$ is a compact subset of $\mathscr{X} \times \mathscr{C}$ and the inequality* $\max_{w \in \mathcal{W}} \mathscr{P}(\pi, w, \bar{u}, \bar{v}, \rho) \geqq \max_{w \in \mathcal{W}} \mathscr{P}(\bar{\pi}, w, \bar{u}, \bar{v}, \rho)$ *holds for any $\rho \geq \bar{\rho}$ and for all $(x, c) \in \mathcal{D}$. Further, if* $\max_{w \in \mathcal{W}} f(\pi, w) = f(\pi, \bar{w})$ *and $\int_\Gamma \bar{\mu}^T f(\pi, \bar{w}) dt$, $\int_\Gamma \bar{v}^T G(\pi, \bar{u}) dt$ and $\int_\Gamma \bar{\gamma}^T (H(\pi, \bar{v}) - x_\alpha) dt$ are convex at $(\bar{x}, \bar{c})$ on $\mathscr{X} \times \mathscr{C}$, then $((\bar{x}, \bar{c}), \bar{v}, \bar{\gamma}, \bar{w}, \bar{u}, \bar{v})$ is a robust saddle-point to the semi-infinite multi-dimensional multi-objective optimization problem* (SMMOPU).

Proof To show that $((\bar{x}, \bar{c}), \bar{v}, \bar{\gamma}, \bar{w}, \bar{u}, \bar{v})$ is a robust saddle-point to the problem (SMMOPU), firstly, we shall show that $(\bar{x}, \bar{c})$ is a weak robust efficient solution to the problem (SMMOPU). From the assumption, $(\bar{x}, \bar{c})$ is a weak robust efficient solution to the penalized multi-dimensional multi-objective optimization problem (SMMOPU)$_{\bar{\rho}}$ and the inequality $\max_{w \in \mathcal{W}} \mathscr{P}(\pi, w, \bar{u}, \bar{v}, \rho) \geqq \max_{w \in \mathcal{W}} \mathscr{P}(\bar{\pi}, w, \bar{u}, \bar{v}, \rho)$ holds for any $\rho \geq \bar{\rho}$ and for all $(x, c) \in \mathcal{D}$. Therefore, proceeding on the lines of Theorem 6.4.0.2, $(\bar{x}, \bar{c})$ is a weak robust efficient solution to the problem (SMMOPU).

Since $(\bar{x}, \bar{c})$ is a weak robust efficient solution to the problem (SMMOPU), there exist $\bar{\mu} \in \mathbb{R}^s$, $\bar{v} \in \mathbb{R}^{\Lambda}_+$ and $\bar{\gamma} \in \mathbb{R}^{pq}$ and the uncertain parameters $\bar{u} \in \mathcal{U}$, $\bar{v} \in \mathcal{V}$, with $\max_{w \in \mathcal{W}} f(\pi, w) = f(\pi, \bar{w})$, such that the robust necessary efficiency conditions (6.2.1)–(6.2.4) are satisfied at $(\bar{x}, \bar{c})$ on $\mathscr{X} \times \mathscr{C}$.

From the robust feasibility of $(\bar{x}, \bar{c})$ to the problem (SMMOPU), together with the necessary efficiency condition (6.2.3), we have

$$\int_\Gamma \nu^T G(\bar{\pi}, u) dt \leqq \int_\Gamma \bar{v}^T G(\bar{\pi}, \bar{u}) dt, \ \forall \nu \in \mathbb{R}^{\Lambda}_+, \forall u \in \mathcal{U}.$$

It follows that

$$\int_\Gamma \left\{ \max_{w \in w} f(\bar{\pi}, w) + \nu^T G(\bar{\pi}, u) e + \gamma^T (H(\bar{\pi}, v) - x_\alpha) e \right\} dt$$
$$\leqq \int_\Gamma \left\{ \max_{w \in w} f(\bar{\pi}, w) + \bar{v}^T G(\bar{\pi}, \bar{u}) e + \bar{\gamma}^T (H(\bar{\pi}, \bar{v}) - x_\alpha) e \right\} dt.$$

By using the definition of Lagrange functional, we get

$$L((\bar{x}, \bar{c}), \nu, \gamma, w, u, v) \leqq L((\bar{x}, \bar{c}), \bar{v}, \bar{\gamma}, w, \bar{u}, \bar{v}), \ \forall \nu \in \mathbb{R}^{\Lambda}_+, \ \forall \gamma \in \mathbb{R}^{pq}, \ \forall u \in \mathcal{U}, \ \forall v \in \mathcal{V}.$$

To prove the second condition of Definition 6.6.0.1, we proceed by contradiction and assume that

$$L((x, c), \bar{v}, \bar{\gamma}, w, \bar{u}, \bar{v}) < L((\bar{x}, \bar{c}), \bar{v}, \bar{\gamma}, w, \bar{u}, \bar{v}), \ \forall (x, c) \in \mathscr{X} \times \mathscr{C}.$$

Again, by considering the definition of Lagrange functional, we get

$$\int_{\Gamma}\Big\{\max_{w\in\mathcal{W}} f(\pi,w)+\bar{\upsilon}^T G(\pi,\bar{u})e+\bar{\gamma}^T(H(\pi,\bar{\upsilon})-x_\alpha)e\Big\}dt$$
$$<\int_{\Gamma}\Big\{\max_{w\in\mathcal{W}} f(\bar{\pi},w)+\bar{\upsilon}^T G(\bar{\pi},\bar{u})e+\bar{\gamma}^T(H(\bar{\pi},\bar{\upsilon})-x_\alpha)e\Big\}dt.$$

Since $\max_{w\in\mathcal{W}} f(\pi,w)=f(\pi,\bar{w})$, we have

$$\int_{\Gamma}\Big\{f(\pi,\bar{w})+\bar{\upsilon}^T G(\pi,\bar{u})e+\bar{\gamma}^T(H(\pi,\bar{\upsilon})-x_\alpha)e\Big\}dt$$
$$<\int_{\Gamma}\Big\{f(\bar{\pi},\bar{w})+\bar{\upsilon}^T G(\bar{\pi},\bar{u})e+\bar{\gamma}^T(H(\bar{\pi},\bar{\upsilon})-x_\alpha)e\Big\}dt. \quad (6.6.2)$$

On the other hand, from the hypotheses $\int_\Gamma \bar{\mu}^T f(\pi,\bar{w})dt$, $\int_\Gamma \bar{\upsilon}^T G(\pi,\bar{u})dt$ and $\int_\Gamma \bar{\gamma}^T(H(\pi,\bar{\upsilon})-x_\alpha)dt$ are convex at $(\bar{x},\bar{c})$ on $\mathscr{X}\times\mathscr{C}$, we get

$$\int_{\Gamma}\Big\{\bar{\mu}^T f(\pi,\bar{w})-\bar{\mu}^T f(\bar{\pi},\bar{w})\Big\}dt \geqq \int_{\Gamma}(x-\bar{x})\bar{\mu}^T f_x(\bar{\pi},\bar{w})dt+\int_{\Gamma}(c-\bar{c})\bar{\mu}^T f_c(\bar{\pi},\bar{w})dt, \quad (6.6.3)$$

$$\int_{\Gamma}\Big\{\bar{\upsilon}^T G(\pi,\bar{u})-\bar{\upsilon}^T G(\bar{\pi},\bar{u})\Big\}dt \geqq \int_{\Gamma}(x-\bar{x})\bar{\upsilon}^T G_x(\bar{\pi},\bar{u})dt+\int_{\Gamma}(c-\bar{c})\bar{\upsilon}^T G_c(\bar{\pi},\bar{u})dt, \quad (6.6.4)$$

and $$\int_{\Gamma}\Big\{\bar{\gamma}^T(H(\pi,\bar{\upsilon})-x_\alpha)-\bar{\gamma}^T(H(\bar{\pi},\bar{\upsilon})-x_\alpha)\Big\}dt$$
$$\geqq \int_{\Gamma}(x-\bar{x})\Big\{\bar{\gamma}^T H_x(\bar{\pi},\bar{\upsilon})+\bar{\gamma}_t\Big\}dt+\int_{\Gamma}(c-\bar{c})\bar{\gamma}^T H_c(\bar{\pi},\bar{\upsilon})dt. \quad (6.6.5)$$

On combining the inequalities (6.6.3)–(6.6.5), together with the robust necessary efficiency conditions (6.2.1) and (6.2.2), we have

$$\int_{\Gamma}\Big\{\bar{\mu}^T f(\pi,\bar{w})+\bar{\upsilon}^T G(\pi,\bar{u})+\bar{\gamma}^T(H(\pi,\bar{\upsilon})-x_\alpha)\Big\}dt$$
$$\geqq \int_{\Gamma}\Big\{\bar{\mu}^T f(\bar{\pi},\bar{w})+\bar{\upsilon}^T G(\bar{\pi},\bar{u})+\bar{\gamma}^T\Big(H(\bar{\pi},\bar{\upsilon})-x_\alpha\Big)\Big\}dt,$$

equivalently,

$$\int_{\Gamma}\Big\{f(\pi,\bar{w})+\bar{\upsilon}^T G(\pi,\bar{u})e+\bar{\gamma}^T(H(\pi,\bar{\upsilon})-x_\alpha)e\Big\}dt$$
$$\geqq \int_{\Gamma}\Big\{f(\bar{\pi},\bar{w})+\bar{\upsilon}^T G(\bar{\pi},\bar{u})e+\bar{\gamma}^T(H(\bar{\pi},\bar{\upsilon})-x_\alpha)e\Big\}dt,$$

which contradicts the inequality (6.6.2). Hence, this completes the proof. □

References

1. M.A. López, G. Still, Semi-infinite programming. Eur. J. Oper. Res. **180**, 491–518 (2007)
2. J.J. Ye, S.Y. Wu, First order optimality conditions for generalized semi-infinite programming problem. J. Optim. Theory Appl. **137**, 419–434 (2008)
3. M.V. Ştefănescu, A. Ştefănescu, On semi-infinite minmax programming with generalized invexity. Optimization **61**, 1307–1319 (2012)
4. A. Jayswal, A.K. Prasad, I. Ahmad, R.P. Agarwal, Duality for semiinfinite programming problems involving (H_p, r)-invex functions. J. Inequal. Appl. **2013**, 200–209 (2013)
5. Şt. Mititelu, Optimality and duality for invex multi-dimensional control problems with mixed constraints. J. Adv. Math. Stud. **2**, 25–34 (2009)
6. S. Treanţă, Higher-order Hamilton dynamics and Hamilton-Jacobi divergence PDE. Comput. Math. Appl. **75**, 547–560 (2018)

Chapter 7
Robust Duality for Multi-dimensional Variational Control Problem with Data Uncertainty

7.1 Introduction

Duality is a powerful tool to study optimization problems. This is because duality principles relate a pair of optimization problems, known as primal and dual problems, in such a way that the existence of solution of one ensures the solution of the other and the extrema of two problems coincide under certain conditions. The duality theorems for linear optimization were conjectured by Neumann [1]. Thereafter, Wolfe [2] introduced a dual model known as "Wolfe dual" for a class of nonlinear convex optimization problems and established the duality results. Later, Mond and Weir [3] introduced the "Mond-Weir dual" whose objective function is simple as compared to the objective function of the Wolfe dual. After that, many researchers obtained duality results for various optimization problems under different assumptions. Mond and Smart [4] studied the control problem and derived the duality results under the invexity assumption. Thereafter, Mititelu [5] formulated Wolfe's dual for the multi-time control problem and established the duality results under the invexity hypotheses. Beck and Ben-Tal [6] investigated the solution of the uncertain optimization problem via duality and shown that the primal worst case is equal to the dual best case. Jeyakumar et al. [7] also studied the robust duality for uncertain optimization problem under generalized convexity. Recently, Treanţă and Mititelu [8], extended the duality theory over multi-dimensional fractional control problem under the assumption of (ρ, b) quasiinvexity.

7.2 Problem Description

Throughout the chapter, we are considering the following notations and working hypotheses:

• $\mathbb{R}^p$, $\mathbb{R}^q$, $\mathbb{R}^m$ and $\mathbb{R}^n$ are Euclidean spaces of dimensions p, q, r and n, respectively.

A. Jayswal et al., *Multi-dimensional Control Problems*, Industrial and Applied Mathematics, https://doi.org/10.1007/978-981-19-6561-6_7

• $\Gamma_{t_0,t_1} \subset \mathbb{R}^p$ is a hyperparallelepiped fixed by the diagonally opposite points $t_0 = (t_0^\alpha)$ and $t_1 = (t_1^\alpha)$, $\alpha = \overline{1,p}$ and the point $t = ((t^\alpha), \alpha = \overline{1,p}) \in \Gamma_{t_0,t_1} \subset \mathbb{R}^p$.
• $\mathcal{X}$ is the space of state functions (piecewise smooth) $x = (x^\tau) : \Gamma_{t_0,t_1} \subset \mathbb{R}^p \to \mathbb{R}^q$ and $\frac{\partial x}{\partial t^\alpha} = x_\alpha$ denotes the partial derivative of x with respect to t^α.
• $\mathscr{C}$ is the space of control functions (piecewise continuous) $c = (c^j) : \Gamma_{t_0,t_1} \subset \mathbb{R}^p \to \mathbb{R}^m$.
• $dt = dt^1 \dots dt^p$ is the volume element on $\mathbb{R}^p \supset \Gamma_{t_0,t_1}$.
• T denotes the transpose of a vector.
• We assume the following convention for inequalities and equalities: for $x, y \in \mathbb{R}^n$, we have

(i) $x < y \Leftrightarrow x_i < y_i, \forall i = \overline{1,n},$
(ii) $x = y \Leftrightarrow x_i = y_i, \forall i = \overline{1,n},$
(iii) $x \leqq y \Leftrightarrow x_i \leq y_i, \forall i = \overline{1,n},$
(iv) $x \leq y \Leftrightarrow x_i \leq y_i, \forall i = \overline{1,n}$ and $x_i < y_i$ for some i.

The multi-dimensional multi-objective optimization problem with data uncertainty in the objective and constraint functionals is defined as follows:

$$\text{(MMOPU)} \quad \min_{(x(\cdot),c(\cdot))} \left\{ \int_{\Gamma_{t_0,t_1}} f(t, x(t), c(t), w)dt \right.$$
$$\left. = \left(\int_{\Gamma_{t_0,t_1}} f_1(t, x(t), c(t), w_1)dt, \dots, \int_{\Gamma_{t_0,t_1}} f_s(t, x(t), c(t), w_s)dt \right) \right\}$$
$$\text{subject to} \quad G(t, x(t), c(t), u) \leqq 0,$$
$$x_\alpha = H(t, x(t), c(t), v),$$
$$x(t_0) = x_0, \; x(t_1) = x_1,$$

where $f = (f_1, \dots, f_s)$; $(f_k) : \Gamma_{t_0,t_1} \times \mathscr{X} \times \mathscr{C} \times \mathcal{W}_k \to \mathbb{R}^s, k = \overline{1,s}, G = (G_1, \dots, G_m)$; $(G_l) : \Gamma_{t_0,t_1} \times \mathscr{X} \times \mathscr{C} \times \mathcal{U}_l \to \mathbb{R}^m$, $l = \overline{1,m}$, $H = (H_\alpha^\tau) : \Gamma_{t_0,t_1} \times \mathscr{X} \times \mathscr{C} \times \mathcal{V}_\alpha^\tau \to \mathbb{R}^{pq}, \alpha = \overline{1,p}, \tau = \overline{1,q}$ are C^∞-class functionals. Consider $w = (w_k)$, $u = (u_l)$ and $v = (v_\alpha^\tau)$ are the uncertain parameters for some convex compact subsets $\mathcal{W} = (\mathcal{W}_k) \subset \mathbb{R}^s$, $\mathcal{U} = (\mathcal{U}_l) \subset \mathbb{R}^m$ and $\mathcal{V} = (\mathcal{V}_\alpha^\tau) \subset \mathbb{R}^{pq}$, respectively. The functions H_α satisfy the closeness conditions (complete integrability conditions) $D_\beta H_\alpha = D_\alpha H_\beta, \alpha, \beta = \overline{1,p}, \alpha \neq \beta$, where the total derivative is denoted by D_β.

The associated robust counterpart of the multi-dimensional multi-objective optimization problem (MMOPU) is defined as:

$$\text{(RMMOPU)} \quad \min_{(x(\cdot),c(\cdot))} \left\{ \int_{\Gamma_{t_0,t_1}} \max_{w \in \mathcal{W}} f(t, x(t), c(t), w)dt \right.$$
$$\left. = \left(\int_{\Gamma_{t_0,t_1}} \max_{w_1 \in \mathcal{W}_1} f_1(t, x(t), c(t), w_1)dt, \dots, \int_{\Gamma_{t_0,t_1}} \max_{w_s \in \mathcal{W}_s} f_s(t, x(t), c(t), w_s)dt \right) \right\}$$
$$\text{subject to} \quad G(t, x(t), c(t), u) \leqq 0, \forall u \in \mathcal{U},$$
$$x_\alpha = H(t, x(t), c(t), v), \forall v \in \mathcal{V},$$
$$x(t_0) = x_0, \; x(t_1) = x_1,$$

where f, G and H are defined as above in the multi-dimensional multi-objective optimization problem (MMOPU).

We denote $\mathcal{D} = \{(x, c) \in \mathcal{X} \times \mathcal{C} : G(t, x(t), c(t), u) \leqq 0, x_\alpha = H(t, x(t), c(t), v), x(t_0) = x_0, x(t_1) = x_1\}$ the set of all feasible solutions to (RMMOPU) and we say that it is the robust feasible solution set to the problem (MMOPU).

From now on, to simplify our presentation, we introduce some notations, as follows: $x = x(t), \hat{x} = \hat{x}(t), \bar{x} = \bar{x}(t), \tilde{x} = \tilde{x}(t), c = c(t), \hat{c} = \hat{c}(t), \bar{c} = \bar{c}(t), \tilde{c} = \tilde{c}(t), \pi = (t, x(t), c(t)), \bar{\pi} = (t, \bar{x}(t), \bar{c}(t)), \hat{\pi} = (t, \hat{x}(t), \hat{c}(t)), \tilde{\pi} = (t, \tilde{x}(t), \tilde{c}(t)), \eta = (t, y(t), z(t)), \bar{\eta} = (t, \bar{y}(t), \bar{z}(t)), \rho = \rho(t), \bar{\rho} = \bar{\rho}(t)$ and $\Gamma = \Gamma_{t_0,t_1}$.

The partial derivatives of Lagrange multiplier γ and the partial derivatives associated with f are defined as

$$\gamma_t = \begin{bmatrix} \frac{\partial \gamma_1^\tau}{\partial t^1} & \cdots & \frac{\partial \gamma_1^\tau}{\partial t^p} \\ \vdots & \ddots & \vdots \\ \frac{\partial \gamma_p^\tau}{\partial t^1} & \cdots & \frac{\partial \gamma_p^\tau}{\partial t^p} \end{bmatrix}, \quad f_x = \begin{bmatrix} \frac{\partial f_1}{\partial x^1} & \cdots & \frac{\partial f_1}{\partial x^q} \\ \vdots & \ddots & \vdots \\ \frac{\partial f_s}{\partial x^1} & \cdots & \frac{\partial f_s}{\partial x^q} \end{bmatrix}, \quad f_c = \begin{bmatrix} \frac{\partial f_1}{\partial c^1} & \cdots & \frac{\partial f_1}{\partial c^r} \\ \vdots & \ddots & \vdots \\ \frac{\partial f_s}{\partial c^1} & \cdots & \frac{\partial f_s}{\partial c^r} \end{bmatrix}.$$

Similarly, we have G_x and G_c using matrices with m rows and H_x, H_c, x_α using matrices with p rows.

In general, for a vector optimization problem, it is not so easy to find a point which optimizes more than one objective simultaneously. Therefore, the notion of an efficient solution is used to determine an optimality in vector optimization problems.

Definition 7.2.0.1 A point $(\bar{x}, \bar{c}) \in \mathcal{D}$ is said to be weak robust efficient solution to the multi-dimensional multi-objective optimization problem (MMOPU), if there does not exist another point $(x, c) \in \mathcal{D}$ such that

$$\int_\Gamma \max_{w \in \mathcal{W}} f(\pi, w)dt < \int_\Gamma \max_{w \in \mathcal{W}} f(\bar{\pi}, w)dt.$$

Definition 7.2.0.2 A functional $F(x, c, \bar{w}) = \int_\Gamma f(t, x(t), x_\sigma(t), c(t), \bar{w})dt$ is said to be convex (strict convex) at $(\bar{x}, \bar{c})$, if the following inequality

$$\begin{aligned} F(x, c, \bar{w}) - F(\bar{x}, \bar{c}, \bar{w}) \geqq (>) & \int_\Gamma (x(t) - \bar{x}(t)) f_x(t, \bar{x}(t), \bar{x}_\sigma(t), \bar{c}(t), \bar{w})dt \\ & + \int_\Gamma (x_\sigma(t) - \bar{x}_\sigma(t)) f_{x_\sigma}(t, \bar{x}(t), \bar{x}_\sigma(t), \bar{c}(t), \bar{w})dt \\ & + \int_\Gamma (c(t) - \bar{c}(t)) f_c(t, \bar{x}(t), \bar{x}_\sigma(t), \bar{c}(t), \bar{w})dt \end{aligned}$$

holds for all $(x(t), c(t))$ on $\mathcal{X} \times \mathcal{C}$.

Following the same lines formulated in Mititelu [5], we establish the following result. Thus, the proof is omitted." The proof can be read (it is free) at the following link: https://www.journal.fairpartners.ro/volume-22009-no-1_3.html

Theorem 7.2.0.1 *Let* $(\bar{x}, \bar{c}) \in \mathcal{D}$ *be a weak robust efficient solution to the problem* (MMOPU). *Further assume that* $\max_{w \in \mathcal{W}} f(\pi, w) = f(\pi, \bar{w})$. *If the constraint conditions (for the existence of the multipliers) hold, then there exist the scalar vector* $\bar{\mu} \in \mathbb{R}^s$, *the piecewise smooth Lagrange multipliers* $\bar{\nu} = (\bar{\nu}_l(t)) \in \mathbb{R}^m_+, l \in \mathbb{R}^m, \bar{\gamma} = (\bar{\gamma}^\tau_\alpha(t)) \in \mathbb{R}^{pq}, \tau = \overline{1, q}, \alpha = \overline{1, p}$, *and the uncertain parameters* $\bar{u} \in \mathcal{U}, \bar{v} \in \mathcal{V}$ *such that* $(\bar{x}, \bar{c})$ *satisfies the following conditions:*

$$\bar{\mu}^T f_x(\bar{\pi}, \bar{w}) + \bar{\nu}^T G_x(\bar{\pi}, \bar{u}) + \bar{\gamma}^T H_x(\bar{\pi}, \bar{v}) + \bar{\gamma}_t = 0, \tag{7.2.1}$$

$$\bar{\mu}^T f_c(\bar{\pi}, \bar{w}) + \bar{\nu}^T G_c(\bar{\pi}, \bar{v}) + \bar{\gamma}^T H_c(\bar{\pi}, \bar{v}) = 0, \tag{7.2.2}$$

$$\bar{\nu}^T G(\bar{\pi}, \bar{u}) = 0, \tag{7.2.3}$$

$$\bar{\mu} \geq 0, \ \bar{\nu} \geqq 0, \tag{7.2.4}$$

hold for all $t \in \Gamma$, *except at discontinuities.*

Remark 7.2.0.1 The conditions (7.2.1)–(7.2.4) are known as the robust necessary efficiency conditions for the multi-dimensional multi-objective optimization problem (MMOPU).

7.3 Wolfe Type Robust Duality

In this section, we formulate Wolfe type robust dual problem for the multi-dimensional multi-objective optimization problem with data uncertainty in the objective and constraint functionals (MMOPU) as follows:

$$\text{(UWRDP)} \quad \max_{(y(\cdot),z(\cdot))} \int_{\Gamma_{t_0,t_1}} \Big\{ f(t, y(t), z(t), w) + \nu^T G(t, y(t), z(t), u)e + \gamma^T (H(t, y(t), z(t), v) - y_\alpha)e \Big\} dt$$

subject to

$$\mu^T f_x(t, y(t), z(t), w) + \nu^T G_x(t, y(t), z(t), u) + \gamma^T H_x(t, y(t), z(t), v) + \gamma_t = 0, \tag{7.3.1}$$

$$\mu^T f_c(t, y(t), z(t), w) + \nu^T G_c(t, y(t), z(t), u) + \gamma^T H_c(t, y(t), z(t), v) = 0, \tag{7.3.2}$$

$$y(t_0) = x_0, \ y(t_1) = x_1, \tag{7.3.3}$$

$$\mu^T > 0, \ e^T \mu = 1, e = (1, \ldots, 1) \in \mathbb{R}^s. \tag{7.3.4}$$

The associated robust counterpart for the problem (UWRDP) is given as:

$$\text{(RUWRDP)} \quad \max_{(y(\cdot),z(\cdot),w,u,v)} \int_{\Gamma_{t_0,t_1}} \Big\{ f(t, y(t), z(t), w) + \nu^T G(t, y(t), z(t), u)e + \gamma^T (H(t, y(t), z(t), v) - y_\alpha)e \Big\} dt$$

$$\begin{aligned}
\text{subject to} \quad & \mu^T f_x(t, y(t), z(t), w) + \nu^T G_x(t, y(t), z(t), u) \\
& \quad + \gamma^T H_x(t, y(t), z(t), v) + \gamma_t = 0, \\
& \mu^T f_c(t, y(t), z(t), w) + \nu^T G_c(t, y(t), z(t), u) \\
& \quad + \gamma^T H_c(t, y(t), z(t), v) = 0, \\
& y(t_0) = x_0, \ y(t_1) = x_1, \\
& \mu^T > 0, \ e^T \mu = 1, e = (1, \ldots, 1) \in \mathbb{R}^s,
\end{aligned}$$

for all $w \in \mathcal{W}, u \in \mathcal{U}, v \in \mathcal{V}$.

We denote $\mathcal{D}_w = \{(y, z) \in \mathscr{X} \times \mathscr{C}$: satisfying conditions (7.3.1)–(7.3.4) $\}$ the set of all feasible solutions to (RUWRDP) and we say that it is the robust feasible solution set to the problem (UWRDP).

Definition 7.3.0.1 A point $(\bar{y}, \bar{z}, \bar{\mu}, \bar{\nu}, \bar{\gamma}, \bar{w}, \bar{u}, \bar{v}) \in \mathcal{D}_w$ is said to be weak robust efficient solution to the Wolfe type robust dual problem (UWRDP), if there does not exist another point $(y, z, \mu, \nu, \gamma, w, u, v) \in \mathcal{D}_w$ such that

$$\int_{\Gamma_{t_0,t_1}} \Big\{ f(\bar{\eta}, \bar{w}) + \nu^T G(\bar{\eta}, \bar{u})e + \gamma^T (H(\bar{\eta}, \bar{v}) - y_\alpha)e \Big\} dt$$
$$< \int_{\Gamma_{t_0,t_1}} \Big\{ f(\eta, \bar{w}) + \nu^T G(\eta, \bar{u})e + \gamma^T (H(\eta, \bar{v}) - y_\alpha)e \Big\} dt.$$

Now, we shall establish the weak duality result for (MMOPU) under convexity assumption which states that the value attained by the objective functional of the dual problem over its feasible set does not exceed the value attained by the objective functional of the primal problem.

Theorem 7.3.0.1 *(Weak Robust Duality) Let* $(\bar{x}, \bar{c})$ *and* $(\bar{y}, \bar{z}, \bar{\mu}, \bar{\nu}, \bar{\gamma}, \bar{w}, \bar{u}, \bar{v})$ *be the robust feasible solutions of* (MMOPU) *and* (UWRDP), *respectively. Assume that* $\max_{w \in \mathcal{W}} f(\bar{\pi}, w) = f(\bar{\pi}, \bar{w})$. *Further, if* $\int_\Gamma \bar{\mu}^T f(\cdot, \bar{w}) dt$, $\int_\Gamma \bar{\nu}^T G(\cdot, \bar{u}) dt$, *and* $\int_\Gamma \bar{\gamma}^T \{H(\cdot, \bar{v}) - y_\alpha\} dt$ *are convex at* $(\bar{y}, \bar{z})$, *then the following inequality cannot hold*

$$\int_\Gamma \max_{w \in \mathcal{W}} f(\bar{\pi}, w) dt < \int_\Gamma \Big\{ f(\bar{\eta}, \bar{w}) + \bar{\nu}^T G(\bar{\eta}, \bar{u})e + \bar{\gamma}^T (H(\bar{\eta}, \bar{v}) - y_\alpha)e \Big\} dt.$$

Proof Contrary to the result, we assume that

$$\int_\Gamma \max_{w \in \mathcal{W}} f(\bar{\pi}, w) dt < \int_\Gamma \Big\{ f(\bar{\eta}, \bar{w}) + \bar{\nu}^T G(\bar{\eta}, \bar{u})e + \bar{\gamma}^T (H(\bar{\eta}, \bar{v}) - y_\alpha)e \Big\} dt.$$

Since $\max_{w \in W} f(\bar{\pi}, w) = f(\bar{\pi}, \bar{w})$, we have

$$\int_{\Gamma} f(\bar{\pi}, \bar{w})dt < \int_{\Gamma} \Big\{ f(\bar{\eta}, \bar{w}) + \bar{v}^T G(\bar{\eta}, \bar{u})e + \bar{\gamma}^T (H(\bar{\eta}, \bar{v}) - y_\alpha)e \Big\} dt.$$

The above inequality together with the robust feasibility of $(\bar{x}, \bar{c})$ to the problem (MMOPU) yields

$$\int_{\Gamma} \Big\{ f(\bar{\pi}, \bar{w}) + \bar{v}^T G(\bar{\pi}, \bar{u})e + \bar{\gamma}^T (H(\bar{\pi}, \bar{v}) - x_\alpha)e \Big\} dt$$
$$< \int_{\Gamma} \Big\{ f(\bar{\eta}, \bar{w}) + \bar{v}^T G(\bar{\eta}, \bar{u})e + \bar{\gamma}^T (H(\bar{\eta}, \bar{v}) - y_\alpha)e \Big\} dt.$$

As $\bar{\mu}^T > 0$ and $\bar{\mu}^T e = 1$, therefore, the above inequality can be written as

$$\int_{\Gamma} \Big\{ \bar{\mu}^T f(\bar{\pi}, \bar{w}) + \bar{v}^T G(\bar{\pi}, \bar{u}) + \bar{\gamma}^T (H(\bar{\pi}, \bar{v}) - x_\alpha) \Big\} dt$$
$$< \int_{\Gamma} \Big\{ \mu^T f(\bar{\eta}, \bar{w}) + \bar{v}^T G(\bar{\eta}, \bar{u}) + \bar{\gamma}^T (H(\bar{\eta}, \bar{v}) - y_\alpha) \Big\} dt. \tag{7.3.5}$$

Since $\int_{\Gamma} \bar{\mu}^T f(\cdot, \bar{w})dt$, $\int_{\Gamma} \bar{v}^T G(\cdot, \bar{u})dt$ and $\int_{\Gamma} \bar{\gamma}^T (H(\cdot, \bar{v}) - y_\alpha)dt$ are convex at $(\bar{y}, \bar{z}) \in \mathcal{D}_w$, we have

$$\int_{\Gamma} \Big\{ \bar{\mu}^T f(\bar{\pi}, \bar{w}) - \bar{\mu}^T f(\bar{\eta}, \bar{w}) \Big\} dt \geqq \int_{\Gamma} (\bar{x} - \bar{y}) \bar{\mu}^T f_x(\bar{\eta}, \bar{w})dt + \int_{\Gamma} (\bar{c} - \bar{z}) \bar{\mu}^T f_c(\bar{\eta}, \bar{w})dt, \tag{7.3.6}$$

$$\int_{\Gamma} \Big\{ \bar{v}^T G(\bar{\pi}, \bar{u}) - \bar{v}^T G(\bar{\eta}, \bar{u}) \Big\} dt \geqq \int_{\Gamma} (\bar{x} - \bar{y}) \bar{v}^T G_x(\bar{\eta}, \bar{u})dt + \int_{\Gamma} (\bar{c} - \bar{z}) \bar{v}^T G_c(\bar{\eta}, \bar{u})dt, \tag{7.3.7}$$

and

$$\int_{\Gamma} \Big\{ \bar{\gamma}^T (H(\bar{\pi}, \bar{v}) - x_\alpha) - \bar{\gamma}^T (H(\bar{\eta}, \bar{v}) - y_\alpha) \Big\} dt$$
$$\geqq \int_{\Gamma} (\bar{x} - \bar{y}) \Big\{ \bar{\gamma}^T H_x(\bar{\eta}, \bar{v}) + \bar{\gamma}_t \Big\} dt + \int_{\Gamma} (\bar{c} - \bar{z}) \bar{\gamma}^T H_c(\bar{\eta}, \bar{v})dt. \tag{7.3.8}$$

On adding the inequalities (7.3.6), (7.3.7) and (7.3.8) along with the robust feasibility of $(\bar{y}, \bar{z}, \bar{\mu}, \bar{v}, \bar{\gamma}, \bar{w}, \bar{u}, \bar{v})$ to the problem (UWRDP), respectively. We have

$$\int_{\Gamma} \left\{ \bar{\mu}^T f(\bar{\pi}, \bar{w}) + \bar{\upsilon}^T G(\bar{\pi}, \bar{u}) + \bar{\gamma}^T (H(\bar{\pi}, \bar{\upsilon}) - x_\alpha) \right\} dt$$
$$\geqq \int_{\Gamma} \left\{ \bar{\mu}^T f(\bar{\eta}, \bar{w}) + \bar{\upsilon}^T G(\bar{\eta}, \bar{u}) + \bar{\gamma}^T (H(\bar{\eta}, \bar{\upsilon}) - y_\alpha) \right\} dt,$$

which contradicts the inequality (7.3.5). This completes the proof. □

We present an example to authenticate Theorem 7.3.0.1.

Example 7.3.0.1 Let $p = 2, q = r = 1, k = 2, \mathcal{W}_1 = [0, 1], \mathcal{W}_2 = [\frac{1}{2}, 1], \mathcal{U} = [0, 1], \mathcal{V}_1^1 = \mathcal{V}_2^1 = [1, 2]$ and Γ_{t_0,t_1} is fixed by the diagonally opposite points $t_0 = (t_0^1, t_0^2)$ and $t_1 = (t_1^1, t_1^2)$ in $\mathbb{R}^2$.

We consider the following multi-dimensional multi-objective optimization problem:

$$\text{(MMOPU1)} \quad \min_{(x(\cdot),c(\cdot))} \int_{\Gamma_{0,\frac{1}{4}}} \left(x^2 w_1 + \frac{5}{4}c, \frac{1}{w_2}c^2 \right) dt^1 dt^2$$
$$\text{subject to} \quad ux(x-3) \leq 0,$$
$$x_{t^1} = v_1^1 - c,$$
$$x_{t^2} = v_2^1 - c,$$
$$x(0,0) = 0, \ x\left(\frac{1}{4}, \frac{1}{4}\right) = \frac{1}{2},$$

where $t = (t^1, t^2) \in \Gamma_{0,\frac{1}{4}}$. Let $(\bar{x}, \bar{c}) = (t^1, t^2)$ be a robust feasible solution to the problem (MMOPU1).

The robust counterpart of the multi-dimensional multi-objective optimization problem (MMOPU1) is defined as:

$$\text{(RMMOPU1)} \quad \min_{(x(\cdot),c(\cdot))} \int_{\Gamma_{0,1}} \left(\max_{w_1 \in \mathcal{W}_1} \left(x^2 w_1 + \frac{5}{4}c \right), \max_{w_2 \in \mathcal{W}_2} \left(\frac{1}{w_2}c^2 \right) \right) dt^1 dt^2$$
$$\text{subject to} \quad ux(x-3) \leq 0, \ \forall u \in \mathcal{U},$$
$$x_{t^1} = v_1^1 - c, \ \forall v_1^1 \in \mathcal{V}_1^1,$$
$$x_{t^2} = v_2^1 - c, \ \forall v_2^1 \in \mathcal{V}_2^1,$$
$$x(0,0) = 0, \ x\left(\frac{1}{4}, \frac{1}{4}\right) = \frac{1}{2},$$

where $t = (t^1, t^2) \in \Gamma_{0,\frac{1}{4}}$.

The Wolfe type robust dual problem associated with the problem (MMOPU1) is defined as follows:

$$\text{(UWRDP1)} \max_{(y(\cdot),z(\cdot))} \int_{\Gamma_{0,\frac{1}{4}}} \Big\{ \Big(y^2 w_1 + \frac{5}{4}z\Big) + \nu u y(y-3) + \gamma_1^1(y_{t^1} - v_1^1 + z) + \gamma_2^1(y_{t^2} - v_2^1 + z), \frac{1}{w_2}z^2 + \nu u y(y-3) + \gamma_1^1(y_{t^1} - v_1^1 + z) + \gamma_2^1(y_{t^2} - v_2^1 + z) \Big\} dt^1 dt^2$$

subject to

$$2\mu_1 w_1 y + \nu u(2y-3) + (\gamma_1^1)_{t^1} + (\gamma_1^1)_{t^2} = 0, \tag{7.3.9}$$

$$\frac{5}{4}\mu_1 + 2\frac{1}{w_2}z\mu_2 - \gamma_1^1 - \gamma_2^1 = 0, \tag{7.3.10}$$

$$y(0,0) = 0, \; y\Big(\frac{1}{4}, \frac{1}{4}\Big) = \frac{1}{2}, \tag{7.3.11}$$

$$\mu^T > 0, \; e^T\mu = 1, \; e = (1,1) \in \mathbb{R}^2. \tag{7.3.12}$$

The robust counterpart to the problem (UWRDP1) is given as:

$$\text{(RUWRDP1)} \max_{(y(\cdot),z(\cdot),w,u,v)} \int_{\Gamma_{0,\frac{1}{4}}} \Big\{ \Big(y^2 w_1 + \frac{5}{4}z\Big) + \nu u y(y-3) + \gamma_1^1(y_{t^1} - v_1^1 + z) + \gamma_2^1(y_{t^2} - v_2^1 + z), \frac{1}{w_2}z^2 + \nu u y(y-3) + \gamma_1^1(y_{t^1} - v_1^1 + z) + \gamma_2^1(y_{t^2} - v_2^1 + z) \Big\} dt^1 dt^2$$

subject to

$$2\mu_1 w_1 y + \nu u(2y-3) + (\gamma_1^1)_{t^1} + (\gamma_1^1)_{t^2} = 0,$$

$$\frac{5}{4}\mu_1 + 2\frac{1}{w_2}z\mu_2 - \gamma_1^1 - \gamma_2^1 = 0,$$

$$y(0,0) = 0, \; y\Big(\frac{1}{4}, \frac{1}{4}\Big) = \frac{1}{2},$$

$$\mu^T > 0, \; e^T\mu = 1, \; e = (1,1) \in \mathbb{R}^2,$$

for all $w \in \mathcal{W}$, $u \in \mathcal{U}$, $v \in \mathcal{V}$.
Note that $\mathcal{D}_w = \{(y,z) \in \mathscr{X} \times \mathscr{C} :$ satisfying conditions (7.3.9)–(7.3.12)$\}$ is the feasible set to the (UWRDP1). Let $\bar{y} = 0, \bar{z} = \frac{1}{2}, \; \bar{\mu} = (\bar{\mu}_1, \bar{\mu}_2) = (1,0), \; \bar{\nu} = 2, \; \bar{\gamma} = (\bar{\gamma}_1^1, \bar{\gamma}_2^1) = (1, \frac{1}{4}), \; \bar{w} = (\bar{w}_1, \bar{w}_2) = (1,1), \; \bar{u} = 1, \; \bar{v} = (\bar{v}_1^1, \bar{v}_2^1) = (1,2)$. Then $(\bar{y}, \bar{z}, \bar{\mu}, \bar{\nu}, \bar{\gamma}, \bar{w}, \bar{u}, \bar{v})$ is a robust feasible solution to (UWRDP1). Further, it can be easily verified that all the involved functionals are convex at $(\bar{y}, \bar{z}) \in \mathcal{D}_w$. Thus, the following inequality

$$\int_{\Gamma} f(\bar{\pi}, \bar{w})dt - \int_{\Gamma} \Big\{ f(\bar{\eta}, \bar{w}) + \bar{v}^T G(\bar{\eta}, \bar{u})e + \bar{\gamma}^T (H(\bar{\eta}, \bar{v}) - y_\alpha)e \Big\} dt$$

$$= \int_{\Gamma_{0,\frac{1}{4}}} \Big(\bar{x}^2 \bar{w}_1 + \frac{5}{4}\bar{c}, \frac{1}{\bar{w}_2}\bar{c}^2 \Big) dt^1 dt^2 - \int_{\Gamma_{0,\frac{1}{4}}} \Big\{ \Big(\bar{y}^2 \bar{w}_1 + \frac{5}{4}\bar{z} \Big) + \bar{v}\bar{u}\bar{y}(\bar{y} - 3)$$

$$+ \bar{\gamma}_1^1 (y_{t^1} - \bar{v}_1^1 + \bar{z}) + \bar{\gamma}_2^1 (y_{t^2} - \bar{v}_2^1 + \bar{z}), \frac{1}{\bar{w}_2}\bar{z}^2 + \bar{v}\bar{u}\bar{y}(\bar{y} - 3)$$

$$+ \bar{\gamma}_1^1 (\bar{y}_{t^1} - \bar{v}_1^1 + \bar{z}) + \bar{\gamma}_2^1 (\bar{y}_{t^2} - \bar{v}_2^1 + \bar{z}) \Big\} dt^1 dt^2$$

$$= \int_{\Gamma_{0,\frac{1}{4}}} \Big(t^1 + t^2 + \frac{5}{4}, 1 \Big) dt^1 dt^2 - \int_{\Gamma_{0,\frac{1}{4}}} \Big(\frac{-5}{4}, \frac{-13}{8} \Big) dt^1 dt^2$$

$$= \Big(\frac{3}{32}, \frac{1}{16} \Big) - \Big(\frac{-5}{64}, \frac{-13}{128} \Big)$$

$$= \Big(\frac{11}{64}, \frac{21}{128} \Big) > 0,$$

which shows that the duality gap is positive. Hence, Theorem 7.3.0.1 (Weak Robust Duality) is verified.

Now, we shall prove the strong robust duality result which states that duality gap is zero.

Theorem 7.3.0.2 (*Strong Robust Duality*) *Let* $(\bar{x}, \bar{c})$ *be a weak robust efficient solution to the problem* (MMOPU). *Assume that* $\max_{w \in \mathcal{W}} f(\bar{\pi}, w) = f(\bar{\pi}, \bar{w})$ *and the constraint conditions (for the existence of multiplier) holds for* (MMOPU). *Then, there exist the scalar vector* $\bar{\mu} \in \mathbb{R}_+^s$, *the piecewise smooth Lagrange multipliers* $\bar{v} = (\bar{v}_l(t)) \in \mathbb{R}_+^m, l = \overline{1, m}$ *and* $\bar{\gamma} = (\bar{\gamma}_\alpha^\tau(t)) \in \mathbb{R}^{pq}, \tau = \overline{1, q}, \alpha = \overline{1, p}$ *and the uncertain parameters* $\bar{u} \in \mathcal{U}, \bar{v} \in \mathcal{V}$ *such that* $(\bar{x}, \bar{c}, \bar{\mu}, \bar{v}, \bar{\gamma}, \bar{w}, \bar{u}, \bar{v})$ *is a robust feasible solution to the problem* (UWRDP). *Further, if the Weak Robust Duality (Theorem 7.3.0.1) holds, then* $(\bar{x}, \bar{c}, \bar{\mu}, \bar{v}, \bar{\gamma}, \bar{w}, \bar{u}, \bar{v})$ *is a weak robust efficient solution to the problem* (UWRDP).

Proof Since $(\bar{x}, \bar{c})$ be a weak robust efficient solution to the problem (MMOPU), therefore, by Theorem 7.2.0.1, there exist the scalar vector $\bar{\mu} \in \mathbb{R}_+^s$, the piecewise smooth Lagrange multiplies $\bar{v} \in \mathbb{R}_+^m, \bar{\gamma} \in \mathbb{R}^{pq}$ and the uncertain parameters $\bar{u} \in \mathcal{U}, \bar{v} \in \mathcal{V}$ such that the conditions (7.2.1)–(7.2.4) are satisfied at $(\bar{x}, \bar{c})$. This shows the robust feasibility of $(\bar{x}, \bar{c}, \bar{\mu}, \bar{v}, \bar{\gamma}, \bar{w}, \bar{u}, \bar{v})$ to the problem (UWRDP) and the corresponding objective values are equal. If $(\bar{x}, \bar{c}, \bar{\mu}, \bar{v}, \bar{\gamma}, \bar{w}, \bar{u}, \bar{v})$ is not a weak efficient solution to the problem (UWRDP), then there exists another point $(y, z, \bar{\mu}, \bar{v}, \bar{\gamma}, \bar{w}, \bar{u}, \bar{v})$ such that

$$\int_{\Gamma} \Big\{ f(\bar{\pi}, \bar{w}) + v^T G(\bar{\pi}, \bar{u})e + \gamma^T (H(\bar{\pi}, \bar{v}) - y_\alpha)e \Big\} dt$$

$$< \int_{\Gamma} \Big\{ f(\eta, \bar{w}) + v^T G(\eta, \bar{u})e + \gamma^T (H(\eta, \bar{v}) - y_\alpha)e \Big\} dt.$$

From the condition (7.2.3), we get

$$\int_{\Gamma} f(\bar{\pi}, \bar{w})dt < \int_{\Gamma} \Big\{f(\eta, \bar{w}) + \nu^T G(\eta, \bar{u})e + \gamma^T (H(\eta, \bar{v}) - y_{\alpha})e\Big\}dt.$$

Since $\max_{w \in W} f(\pi, w) = f(\pi, \bar{w})$, we have

$$\int_{\Gamma} \max_{w \in W} f(\bar{\pi}, w)dt < \int_{\Gamma} \Big\{f(\eta, \bar{w}) + \nu^T G(\eta, \bar{u})e + \gamma^T (H(\eta, \bar{v}) - y_{\alpha})e\Big\}dt,$$

which contradict the Weak Robust Duality (Theorem 7.3.0.1). Hence, $(\bar{x}, \bar{c}, \bar{\mu}, \bar{\nu}, \bar{\gamma}, \bar{w}, \bar{u}, \bar{v})$ is a weak efficient solution to the problem (UWRDP). □

Theorem 7.3.0.3 (*Strict Converse Robust Duality*) *Let* $(\bar{x}, \bar{c}, \bar{\mu}, \bar{\nu}, \bar{\gamma}, \bar{w}, \bar{u}, \bar{v})$ *be a robust feasible solution to the problem* (UWRDP). *Assume that* $\max_{w \in W} f(\pi, w) = f(\pi, \bar{w})$ *and* $\int_{\Gamma} \bar{\mu}^T f(\cdot, \bar{w})dt$, $\int_{\Gamma} \bar{\nu}^T G(\cdot, \bar{u})dt$ *and* $\int_{\Gamma}\{\bar{\gamma}^T H(\cdot, \bar{v}) - y_{\alpha}\}dt$ *are strict convex at* $(\bar{y}, \bar{z})$. *If* $(\bar{x}, \bar{c}) \in \mathcal{D}$ *such that* $\int_{\Gamma} f(\bar{\pi}, \bar{w})dt = \int_{\Gamma} f(\bar{\eta}, \bar{w})dt$, *then* $(\bar{x}, \bar{c})$ *is a weak robust efficient solution to the problem* (MMOPU).

Proof Since $(\bar{x}, \bar{c}, \bar{\mu}, \bar{\nu}, \bar{\gamma}, \bar{w}, \bar{u}, \bar{v})$ is a robust feasible solution to the problem (UWRDP), on multiplying the inequality 8d1) and (7.2.2) by $(\hat{x} - \bar{y})$ and $(\hat{c} - \bar{z})$, respectively, and then integrate them, we get

$$\int_{\Gamma} (\hat{x} - \bar{y})\Big\{\bar{\mu}^T f_x(\bar{\eta}, \bar{w}) + \bar{\nu}^T G_x(\bar{\eta}, \bar{u}) + \bar{\gamma}^T H_x(\bar{\eta}, \bar{v}) + \bar{\gamma}_t\Big\}dt$$
$$+ \int_{\Gamma} (\hat{c} - \bar{z})\Big\{\bar{\mu}^T f_c(\bar{\eta}, \bar{w}) + \bar{\nu}^T G_c(\bar{\eta}, \bar{u}) + \bar{\gamma}^T H_c(\bar{\eta}, \bar{v})\Big\}dt = 0. \tag{7.3.13}$$

We proceed by contradiction and assume that $(\bar{x}, \bar{c})$ is not a weak robust efficient solution to the problem (MMOPU), then there exists $(\hat{x}, \hat{c}) \in \mathcal{D}$ such that

$$\int_{\Gamma} \max_{w \in W} f(\hat{\pi}, w)dt < \int_{\Gamma} \max_{w \in W} f(\bar{\pi}, w)dt.$$

Since $\max_{w \in W} f(\pi, w) = f(\pi, \bar{w})$, therefore

$$\int_{\Gamma} f(\hat{\pi}, \bar{w})dt < \int_{\Gamma} f(\bar{\pi}, \bar{w})dt.$$

By assumption $\int_{\Gamma} f(\bar{\pi}, \bar{w})dt = \int_{\Gamma} f(\bar{\eta}, \bar{w})dt$. Therefore, the above inequality yields

$$\int_{\Gamma} f(\hat{\pi}, \bar{w})dt < \int_{\Gamma} f(\bar{\eta}, \bar{w})dt.$$

Since $\bar{\mu} > 0 \in \mathbb{R}^s_+$, therefore

$$\int_\Gamma \bar{\mu}^T f(\hat{\pi}, \bar{w})dt < \int_\Gamma \bar{\mu}^T f(\bar{\eta}, \bar{w})dt. \tag{7.3.14}$$

On the other hand, from the assumption $\int_\Gamma \bar{\mu}^T f(\cdot, \bar{w})dt$ is strict convex at $(\bar{y}, \bar{z})$, we have

$$\int_\Gamma \Big\{\bar{\mu}^T f(\hat{\pi}, \bar{w}) - \bar{\mu}^T f(\bar{\eta}, \bar{w})\Big\}dt > \int_\Gamma (\hat{x} - \bar{y})\bar{\mu}^T f_x(\bar{\eta}, \bar{w})dt + \int_\Gamma (\hat{c} - \bar{z})\bar{\mu}^T f_c(\bar{\eta}, \bar{w})dt,$$

which together with the inequality (7.3.14), gives

$$\int_\Gamma (\hat{x} - \bar{y})\bar{\mu}^T f_x(\bar{\eta}, \bar{w})dt + \int_\Gamma (\hat{c} - \bar{z})\bar{\mu}^T f_c(\bar{\eta}, \bar{w})dt < 0. \tag{7.3.15}$$

Again, by assumption $\int_\Gamma \bar{\nu}^T G(\cdot, \bar{u})dt$ is convex at $(\bar{y}, \bar{z})$, we get

$$\int_\Gamma \Big\{\bar{\nu}^T G(\hat{\pi}, \bar{u}) - \bar{\nu}^T G(\bar{\eta}, \bar{u})\Big\}dt > \int_\Gamma (\hat{x} - \bar{y})\bar{\nu}^T G_x(\bar{\eta}, \bar{u})dt + \int_\Gamma (\hat{c} - \bar{z})\bar{\nu}^T G_c(\bar{\eta}, \bar{u})dt. \tag{7.3.16}$$

Since $(\hat{x}, \hat{c})$ and $(\bar{x}, \bar{c}, \bar{\nu}, \bar{\gamma}, \bar{w}, \bar{u}, \bar{v})$ are the robust feasible solutions to the problem (MMOPU) and (UWRDP), respectively, we obtain

$$\int_\Gamma \bar{\nu}^T G(\hat{\pi}, \bar{u})dt - \int_\Gamma \bar{\nu}^T G(\bar{\eta}, \bar{u})dt \leqq 0,$$

which along with the inequality (7.3.16), gives

$$\int_\Gamma (\hat{x} - \bar{y})\bar{\nu}^T G_x(\bar{\eta}, \bar{u})dt + \int_\Gamma (\hat{c} - \bar{z})\bar{\nu}^T G_c(\bar{\eta}, \bar{u})dt < 0. \tag{7.3.17}$$

Similarly, $\int_\Gamma \{\bar{\gamma}^T H(\cdot, \bar{v}) - y_\alpha\}dt$ is also convex at $(\bar{y}, \bar{z})$. The robust feasible solutions $(\hat{x}, \hat{c})$ and $(\bar{x}, \bar{c}, \bar{\mu}, \bar{\nu}, \bar{\gamma}, \bar{w}, \bar{u}, \bar{v})$ to the problem (MMOPU) and (UWRDP), respectively, yields

$$\int_\Gamma (\hat{x} - \bar{y})\Big\{\bar{\gamma}^T H_x(\bar{\eta}, \bar{v}) + \bar{\gamma}_t\Big\}dt + \int_\Gamma (\hat{c} - \bar{z})\bar{\gamma}^T H_c(\bar{\eta}, \bar{v})dt < 0. \tag{7.3.18}$$

On adding the inequalities (7.3.15), (7.3.17) and (7.3.18), we obtain the following inequality

$$\int_{\Gamma}(\hat{x}-\bar{y})\left\{\bar{\mu}^T f_x(\bar{\eta},\bar{w})+\bar{\nu}^T G_x(\bar{\eta},\bar{u})+\bar{\gamma}^T H_x(\bar{\eta},\bar{v})+\bar{\gamma}_t\right\}dt$$
$$+\int_{\Gamma}(\hat{c}-\bar{z})\left\{\bar{\mu}^T f_c(\bar{\eta},\bar{w})+\bar{\nu}^T G_c(\bar{\eta},\bar{u})+\bar{\gamma}^T H_c(\bar{\eta},\bar{v})\right\}dt<0,$$

which contradicts the inequality (7.3.13). This completes the proof. □

7.4 Mond-Weir Type Robust Dual Problem

In this section, we formulate the Mond-Weir type robust dual problem for the considered multi-dimensional multi-objective optimization problem with data uncertainty in the objective and constraint functionals (MMOPU) as follows:

$$\text{(UMWRDP)}\quad \max_{(y(\cdot),z(\cdot))}\int_{\Gamma_{t_0,t_1}} f(t,y(t),z(t),w)dt$$

subject to
$$\mu^T f_x(t,y(t),z(t),w)+\nu^T G_x(t,y(t),z(t),u)+\gamma^T H_x(t,y(t),z(t),v)+\gamma_t=0, \quad (7.4.1)$$
$$\mu^T f_c(t,y(t),z(t),w)+\nu^T G_c(t,y(t),z(t),u)+\gamma^T H_c(t,y(t),z(t),v)=0, \quad (7.4.2)$$
$$\bar{\nu}^T G(t,y(t),z(t),u)\geqq 0, \quad (7.4.3)$$
$$y_\alpha = H(t,y(t),z(t),v), \quad (7.4.4)$$
$$y(t_0)=x_0,\ y(t_1)=x_1, \quad (7.4.5)$$
$$\mu\in\mathbb{R}^s_+,\ \mu^T e=1,\nu\in\mathbb{R}^m_+,\gamma\in\mathbb{R}^{pq}. \quad (7.4.6)$$

The associated robust counterpart to the problem (UMWRDP) is given as follows:

$$\text{(RUMWRDP)}\quad \max_{(y(\cdot),z(\cdot),w)}\int_{\Gamma_{t_0,t_1}} f(t,y(t),z(t),w)dt$$

subject to
$$\mu^T f_x(t,y(t),z(t),w)+\nu^T G_x(t,y(t),z(t),u)+\gamma^T H_x(t,y(t),z(t),v)+\gamma_t=0,$$
$$\mu^T f_c(t,y(t),z(t),w)+\nu^T G_c(t,y(t),z(t),u)+\gamma^T H_c(t,y(t),z(t),v)=0,$$
$$\nu^T G(t,y(t),z(t),u)\geqq 0,$$
$$y_\alpha = H(t,y(t),z(t),v),$$
$$y(t_0)=x_0,\ y(t_1)=x_1,$$
$$\mu^T e=1,\mu\in\mathbb{R}^s_+,\nu\in\mathbb{R}^m_+,\gamma\in\mathbb{R}^{pq},$$

for all $w\in\mathcal{W},u\in\mathcal{U},v\in\mathcal{V}$.

We denote $\mathcal{D}_{mw} = \{(y, z) \in \mathscr{X} \times \mathscr{C} :$ satisfying conditions (7.4.1)–(7.4.6)$\}$ the set of all feasible solutions to (RUWRDP) and we say that it is the robust feasible solution set to the problem (UWRDP).

Now, we establish the robust duality results (weak and strong robust duality results) for (MMOPU) and (UMWRDP) under the convexity hypotheses.

Theorem 7.4.0.1 *(Weak Robust Duality) Let* $(\bar{x}, \bar{c})$ *and* $(\bar{y}, \bar{z}, \bar{\mu}, \bar{\nu}, \bar{\gamma}, \bar{w}, \bar{u}, \bar{v})$ *be the robust feasible solutions to the problem* (MMOPU) *and* (UMWRDP)*, respectively. Assume that* $\max_{w\in\mathcal{W}} f(\bar{\pi}, w) = f(\bar{\pi}, \bar{w})$*. Further, if* $\int_\Gamma \bar{\mu}^T f(\cdot, \bar{w})dt$*,* $\int_\Gamma \bar{\nu}^T G(\cdot, \bar{u})dt$*, and* $\int_\Gamma \bar{\gamma}^T\{H(\cdot, \bar{v}) - y_\alpha\}dt$ *are convex at* $(\bar{y}, \bar{z})$*, then the following inequality cannot hold*

$$\int_\Gamma \max_{w\in\mathcal{W}} f(\bar{\pi}, w)dt < \int_\Gamma f(\bar{\eta}, \bar{w})dt.$$

Proof Contrary to the result, we assume that

$$\int_\Gamma \max_{w\in\mathcal{W}} f(\bar{\pi}, w)dt < \int_\Gamma f(\bar{\eta}, \bar{w})dt.$$

Since $\max_{w\in\mathcal{W}} f(\bar{\pi}, w) = f(\bar{\pi}, \bar{w})$, we have

$$\int_\Gamma f(\bar{\pi}, \bar{w})dt < \int_\Gamma f(\bar{\eta}, \bar{w})dt. \tag{7.4.7}$$

By assumptions $\int_\Gamma \bar{\mu}^T f(\cdot, \bar{w})dt$, $\int_\Gamma \bar{\nu}^T G(\cdot, \bar{u})dt$ and $\int_\Gamma \bar{\gamma}^T (H(\cdot, \bar{v}) - y_\alpha)dt$ are convex at $(\bar{y}, \bar{z})$. Then, we have

$$\int_\Gamma \left\{\bar{\mu}^T f(\bar{\pi}, \bar{w}) - \bar{\mu}^T f(\bar{\eta}, \bar{w})\right\}dt \geqq \int_\Gamma (\bar{x} - \bar{y})\bar{\mu}^T f_x(\bar{\eta}, \bar{w})dt + \int_\Gamma (\bar{c} - \bar{z})\bar{\mu}^T f_c(\bar{\eta}, \bar{w})dt, \tag{7.4.8}$$

$$\int_\Gamma \left\{\bar{\nu}^T G(\bar{\pi}, \bar{u}) - \bar{\nu}^T G(\bar{\eta}, \bar{u})\right\}dt \geqq \int_\Gamma (\bar{x} - \bar{y})\bar{\nu}^T G_x(\bar{\eta}, \bar{u})dt + \int_\Gamma (\bar{c} - \bar{z})\bar{\nu}^T G_c(\bar{\eta}, \bar{u})dt, \tag{7.4.9}$$

and

$$\int_\Gamma \left\{(\bar{\gamma}^T H(\bar{\pi}, \bar{v}) - x_\alpha) - (\bar{\gamma}^T H(\bar{\eta}, \bar{v}) - y_\alpha)\right\}dt$$
$$\geqq \int_\Gamma (\bar{x} - \bar{y})\left\{\bar{\gamma}^T H_x(\bar{\eta}, \bar{v}) + \dot{\bar{\gamma}}_t\right\}dt + \int_\Gamma (\bar{c} - \bar{z})\bar{\gamma}^T H_c(\bar{\eta}, \bar{v})dt. \tag{7.4.10}$$

On adding the inequalities (7.4.8), (7.4.9) and (7.4.10) along with the robust feasibility of $(\bar{x}, \bar{c})$ and $(\bar{y}, \bar{z}, \bar{\mu}, \bar{\nu}, \bar{\gamma}, \bar{w}, \bar{u}, \bar{v})$ to the problem (MMOPU) and (UMWRDP), respectively, we have

$$\int_{\Gamma} f(\bar{\pi}, \bar{w})dt \geqq \int_{\Gamma} f(\bar{\eta}, \bar{w})dt,$$

which contradicts the inequality (7.4.7). This completes the proof. □

We present an example to authenticate Theorem 7.4.0.1.

Example 7.4.0.1 Let $p = 2, q = r = 1, k = 2, \mathcal{W}_1 = \mathcal{W}_2 = [0, 1], \mathcal{U} = [3, 4]$, $\mathcal{V}_1^1 = \mathcal{V}_2^1 = [0, 2]$ and Γ_{t_0,t_1} is fixed by the diagonally opposite points $t_0 = (t_0^1, t_0^2)$ and $t_1 = (t_1^1, t_1^2)$ in $\mathbb{R}^2$.

We consider the following multi-dimensional multi-objective optimization problem:

$$\text{(MMOPU2)} \quad \min_{(x(\cdot),c(\cdot))} \int_{\Gamma_{0,1}} \left(x^2 + c + w_1, x + w_2c^2 + 1\right) dt^1dt^2$$
$$\text{subject to} \quad u + x(x-4) \leq 0,$$
$$x_{t^1} = 2 - v_1^1 c,$$
$$x_{t^2} = 2 - v_2^1 c,$$
$$x(0,0) = 1, \; x(1,1) = 2,$$

where $t = (t^1, t^2) \in \Gamma_{0,1}$.
The robust counterpart of the multi-dimensional multi-objective optimization problem (MMOPU2) is defined as follows:

$$\text{(RMMOPU2)} \quad \min_{(x(\cdot),c(\cdot))} \int_{\Gamma_{0,1}} \left(\max_{w_1\in\mathcal{W}_1} (x^2 + c + w_1), \max_{w_2\in\mathcal{W}_2} (x + w_2c^2 + 1)\right) dt^1dt^2$$
$$\text{subject to} \quad ux(x-3) \leq 0, \; \forall u \in \mathcal{U},$$
$$x_{t^1} = 2 - v_1^1 c, \; \forall v_1^1 \in \mathcal{V}_1^1,$$
$$x_{t^2} = 2 - v_2^1 c, \; \forall v_2^1 \in \mathcal{V}_2^1,$$
$$x(0,0) = 1, \; x(1,1) = 2,$$

where $t = (t^1, t^2) \in \Gamma_{0,1}$. Let $(\bar{x}, \bar{c}) = (\frac{1}{2}(t^1 + t^2) + 1, \frac{3}{2})$ be a robust feasible solution to the problem (MMOPU2).

The Mond-Weir type robust dual problem associated with the problem (MMOPU2) is defined as follows:

$$\text{(UMWRDP2)} \quad \max_{(y(\cdot),z(\cdot))} \int_{\Gamma_{0,1}} \left(y^2 + z + w_1, y + w_2z^2 + 1\right) dt^1dt^2$$
$$\text{subject to} \quad 2\mu_1 y + \mu_2 + \nu(2y-4) + (\gamma_1^1)_{t^1} + (\gamma_1^1)_{t^2} = 0, \tag{7.4.11}$$
$$\mu_1 + 2w_2z\mu_2 - v_1^1\gamma_1^1 - v_2^1\gamma_2^1 = 0, \tag{7.4.12}$$
$$\nu(u + y(y-4)) \geqq 0, \tag{7.4.13}$$
$$y_{t^1} = 2 - v_1^1 c, \tag{7.4.14}$$

$$y_{t^2} = 2 - v_2^1 c, \tag{7.4.15}$$
$$y(0,0) = 1, \ \ y(1,1) = 2. \tag{7.4.16}$$

The robust counterpart to the problem (UMWRDP2) is given as follows:

$$\text{(RUMWRDP2)} \quad \max_{(y(\cdot),z(\cdot),w)} \int_{\Gamma_{0,1}} \left(y^2 + z + w_1, y + w_2 z^2 + 1 \right) dt^1 dt^2$$

$$\begin{aligned} \text{subject to} \quad & 2\mu_1 y + \mu_2 + \nu(2y-4) + (\gamma_1^1)_{t^1} + (\gamma_1^1)_{t^2} = 0, \\ & \mu_1 + 2w_2 z \mu_2 - v_1^1 \gamma_1^1 - v_2^1 \gamma_2^1 = 0, \\ & \nu(u + y(y-4)) \geqq 0, \\ & y_{t^1} = 2 - v_1^1 c, \\ & y_{t^2} = 2 - v_2^1 c, \\ & y(0,0) = 1, \ \ y(1,1) = 2, \end{aligned}$$

for all $w \in \mathcal{W}$, $u \in \mathcal{U}$, $v \in \mathcal{V}$.
Note that $\mathcal{D}_{mw} = \{(y,z) \in \mathscr{X} \times \mathscr{C} :$ satisfying conditions (7.4.11)–(7.4.16)$\}$ is the feasible set to the (UMWRDP2). Let $\bar{y} = 1, \bar{z} = 1, \bar{\mu} = (\bar{\mu}_1, \bar{\mu}_2) = (1,0)$, $\bar{\nu} = 1$, $\bar{\gamma} = (\bar{\gamma}_1^1, \bar{\gamma}_2^1) = (\frac{1}{4}, \frac{1}{4})$, $\bar{w} = (\bar{w}_1, \bar{w}_2) = (1,1)$, $\bar{u} = 4$, $\bar{v} = (\bar{v}_1^1, \bar{v}_2^1) = (2,2)$. Then $(\bar{y}, \bar{z}, \bar{\mu}, \bar{\nu}, \bar{\gamma}, \bar{w}, \bar{u}, \bar{v})$ is a robust feasible solution to (UMWRDP2). Further, it can be easily verified that all the involved functionals are convex at $(\bar{y}, \bar{z}) \in \mathcal{D}_{mw}$. Thus, the following inequality

$$\begin{aligned} & \int_\Gamma f(\bar{\pi}, \bar{w}) dt - \int_\Gamma f(\bar{\eta}, \bar{w}) dt \\ & = \int_{\Gamma_{0,1}} \left(\bar{x}^2 + \bar{c} + \bar{w}_1, \bar{x} + \bar{w}_2 \bar{c}^2 + 1 \right) dt^1 dt^2 - \int_{\Gamma_{0,1}} \left(\bar{y}^2 + \bar{z} + \bar{w}_1, \bar{y} + \bar{w}_2 \bar{z}^2 + 1 \right) dt^1 dt^2 \\ & = \int_{\Gamma_{0,1}} \left(\left(\frac{1}{2}(t^1 + t^2) + 1 \right)^2 + \frac{3}{2} + 1, \frac{1}{2}(t^1 + t^2) + \left(\frac{3}{2} \right)^2 + 2 \right) dt^1 dt^2 - \int_{\Gamma_{0,1}} (3,3) dt^1 dt^2 \\ & = \left(\frac{115}{24}, \frac{19}{4} \right) - (3,3) \\ & = \left(\frac{43}{24}, \frac{7}{4} \right) > 0, \end{aligned}$$

which shows that the duality gap is positive. Hence, Theorem 7.4.0.1 (Weak Robust Duality) is verified.

Theorem 7.4.0.2 *(Strong Robust Duality) Let* $(\bar{x}, \bar{c})$ *be a weak robust efficient solution to the problem* (MMOPU). *Assume that* $\max_{w \in \mathcal{W}} f(\bar{\pi}, w) = f(\bar{\pi}, \bar{w})$ *and the constraint conditions (for the existence of multiplier) holds for* (MMOPU). *Then, there exist the scalar vector* $\bar{\mu} \in \mathbb{R}_+^s$, *the piecewise smooth Lagrange multipliers* $\bar{\nu} = (\bar{\nu}_l(t)) \in \mathbb{R}_+^m, l = \overline{1,m}$ *and* $\bar{\gamma} = (\bar{\gamma}_\alpha^\tau(t)) \in \mathbb{R}^{pq}, \tau = \overline{1,q}, \alpha = \overline{1,p}$ *and the uncertain parameters* $\bar{u} \in \mathcal{U}, \bar{v} \in \mathcal{V}$ *such that* $(\bar{x}, \bar{c}, \bar{\mu}, \bar{\nu}, \bar{\gamma}, \bar{w}, \bar{u}, \bar{v})$ *is a robust feasible solution to the problem* (UMWRDP). *Further, if the Weak Robust Duality (The-*

orem 7.4.0.1) holds, then $(\bar{x}, \bar{c}, \bar{\mu}, \bar{\nu}, \bar{\gamma}, \bar{w}, \bar{u}, \bar{v})$ *is a weak robust efficient solution to the problem* (UMWRDP).

Proof Since $(\bar{x}, \bar{c})$ be a weak robust efficient solution to the problem (MMOPU), therefore, by Theorem 7.2.0.1, there exist the scalar vector $\bar{\mu} \in \mathbb{R}^s_+$, the piecewise smooth Lagrange multipliers $\bar{\nu} \in \mathbb{R}^m_+$, $\bar{\gamma} \in \mathbb{R}^{pq}$ and the uncertain $\bar{u} \in \mathcal{U}$, $\bar{v} \in \mathcal{V}$ such that the conditions (7.2.1)–(7.2.4) are satisfied at $(\bar{x}, \bar{c})$. This shows the robust feasibility of $(\bar{x}, \bar{c}, \bar{\mu}, \bar{\nu}, \bar{\gamma}, \bar{w}, \bar{u}, \bar{v})$ to the problem (UMWRDP) and the corresponding objective values are equal. If $(\bar{x}, \bar{c}, \bar{\mu}, \bar{\nu}, \bar{\gamma}, \bar{w}, \bar{u}, \bar{v})$ is not a weak efficient solution to the problem (UMRDP), then there exists another point $(y, z, \bar{\mu}, \bar{\nu}, \bar{\gamma}, \bar{w}, \bar{u}, \bar{v})$ such that

$$\int_{\Gamma} \max_{w \in \mathcal{W}} f(\bar{\pi}, w) dt < \int_{\Gamma} f(\bar{\eta}, \bar{w}) dt,$$

which contradict the Weak Robust Duality (Theorem 7.4.0.1). Hence, $(\bar{x}, \bar{c}, \bar{\mu}, \bar{\nu}, \bar{\gamma}, \bar{w}, \bar{u}, \bar{v})$ is a weak efficient solution to the problem (UMWRDP). □

7.5 Mixed Type Robust Dual Problem

In this section, we formulate the mixed type robust dual problem for the considered multi-dimensional multi-objective optimization problem with data uncertainty in the objective and constraint functionals (MMOPU) as follows:

$$\text{(UMRDP)} \quad \max_{(y(\cdot), z(\cdot))} \int_{\Gamma_{t_0,t_1}} \Big\{ f(t, y(t), z(t), w) + \nu^T G(t, y(t), z(t), u)e + \gamma^T (H(t, y(t), z(t), v) - y_\alpha)e \Big\} dt$$

subject to

$$\mu^T f_x(t, y(t), z(t), w) + \nu^T G_x(t, y(t), z(t), u) + \gamma^T H_x(t, y(t), z(t), v) + \gamma_t = 0, \quad (7.5.1)$$

$$\mu^T f_c(t, y(t), z(t), w) + \nu^T G_c(t, y(t), z(t), u) + \gamma^T H_c(t, y(t), z(t), v) = 0, \quad (7.5.2)$$

$$\bar{\nu}^T G(t, y(t), z(t), u) \geq 0, \quad (7.5.3)$$

$$y_\alpha = H(t, y(t), z(t), v), \quad (7.5.4)$$

$$y(t_0) = x_0, \; y(t_1) = x_1, \quad (7.5.5)$$

$$\mu^T e = 1, \; \mu \in \mathbb{R}^s_+, \; \nu \in \mathbb{R}^m_+, \gamma \in \mathbb{R}^{pq}. \quad (7.5.6)$$

The associated robust counterpart to the problem (UMRDP) is given as follows:

$$\text{(RUMRDP)} \quad \max_{(y(\cdot),z(\cdot),w,u,v)} \int_{\Gamma_{t_0,t_1}} \Big\{f(t,y(t),z(t),w) + \nu^T G(t,y(t),z(t),u)e + \gamma^T(H(t,y(t),z(t),v) - y_\alpha)e\Big\}dt$$

$$\text{subject to} \quad \mu^T f_x(t,y(t),z(t),w) + \nu^T G_x(t,y(t),z(t),u) + \gamma^T H_x(t,y(t),z(t),v) + \gamma_t = 0,$$

$$\mu^T f_c(t,y(t),z(t),w) + \nu^T G_c(t,y(t),z(t),u) + \gamma^T H_c(t,y(t),z(t),v) = 0,$$

$$\nu^T G(t,y(t),z(t),u) \geq 0,$$

$$y_\alpha = H(t,y(t),z(t),v),$$

$$y(t_0) = x_0,\ y(t_1) = x_1,$$

$$\mu^T e = 1,\ \mu \in \mathbb{R}^s_+,\ \nu \in \mathbb{R}^m_+, \gamma \in \mathbb{R}^{pq},$$

for all $w \in \mathcal{W}, u \in \mathcal{U}, v \in \mathcal{V}$.

We denote $\mathcal{D}_m = \{(y,z) \in \mathscr{X} \times \mathscr{C} :$ satisfying conditions (7.5.1)–(7.5.6)$\}$ the set of all feasible solutions to (RUWRDP) and we say that it is the robust feasible solution set to the problem (UWRDP).

Theorem 7.5.0.1 (*Weak Robust Duality*) *Let* $(\bar{x}, \bar{c})$ *and* $(\bar{y}, \bar{z}, \bar{\mu}, \bar{\nu}, \bar{\gamma}, \bar{w}, \bar{u}, \bar{v})$ *be the robust feasible solutions to the problem* (MMOPU) *and* (UMRDP), *respectively. Assume that* $\max_{w\in\mathcal{W}} f(\bar{\pi}, w) = f(\bar{\pi}, \bar{w})$. *Further, if* $\int_\Gamma \bar{\mu}^T f(\cdot, \bar{w})dt$, $\int_\Gamma \bar{\nu}^T G(\cdot, \bar{u})dt$, *and* $\int_\Gamma \bar{\gamma}^T\{H(\cdot, \bar{v}) - y_\alpha\}dt$ *are convex at* $(\bar{y}, \bar{z})$, *then the following inequality cannot hold*

$$\int_\Gamma \max_{w\in\mathcal{W}} f(\bar{\pi}, w)dt < \int_\Gamma \Big\{f(\bar{\eta}, \bar{w}) + \bar{\nu}^T G(\bar{\eta}, \bar{u})e + \bar{\gamma}^T(H(\bar{\eta}, \bar{v}) - y_\alpha)e\Big\}dt.$$

Proof Contrary to the result, we assume that

$$\int_\Gamma \max_{w\in\mathcal{W}} f(\bar{\pi}, w)dt < \int_\Gamma \Big\{f(\bar{\eta}, \bar{w}) + \bar{\nu}^T G(\bar{\eta}, \bar{u})e + \bar{\gamma}^T(H(\bar{\eta}, \bar{v}) - y_\alpha)e\Big\}dt.$$

Since $\max_{w\in\mathcal{W}} f(\bar{\pi}, w) = f(\bar{\pi}, \bar{w})$, we have

$$\int_\Gamma f(\bar{\pi}, \bar{w})dt < \int_\Gamma \Big\{f(\bar{\eta}, \bar{w}) + \bar{\nu}^T G(\bar{\eta}, \bar{u})e + \bar{\gamma}^T(H(\bar{\eta}, \bar{v}) - y_\alpha)e\Big\}dt.$$

The above inequality together with the robust feasibility of $(\bar{x}, \bar{c})$ to the problem (MMOPU), yields

$$\int_{\Gamma}\Big\{f(\bar{\pi},\bar{w})+\bar{\upsilon}^T G(\bar{\pi},\bar{u})e+\bar{\gamma}^T(H(\bar{\pi},\bar{\upsilon})-x_{\alpha})e\Big\}dt$$
$$<\int_{\Gamma}\Big\{f(\bar{\eta},\bar{w})+\bar{\upsilon}^T G(\bar{\eta},\bar{u})e+\bar{\gamma}^T(H(\bar{\eta},\bar{\upsilon})-y_{\alpha})e\Big\}dt.$$

As $\bar{\mu}^T > 0$ and $\bar{\mu}^T e = 1$, therefore, the above inequality can be written as

$$\int_{\Gamma}\bar{\mu}^T f(\bar{\pi},\bar{w})+\bar{\upsilon}^T G(\bar{\pi},\bar{u})+\bar{\gamma}^T(H(\bar{\pi},\bar{\upsilon})-x_{\alpha})\Big\}dt$$
$$<\int_{\Gamma}\Big\{\bar{\mu}^T f(\bar{\eta},\bar{w})+\bar{\upsilon}^T G(\bar{\eta},\bar{u})+\bar{\gamma}^T(H(\bar{\eta},\bar{\upsilon})-y_{\alpha})\Big\}dt. \tag{7.5.7}$$

Since $\int_{\Gamma}\bar{\mu}^T f(\cdot,\bar{w})dt$, $\int_{\Gamma}\bar{\upsilon}^T G(\cdot,\bar{u})dt$ and $\int_{\Gamma}\bar{\gamma}^T(H(\cdot,\bar{\upsilon})-y_{\alpha})dt$ are convex at $(\bar{y},\bar{z})$, we have

$$\int_{\Gamma}\Big\{\bar{\mu}^T f(\bar{\pi},\bar{w})-\bar{\mu}^T f(\bar{\eta},\bar{w})\Big\}dt \geqq \int_{\Gamma}(\bar{x}-\bar{y})\bar{\mu}^T f_x(\bar{\eta},\bar{w})dt+\int_{\Gamma}(\bar{c}-\bar{z})\bar{\mu}^T f_c(\bar{\eta},\bar{w})dt, \tag{7.5.8}$$

$$\int_{\Gamma}\Big\{\bar{\upsilon}^T G(\bar{\pi},\bar{u})-\bar{\upsilon}^T G(\bar{\eta},\bar{u})\Big\}dt \geqq \int_{\Gamma}(\bar{x}-\bar{y})\bar{\upsilon}^T G_x(\bar{\eta},\bar{u})dt+\int_{\Gamma}(\bar{c}-\bar{z})\bar{\upsilon}^T G_c(\bar{\eta},\bar{u})dt, \tag{7.5.9}$$

and

$$\int_{\Gamma}\Big\{\bar{\gamma}^T(H(\bar{\pi},\bar{\upsilon})-x_{\alpha})-\bar{\gamma}^T(H(\bar{\eta},\bar{\upsilon})-y_{\alpha})\Big\}dt$$
$$\geqq \int_{\Gamma}(\bar{x}-\bar{y})\Big\{\bar{\gamma}^T H_x(\bar{\eta},\bar{\upsilon})+\gamma_t\Big\}dt+\int_{\Gamma}(\bar{c}-\bar{z})\bar{\gamma}^T H_c(\bar{\eta},\bar{\upsilon})dt. \tag{7.5.10}$$

On adding the inequalities (7.5.8), (7.5.9) and (7.5.10) along with the robust feasibility of $(\bar{x},\bar{c})$ and $(\bar{y},\bar{z},\bar{\mu},\bar{\upsilon},\bar{\gamma},\bar{w},\bar{u},\bar{\upsilon})$ to the problem (MMOPU) and (UMRDP), respectively. we have

$$\int_{\Gamma}\Big\{\bar{\mu}^T f(\bar{\pi},\bar{w})+\bar{\upsilon}^T G(\bar{\pi},\bar{u})+\bar{\gamma}^T(H(\bar{\pi},\bar{\upsilon})-x_{\alpha})\Big\}dt$$
$$\geqq \int_{\Gamma}\Big\{\bar{\mu}^T f(\bar{\eta},\bar{w})+\bar{\upsilon}^T G(\bar{\eta},\bar{u})+\bar{\gamma}^T(H(\bar{\eta},\bar{\upsilon})-y_{\alpha})\Big\}dt,$$

which contradicts the inequality (7.5.7). This completes the proof. □

We are considering the Example 7.4.0.1 to authenticate Theorem 7.5.0.1.

Example 7.5.0.1 The mixed type robust dual problem associated with the considered problem (MMOPU2) is given as follows:

$$\text{(UMRDP2)} \max_{(y(\cdot),z(\cdot))} \int_{\Gamma_{0,1}} \Big\{\Big(y^2+z+w_1\Big)+\nu(u+y(y-4))+\gamma_1^1(y_{t^1}-2+v_1^1 z) \\ +\gamma_2^1(y_{t^2}-2+v_2^1 z),\, x+w_2c^2+1+\nu(u+y(y-4)) \\ +\gamma_1^1(y_{t^1}-2+v_1^1 z)+\gamma_2^1(y_{t^2}-2+v_2^1 z)\Big\}\, dt^1 dt^2$$

$$\text{subject to} \quad 2\mu_1 y+\mu_2+\nu(2y-4)+(\gamma_1^1)_{t^1}+(\gamma_1^1)_{t^2}=0, \tag{7.5.11}$$

$$\mu_1+2w_2 z\mu_2-v_1^1\gamma_1^1-v_2^1\gamma_2^1=0, \tag{7.5.12}$$

$$\nu(u+y(y-4)) \geqq 0, \tag{7.5.13}$$

$$y_{t^1}=2-v_1^1 z, \tag{7.5.14}$$

$$y_{t^2}=2-v_2^1 z, \tag{7.5.15}$$

$$y(0,0)=1,\ y(1,1)=2. \tag{7.5.16}$$

The robust counterpart to the problem (UMRDP2) is given as follows:

$$\text{(RUMRDP2)} \max_{(y(\cdot),z(\cdot),w,u,v)} \int_{\Gamma_{0,1}} \Big\{\Big(y^2+z+w_1\Big)+\nu(u+y(y-4))+\gamma_1^1(y_{t^1}-2+v_1^1 z) \\ +\gamma_2^1(y_{t^2}-2+v_2^1 z),\, x+w_2c^2+1+\nu(u+y(y-4)) \\ +\gamma_1^1(y_{t^1}-2+v_1^1 z)+\gamma_2^1(y_{t^2}-2+v_2^1 z)\Big\}\, dt^1 dt^2$$

$$\text{subject to} \quad 2\mu_1 y+\mu_2+\nu(2y-4)+(\gamma_1^1)_{t^1}+(\gamma_1^1)_{t^2}=0,$$

$$\mu_1+2w_2 z\mu_2-v_1^1\gamma_1^1-v_2^1\gamma_2^1=0,$$

$$\nu(u+y(y-4)) \geqq 0,$$

$$y_{t^1}=2-v_1^1 z,$$

$$y_{t^2}=2-v_2^1 z,$$

$$y(0,0)=1,\ y(1,1)=2,$$

for all $w \in \mathcal{W}$, $u \in \mathcal{U}$, $v \in \mathcal{V}$.
Note that $\mathcal{D}_m = \{(y,z) \in \mathscr{X} \times \mathscr{C} :$ satisfying conditions (7.5.11)–(7.5.16)$\}$ is the feasible set to the (UMRDP2). Let $(\bar{y}, \bar{z}, \bar{\mu}, \bar{\nu}, \bar{\gamma}, \bar{w}, \bar{u}, \bar{v})$ be a robust feasible solution to (UMRDP2). Thus, the following inequality

$$\int_{\Gamma} f(\bar{\pi}, \bar{w})dt - \int_{\Gamma} \left\{f(\bar{\eta}, \bar{w}) + \bar{v}^T G(\bar{\eta}, \bar{u})e + \bar{\gamma}^T (H(\bar{\eta}, \bar{v}) - y_\alpha)e\right\}dt$$
$$= \int_{\Gamma_{0,1}} \left\{\bar{x}^2 + \bar{c} + \bar{w}_1,\ \bar{x} + \bar{w}_2\bar{c}^2 + 1\right\}dt^1dt^2 - \int_{\Gamma_{0,1}} \left\{\bar{y}^2 + \bar{z} + \bar{w}_1 + \bar{v}(\bar{u} + \bar{y}(\bar{y} - 4))\right.$$
$$+ \bar{\gamma}_1^1(y_{t^1} - 2 + \bar{v}_1^1\bar{z}) + \bar{\gamma}_2^1(y_{t^2} - 2 + \bar{v}_2^1\bar{z}),\ \bar{y} + \bar{w}_2\bar{z}^2 + 1 + \bar{v}(\bar{u} + \bar{y}(\bar{y} - 4))$$
$$\left. + \bar{\gamma}_1^1(y_{t^1} - 2 + \bar{v}_1^1\bar{z}) + \bar{\gamma}_2^1(y_{t^2} - 2 + \bar{v}_2^1\bar{z})\right\} dt^1dt^2$$
$$= \int_{\Gamma_{0,1}} \left(\left(\frac{1}{2}(t^1 + t^2) + 1\right)^2 + \frac{3}{2} + 1,\ \frac{1}{2}(t^1 + t^2) + \left(\frac{3}{2}\right)^2 + 2\right)dt^1dt^2 - \int_{\Gamma_{0,1}} (4, 4)dt^1dt^2$$
$$= \left(\frac{115}{24}, \frac{19}{4}\right) - (4, 4)$$
$$= \left(\frac{19}{24}, \frac{3}{4}\right) > 0,$$

which shows that the duality gap is positive. Hence, Theorem 7.5.0.1 (Weak Robust Duality) is verified.

Next, we shall prove the strong robust duality result which states that duality gap is zero.

Theorem 7.5.0.2 (*Strong Robust Duality*) *Let* $(\bar{x}, \bar{c})$ *be a weak robust efficient solution to the problem* (MMOPU). *Assume that* $\max_{w \in \mathcal{W}} f(\bar{\pi}, w) = f(\bar{\pi}, \bar{w})$ *and the constraint conditions (for the existence of multiplier) holds for* (MMOPU). *Then, there exist the scalar vector* $\bar{\mu} \in \mathbb{R}^s_+$, *the piecewise smooth Lagrange multipliers* $\bar{v} = (\bar{v}_l(t)) \in \mathbb{R}^m_+$, $l = \overline{1, m}$ *and* $\bar{\gamma} = (\bar{\gamma}_\alpha^\tau(t)) \in \mathbb{R}^{pq}$, $\tau = \overline{1, q}$, $\alpha = \overline{1, p}$ *and the uncertain parameters* $\bar{u} \in \mathcal{U}$, $\bar{v} \in \mathcal{V}$ *such that* $(\bar{x}, \bar{c}, \bar{\mu}, \bar{v}, \bar{\gamma}, \bar{w}, \bar{u}, \bar{v})$ *is a robust feasible solution to the problem* (UMRDP). *Further, if the Weak Robust Duality* (*Theorem 7.5.0.1*) *holds, then* $(\bar{x}, \bar{c}, \bar{\mu}, \bar{v}, \bar{\gamma}, \bar{w}, \bar{u}, \bar{v})$ *is a weak robust efficient solution to the problem* (UMRDP).

Proof Since $(\bar{x}, \bar{c})$ be a weak robust efficient solution to the problem (MMOPU). Therefore, by Theorem 7.2.0.1, there exist the scalar vector $\bar{\mu} \in \mathbb{R}^s_+$, the piecewise smooth Lagrange multiplies $\bar{v} \in \mathbb{R}^m_+$, $\bar{\gamma} \in \mathbb{R}^{pq}$ and the uncertain $\bar{u} \in \mathcal{U}$, $\bar{v} \in \mathcal{V}$ such that the conditions (7.2.1)–(7.2.4) are satisfied at $(\bar{x}, \bar{c})$. This shows the robust feasibility of $(\bar{x}, \bar{c}, \bar{\mu}, \bar{v}, \bar{\gamma}, \bar{w}, \bar{u}, \bar{v})$ to the problem (UMRDP) and the corresponding objective values are equal. If $(\bar{x}, \bar{c}, \bar{\mu}, \bar{v}, \bar{\gamma}, \bar{w}, \bar{u}, \bar{v})$ is not a weak efficient solution to the problem (UMRDP), then there exists another point $(y, z, \bar{\mu}, \bar{v}, \bar{\gamma}, \bar{w}, \bar{u}, \bar{v})$ such that

$$\int_{\Gamma} \max_{w \in \mathcal{W}} f(\bar{\pi}, w)dt < \int_{\Gamma} f(\bar{\eta}, \bar{w})dt,$$

which contradict the Weak Robust Duality (Theorem 7.5.0.1). Hence, $(\bar{x}, \bar{c}, \bar{\mu}, \bar{v}, \bar{\gamma}, \bar{w}, \bar{u}, \bar{v})$ is a weak efficient solution to the problem (UMRDP). □

References

1. J.V. Neumann, *On a Maximization Problem* (Institute for Advanced Study, Princeton, New Jersey, 1947)
2. P. Wolfe, A duality theorem for nonlinear programming. Quart. Appl. Math. **19**, 239–244 (1961)
3. B. Mond, I. Smart, Duality and sufficiency in control problems with invexity. J. Math. Anal. Appl. **136**, 325–333 (1988)
4. B. Mond, T. Weir, Generalized concavity and duality, in *Generalized Concavity in Optimization and Economics*, ed. by S. Schaible, W.T. Ziemba (Academic, New York, 1981), pp. 263–279
5. Şt. Mititelu, Optimality and duality for invex multi-dimensional control problems with mixed constraints. J. Adv. Math. Stud. **2**, 25–34 (2009)
6. A. Beck, A. Ben-Tal, Duality in robust optimization: primal worst equals dual best. Oper. Res. Lett. **37**, 1–6 (2009)
7. V. Jeyakumar, G. Li, G.M. Lee, Robust duality for generalized convex programming problems under data uncertainty. Nonlinear Anal. **75**, 1362–1373 (2012)
8. S. Treanţă, Şt. Mititelu, Duality with (ρ, b)-quasiinvexity for multidimensional vector fractional control problems. J. Inform. Optim. Sci. **40**, 1429–1445 (2019)

Chapter 8
On a Class of Second-Order PDE&PDI Constrained Robust Optimization Problems

8.1 Introduction

Over the last decade, the modified objective function method and saddle-point efficiency criteria have been intensively investigated by many researchers (see, for instance, Jayswal et al. [1, 2], Antczak et al. [3], Jayswal and Preeti [4], Sun et al. [5], Treanţă [6–9], Preeti et al. [10]). More concretely, new auxiliary optimization problems have been constructed for solving the original optimization problems. These auxiliary problems are easier to study than the original ones. For various aspects on interval and robust optimization problems, including efficiency conditions, Lagrange multiplier characterizations, robust algorithms, the reader is directed, for instance, to Liu and Yuan [11], Jeyakumar et al. [12], Wei et al. [13], and Treanţă [14].

In this chapter, motivated and inspired by the above mentioned works, by using vector multiple integral cost functionals and the notion of convexity associated with a multi-time uncertain controlled second-order Lagrangian, we develop a new mathematical framework on multi-dimensional multi-objective variational control problems with mixed constraints implying second-order partial differential equations and inequations. More specifically, we start with a multi-dimensional multi-objective optimization problem with uncertainty in the objective and constraint functionals, denoted by (MOP), and introduce a modified optimization problem for which we formulate and prove conditions of efficiency. As can be easily noticed, the modified variational control problem is simpler and easier to solve than the original optimization problem. Furthermore, some connections between a robust saddle-point for a Lagrange functional associated with the modified optimization problem and a weak robust efficient solution in the considered variational control problem (MOP) are presented.

A. Jayswal et al., *Multi-dimensional Control Problems*, Industrial and Applied Mathematics, https://doi.org/10.1007/978-981-19-6561-6_8

8.2 Problem Description

Throughout the chapter, we are considering the following notations and working hypotheses:

• $\mathbb{R}^p$, $\mathbb{R}^q$, $\mathbb{R}^r$ and $\mathbb{R}^n$ are Euclidean spaces of dimensions p, q, r and n, respectively;

• $\Gamma_{t_0,t_1} \subset \mathbb{R}^p$ is a hyperparallelepiped fixed by the diagonally opposite points $t_0 = (t_0^\alpha)$ and $t_1 = (t_1^\alpha)$, $\alpha = \overline{1,p}$, and the point $t = (t^\alpha) \in \Gamma_{t_0,t_1}$ is a *multi-parameter of evolution*;

• $\mathscr{X}$ is the space of C^4-class state functions $x = (x^\tau) : \Gamma_{t_0,t_1} \to \mathbb{R}^q$ and $x_\alpha := \frac{\partial x}{\partial t^\alpha}$, $x_{\alpha\beta} := \frac{\partial^2 x}{\partial t^\alpha \partial t^\beta}$ denote the *partial speed* and *partial acceleration*, respectively;

• $\mathscr{C}$ is the space of C^1-class control functions $c = (c^j) : \Gamma_{t_0,t_1} \to \mathbb{R}^r$;

• $dt = dt^1 \ldots dt^p$ is the volume element on $\mathbb{R}^p \supset \Gamma_{t_0,t_1}$.

• T denotes the transpose of a vector;

• for two vectors $x, y \in \mathbb{R}^n$, we use the following convention for inequalities and equalities:

$$
\begin{aligned}
&(i)\ x < y \Leftrightarrow x_i < y_i,\ \forall i = \overline{1,n},\\
&(ii)\ x = y \Leftrightarrow x_i = y_i,\ \forall i = \overline{1,n},\\
&(iii)\ x \leqq y \Leftrightarrow x_i \leq y_i,\ \forall i = \overline{1,n},\\
&(iv)\ x \leq y \Leftrightarrow x_i \leq y_i,\ \forall i = \overline{1,n} \text{ and } x_i < y_i \text{ for some } i.
\end{aligned}
$$

The second-order PDE&PDI constrained controlled multi-objective optimization problem with data uncertainty in the objective and constraint functionals is formulated as follows:

$$
\begin{aligned}
\text{(MOP)}\quad & \min_{(x(\cdot),c(\cdot))} \left\{ \int_{\Gamma_{t_0,t_1}} f(t, x(t), x_\sigma(t), x_{\alpha\beta}(t), c(t), w)dt \right.\\
&= \left(\int_{\Gamma_{t_0,t_1}} f_1(t, x(t), x_\sigma(t), x_{\alpha\beta}(t), c(t), w_1)dt, \ldots, \right.\\
&\left.\left. \int_{\Gamma_{t_0,t_1}} f_s(t, x(t), x_\sigma(t), x_{\alpha\beta}(t), c(t), w_s)dt \right) \right\}\\
\text{subject to}\quad & g(t, x(t), x_\sigma(t), x_{\alpha\beta}(t), c(t), u) \leqq 0, \quad t \in \Gamma_{t_0,t_1},\\
& h(t, x(t), x_\sigma(t), x_{\alpha\beta}(t), c(t), v) = 0, \quad t \in \Gamma_{t_0,t_1},\\
& x(t_0) = x_0,\ x(t_1) = x_1,\ x_\sigma(t_0) = x_{\sigma 0},\ x_\sigma(t_1) = x_{\sigma 1},\\
\text{or}\quad & x(t)|_{\partial\Gamma_{t_0,t_1}} = \text{given}, \quad x_\sigma(t)|_{\partial\Gamma_{t_0,t_1}} = \text{given},
\end{aligned}
$$

where $f = (f_1, \ldots, f_s) = (f_k) : J^2\left(\Gamma_{t_0,t_1}, \mathbb{R}^q\right) \times \mathscr{C} \times \mathcal{W}_k \to \mathbb{R}^s$, $k = \overline{1,s}$, $g = (g_1, \ldots, g_m) = (g_l) : J^2\left(\Gamma_{t_0,t_1}, \mathbb{R}^q\right) \times \mathscr{C} \times \mathcal{U}_l \to \mathbb{R}^m$, $l = \overline{1,m}$, $h = (h_1, \ldots, h_n) = (h_\zeta) : J^2\left(\Gamma_{t_0,t_1}, \mathbb{R}^q\right) \times \mathscr{C} \times \mathcal{V}_\zeta \to \mathbb{R}^n$, $\zeta = \overline{1,n}$, are C^3-class functionals, $w = (w_k)$, $u = (u_l)$ and $v = (v_\zeta)$ are the uncertain parameters for some convex compact subsets $\mathcal{W} = (\mathcal{W}_k) \subset \mathbb{R}^s$, $\mathcal{U} = (\mathcal{U}_l) \subset \mathbb{R}^m$ and $\mathcal{V} = (\mathcal{V}_\zeta) \subset \mathbb{R}^n$, respectively, and $J^2\left(\Gamma_{t_0,t_1}, \mathbb{R}^q\right)$ denotes the second-order jet bundle associated to Γ_{t_0,t_1} and $\mathbb{R}^q$.

The associated robust counterpart of the aforementioned multi-objective optimization problem (MOP) is defined as:

$$
\begin{aligned}
\text{(RMOP)} \quad & \min_{(x(\cdot),c(\cdot))} \left\{ \int_{\Gamma_{t_0,t_1}} \max_{w \in \mathcal{W}} f(t, x(t), x_\sigma(t), x_{\alpha\beta}(t), c(t), w)dt \right. \\
& = \left(\int_{\Gamma_{t_0,t_1}} \max_{w_1 \in \mathcal{W}_1} f_1(t, x(t), x_\sigma(t), x_{\alpha\beta}(t), c(t), w_1)dt, \ldots, \right. \\
& \left. \left. \int_{\Gamma_{t_0,t_1}} \max_{w_s \in \mathcal{W}_s} f_s(t, x(t), x_\sigma(t), x_{\alpha\beta}(t), c(t), w_s)dt \right) \right\} \\
\text{subject to} \quad & g(t, x(t), x_\sigma(t), x_{\alpha\beta}(t), c(t), u) \leqq 0, \quad t \in \Gamma_{t_0,t_1}, \ \forall u \in \mathcal{U}, \\
& h(t, x(t), x_\sigma(t), x_{\alpha\beta}(t), c(t), v) = 0, \quad t \in \Gamma_{t_0,t_1}, \ \forall v \in \mathcal{V}, \\
& x(t_0) = x_0, \ x(t_1) = x_1, \ x_\sigma(t_0) = x_{\sigma 0}, \ x_\sigma(t_1) = x_{\sigma 1}, \\
\text{or} \quad & x(t)|_{\partial \Gamma_{t_0,t_1}} = \text{given}, \quad x_\sigma(t)|_{\partial \Gamma_{t_0,t_1}} = \text{given}.
\end{aligned}
$$

Further, denote by

$$
\begin{aligned}
\mathcal{D} = \{(x, c) \in \mathscr{X} \times \mathscr{C} : \ & g(t, x(t), x_\sigma(t), x_{\alpha\beta}(t), c(t), u) \leqq 0, \\
& h(t, x(t), x_\sigma(t), x_{\alpha\beta}(t), c(t), v) = 0, \ x(t_0) = x_0, \ x(t_1) = x_1, \\
& x_\sigma(t_0) = x_{\sigma 0}, \ x_\sigma(t_1) = x_{\sigma 1}, \ t \in \Gamma_{t_0,t_1}, u \in \mathcal{U}, \ v \in \mathcal{V}\}
\end{aligned}
$$

the set of all feasible solutions in (RMOP) and we say that it is the robust feasible solution set to the problem (MOP).

From now on, to simplify our presentation, we introduce some notations as follows: $\pi = (t, x(t), x_\sigma(t), x_{\alpha\beta}(t), c(t))$ and $\Gamma = \Gamma_{t_0,t_1}$.

The first-order partial derivatives associated with f are defined as

$$
f_x = \begin{bmatrix} \frac{\partial f_1}{\partial x^1} & \cdots & \frac{\partial f_1}{\partial x^q} \\ \vdots & \ddots & \vdots \\ \frac{\partial f_s}{\partial x^1} & \cdots & \frac{\partial f_s}{\partial x^q} \end{bmatrix}, \quad f_c = \begin{bmatrix} \frac{\partial f_1}{\partial c^1} & \cdots & \frac{\partial f_1}{\partial c^r} \\ \vdots & \ddots & \vdots \\ \frac{\partial f_s}{\partial c^1} & \cdots & \frac{\partial f_s}{\partial c^r} \end{bmatrix}.
$$

Similarly, we have g_x and g_c using matrices with m rows and h_x and h_c using matrices with n rows.

In the following, we introduce the notion of an efficient solution for the considered class of constrained optimization problems. This concept will be used to formulate the associated efficiency conditions.

Definition 8.2.0.1 A robust feasible solution $(\bar{x}, \bar{c}) \in \mathcal{D}$ is said to be a weak robust efficient solution to the multi-dimensional multi-objective optimization problem (MOP) if there does not exist another point $(x, c) \in \mathcal{D}$ such that

$$\int_{\Gamma} \max_{w \in \mathcal{W}} f(\pi, w)dt < \int_{\Gamma} \max_{w \in \mathcal{W}} f(\bar{\pi}, w)dt.$$

To formulate the concept of convexity and the robust necessary efficiency conditions associated with the aforementioned controlled optimization problem, we shall introduce the Saunders's multi-index notation (see Saunders [15]). A *multi-index* is an m-tuple U of natural numbers. The components of U are denoted $U(\alpha)$, where α is an ordinary index, $1 \leq \alpha \leq m$. The multi-index 1_α is defined by $1_\alpha(\alpha) = 1,\ 1_\alpha(\beta) = 0$ for $\alpha \neq \beta$. The addition and the substraction of the multi-indexes are defined componentwise (although the result of a substraction might not be a multi-index): $(U \pm V)(\alpha) = U(\alpha) \pm V(\alpha)$. The length of a multi-index is $\mid U \mid = \sum_{\alpha=1}^{m} U(\alpha)$, and its factorial is $U! = \prod_{\alpha=1}^{m}(U(\alpha))!$. The number of distinct indices represented by $\{\alpha_1, \alpha_2, ..., \alpha_k\},\ \alpha_j \in \{1, 2, ..., m\},\ j = \overline{1, k}$, is

$$n(\alpha_1, \alpha_2, ..., \alpha_k) = \frac{\mid 1_{\alpha_1} + 1_{\alpha_2} + ... + 1_{\alpha_k} \mid !}{(1_{\alpha_1} + 1_{\alpha_2} + ... + 1_{\alpha_k})!}.$$

Definition 8.2.0.2 A controlled multiple integral functional (uncertain multi-time controlled second-order Lagrangian)

$$F(x, c, \bar{w}) = \int_{\Gamma} f(t, x(t), x_\sigma(t), x_{\alpha\beta}(t), c(t), \bar{w})dt = \int_{\Gamma} f(\pi, \bar{w})dt$$

is said to be *convex* at $(\bar{x}, \bar{c}) \in \mathscr{X} \times \mathscr{C}$ if the following inequality

$$F(x, c, \bar{w}) - F(\bar{x}, \bar{c}, \bar{w}) \geqq \int_{\Gamma} [x(t) - \bar{x}(t)] f_x(\bar{\pi}, \bar{w})dt + \int_{\Gamma} [x_\sigma(t) - \bar{x}_\sigma(t)] f_{x_\sigma}(\bar{\pi}, \bar{w})dt$$
$$+ \frac{1}{n(\alpha, \beta)} \int_{\Gamma} [x_{\alpha\beta}(t) - \bar{x}_{\alpha\beta}(t)] f_{x_{\alpha\beta}}(\bar{\pi}, \bar{w})dt + \int_{\Gamma} [c(t) - \bar{c}(t)] f_c(\bar{\pi}, \bar{w})dt$$

holds for all $(x(t), c(t)) \in \mathscr{X} \times \mathscr{C}$.

Next, we describe the robust Kuhn-Tucker (KT) point associated with (MOP), in accordance to Treanţă and Arana-Jiménez [16] and following Mititelu [17] as:

Definition 8.2.0.3 Let $\max_{w \in \mathcal{W}} f(\pi, w) = f(\pi, \bar{w})$. A robust feasible solution $(\bar{x}, \bar{c})$ is said to be a *robust KT point* to the multi-dimensional multi-objective optimization problem (MOP) if there exist the piecewise smooth Lagrange multipliers $\bar{\nu} = (\bar{\nu}_l(t)) \in \mathbb{R}^m_+$, $\bar{\gamma} = (\bar{\gamma}_\varsigma(t)) \in \mathbb{R}^n$, and the uncertain parameters $\bar{u} \in \mathcal{U}$ and $\bar{v} \in \mathcal{V}$ such that the following conditions

$$e^T f_x(\bar{\pi}, \bar{w}) + \bar{\nu}^T g_x(\bar{\pi}, \bar{u}) + \bar{\gamma}^T h_x(\bar{\pi}, \bar{v}) - D_\sigma\left[e^T f_{x_\sigma}(\bar{\pi}, \bar{w}) + \bar{\nu}^T g_{x_\sigma}(\bar{\pi}, \bar{u}) + \bar{\gamma}^T h_{x_\sigma}(\bar{\pi}, \bar{v})\right]$$
$$+ \frac{1}{n(\alpha, \beta)} D^2_{\alpha\beta}\left[e^T f_{x_{\alpha\beta}}(\bar{\pi}, \bar{w}) + \bar{\nu}^T g_{x_{\alpha\beta}}(\bar{\pi}, \bar{u}) + \bar{\gamma}^T h_{x_{\alpha\beta}}(\bar{\pi}, \bar{v})\right] = 0,$$
$$e^T f_c(\bar{\pi}, \bar{w}) + \bar{\nu}^T g_c(\bar{\pi}, \bar{u}) + \bar{\gamma}^T h_c(\bar{\pi}, \bar{v}) = 0,$$
$$\bar{\nu}^T g(\bar{\pi}, \bar{u}) = 0,\ \bar{\nu} \geqq 0,$$

hold for all $t \in \Gamma$, except at discontinuities, where $e^T = (1, \ldots, 1) \in \mathbb{R}^s$.

In accordance with Treanţă [18] (see Lemma 3.2), we formulate the following auxiliary lemma.

Lemma 8.2.0.1 *Under the constraint conditions, a robust feasible solution* $(\bar{x}, \bar{c})$ *is a weak robust efficient solution to the problem* (MOP) *if and only if* $(\bar{x}, \bar{c})$ *is an robust optimal solution of each scalar problem* $(\text{MOP})_k(\bar{x}, \bar{c})$, $k = \overline{1, s}$, *where*

$$
\begin{aligned}
(\text{MOP})_k(\bar{x}, \bar{c}) \quad & \min_{(x(\cdot), c(\cdot))} \int_\Gamma f_k(\pi, w_k)dt \\
\textit{subject to} \quad & g(\pi, u) \leqq 0, \quad t \in \Gamma_{t_0, t_1}, \\
& h(\pi, v) = 0, \quad t \in \Gamma_{t_0, t_1}, \\
& \int_\Gamma f_\epsilon(\pi, w_\epsilon)dt \leq \int_\Gamma f_\epsilon(\bar{\pi}, w_\epsilon)dt, \quad \epsilon = \overline{1, s}, \ \epsilon \neq k, \\
& x(t_0) = x_0, \ x(t_1) = x_1, \ x_\sigma(t_0) = x_{\sigma 0}, \ x_\sigma(t_1) = x_{\sigma 1}, \\
\textit{or} \quad & x(t)|_{\partial \Gamma} = \textit{given}, \quad x_\sigma(t)|_{\partial \Gamma} = \textit{given}.
\end{aligned}
$$

Definition 8.2.0.4 A robust feasible solution $(\bar{x}, \bar{c})$ is said to be a *normal weak robust efficient solution* to the problem (MOP) if it is normal robust optimal solution for each scalar problem $(\text{MOP})_k(\bar{x}, \bar{c})$, $k = \overline{1, s}$, that is, we consider $\mu_k = 1$ in the following Lagrange function $L = \mu^T f + \nu^T g + \gamma^T h$, $\mu = (\mu_\epsilon)$, $\epsilon = \overline{1, s}$.

Theorem 8.2.0.1 *Under the constraint conditions, let* $(\bar{x}, \bar{c}) \in \mathcal{D}$ *be a normal weak robust efficient solution to the problem* (MOP), *and* $\max_{w \in \mathcal{W}} f(\pi, w) = f(\pi, \bar{w})$. *Then,* $(\bar{x}, \bar{c})$ *is a robust KT point to the problem* (MOP).

Proof Assume that the constraint conditions (for the existence of multipliers) hold and $\max_{w \in \mathcal{W}} f(\pi, w) = f(\pi, \bar{w})$. Now, on the line of Mititelu and Treanţă [19], if $(\bar{x}, \bar{c})$ is a weak robust efficient solution to the problem (MOP), then there exist the scalar vector $\bar{\mu} \in \mathbb{R}^s$, the piecewise smooth Lagrange multipliers $\bar{\nu} = (\bar{\nu}_l(t)) \in \mathbb{R}^m_+$, $\bar{\gamma} = (\bar{\gamma}_\varsigma(t)) \in \mathbb{R}^n$, and the uncertain parameters $\bar{u} \in \mathcal{U}$ and $\bar{v} \in \mathcal{V}$ such that the following conditions

$$
\begin{aligned}
& \bar{\mu}^T f_x(\bar{\pi}, \bar{w}) + \bar{\nu}^T g_x(\bar{\pi}, \bar{u}) + \bar{\gamma}^T h_x(\bar{\pi}, \bar{v}) - D_\sigma\left[\bar{\mu}^T f_{x_\sigma}(\bar{\pi}, \bar{w}) + \bar{\nu}^T g_{x_\sigma}(\bar{\pi}, \bar{u}) + \bar{\gamma}^T h_{x_\sigma}(\bar{\pi}, \bar{v})\right] \\
& \quad + \frac{1}{n(\alpha, \beta)} D^2_{\alpha\beta}\left[\bar{\mu}^T f_{x_{\alpha\beta}}(\bar{\pi}, \bar{w}) + \bar{\nu}^T g_{x_{\alpha\beta}}(\bar{\pi}, \bar{u}) + \bar{\gamma}^T h_{x_{\alpha\beta}}(\bar{\pi}, \bar{v})\right] = 0, \qquad (8.2.1)
\end{aligned}
$$

$$\bar{\mu}^T f_c(\bar{\pi}, \bar{w}) + \bar{\nu}^T g_c(\bar{\pi}, \bar{u}) + \bar{\gamma}^T h_c(\bar{\pi}, \bar{v}) = 0, \qquad (8.2.2)$$

$$\bar{\nu}^T g(\bar{\pi}, \bar{u}) = 0, \ \bar{\nu} \geqq 0, \qquad (8.2.3)$$

$$\bar{\mu} \geq 0, \qquad (8.2.4)$$

hold for all $t \in \Gamma$, except at discontinuities. Since the weak robust efficient solution $(\bar{x}, \bar{c})$ to the problem (MOP) is a normal weak robust efficient solution, therefore, we obtain $\bar{\mu}^T = e^T$, and the proof is now complete. □

Remark 8.2.0.1 The conditions (8.2.1)–(8.2.4) are known as *robust necessary efficiency conditions* for the multi-dimensional multi-objective optimization problem (MOP).

8.3 Robust Sufficient Efficiency Conditions

In this section, we derive the robust sufficient efficiency conditions for the multi-dimensional multi-objective optimization problem (MOP). More precisely, we establish that any robust feasible solution to the problem (MOP) is a weak robust efficient solution to the considered problem (MOP) if the involved functionals are convex.

Theorem 8.3.0.1 *Let* $(\bar{x}, \bar{c}) \in \mathcal{D}$ *be a robust feasible solution to the multi-dimensional multi-objective optimization problem* (MOP) *such that the robust necessary efficiency conditions* (8.2.1)–(8.2.4) *are satisfied. Further, assume that* $\int_\Gamma \bar{\mu}^T f(\pi, \bar{w})dt$, $\int_\Gamma \bar{\nu}^T g(\pi, \bar{u})dt$, *and* $\int_\Gamma \bar{\gamma}^T h(\pi, \bar{v})dt$ *are convex at* $(\bar{x}, \bar{c})$. *Then,* $(\bar{x}, \bar{c})$ *is a weak robust efficient solution to the multi-dimensional multi-objective optimization problem* (MOP).

Proof Contrary to the result, we assume that $(\bar{x}, \bar{c})$ is not a weak robust efficient solution to the multi-dimensional multi-objective optimization problem (MOP). Then, there exists $(\hat{x}, \hat{c}) \in \mathcal{D}$ such that

$$\int_\Gamma \max_{w \in \mathcal{W}} f(\hat{\pi}, w)dt < \int_\Gamma \max_{w \in \mathcal{W}} f(\bar{\pi}, w)dt.$$

By assumption, $\max_{w \in \mathcal{W}} f(\pi, w) = f(\pi, \bar{w})$, we get

$$\int_\Gamma f(\hat{\pi}, \bar{w})dt < \int_\Gamma f(\bar{\pi}, \bar{w})dt. \tag{8.3.1}$$

Since $(\bar{x}, \bar{c})$ satisfies the conditions (8.2.1)–(8.2.4), therefore, multiplying equations (8.2.1) and (8.2.2) by $(\hat{x} - \bar{x})$ and $(\hat{c} - \bar{c})$, respectively, and integrating them, we get

$$\begin{aligned}
&\int_\Gamma (\hat{x} - \bar{x})\Big\{\bar{\mu}^T f_x(\bar{\pi}, \bar{w}) + \bar{\nu}^T g_x(\bar{\pi}, \bar{u}) + \bar{\gamma}^T h_x(\bar{\pi}, \bar{v}) \\
&\quad - D_\sigma\Big[\bar{\mu}^T f_{x_\sigma}(\bar{\pi}, \bar{w}) + \bar{\nu}^T g_{x_\sigma}(\bar{\pi}, \bar{u}) + \bar{\gamma}^T h_{x_\sigma}(\bar{\pi}, \bar{v})\Big] \\
&\quad + \frac{1}{n(\alpha, \beta)} D^2_{\alpha\beta}\Big[\bar{\mu}^T f_{x_{\alpha\beta}}(\bar{\pi}, \bar{w}) + \bar{\nu}^T g_{x_{\alpha\beta}}(\bar{\pi}, \bar{u}) + \bar{\gamma}^T h_{x_{\alpha\beta}}(\bar{\pi}, \bar{v})\Big]\Big\}dt \\
&\quad + \int_\Gamma (\hat{c} - \bar{c})\{\bar{\mu}^T f_c(\bar{\pi}, \bar{w}) + \bar{\nu}^T g_c(\bar{\pi}, \bar{u}) + \bar{\gamma}^T h_c(\bar{\pi}, \bar{v})\}dt
\end{aligned}$$

$$= \int_{\Gamma} \Big[(\hat{x} - \bar{x})\{\bar{\mu}^T f_x(\bar{\pi}, \bar{w}) + \bar{\nu}^T g_x(\bar{\pi}, \bar{u}) + \bar{\gamma}^T h_x(\bar{\pi}, \bar{\upsilon})\}$$

$$+ (\hat{x}_\sigma - \bar{x}_\sigma)\{\bar{\mu}^T f_{x_\sigma}(\bar{\pi}, \bar{w}) + \bar{\nu}^T g_{x_\sigma}(\bar{\pi}, \bar{u}) + \bar{\gamma}^T h_{x_\sigma}(\bar{\pi}, \bar{\upsilon})\}$$

$$+ \frac{1}{n(\alpha, \beta)}(\hat{x}_{\alpha\beta} - \bar{x}_{\alpha\beta})\{\bar{\mu}^T f_{x_{\alpha\beta}}(\bar{\pi}, \bar{w}) + \bar{\nu}^T g_{x_{\alpha\beta}}(\bar{\pi}, \bar{u}) + \bar{\gamma}^T h_{x_{\alpha\beta}}(\bar{\pi}, \bar{\upsilon})\}\Big] dt$$

$$+ \int_{\Gamma} (\hat{c} - \bar{c})\{\bar{\mu}^T f_c(\bar{\pi}, \bar{w}) + \bar{\nu}^T g_c(\bar{\pi}, \bar{u}) + \bar{\gamma}^T h_c(\bar{\pi}, \bar{\upsilon})\} dt = 0, \qquad (8.3.2)$$

where we used the formula of integration by parts, the divergence formula and the boundary conditions formulated in the considered problem.

On the other hand, since $\int_{\Gamma} \bar{\mu}^T f(\pi, \bar{w}) dt$ is convex at $(\bar{x}, \bar{c})$, we have

$$\int_{\Gamma} \Big\{ \bar{\mu}^T f(\hat{\pi}, \bar{w}) - \bar{\mu}^T f(\bar{\pi}, \bar{w}) \Big\} dt \geq \int_{\Gamma} (\hat{x} - \bar{x})\bar{\mu}^T f_x(\bar{\pi}, \bar{w}) dt + \int_{\Gamma} (\hat{x}_\sigma - \bar{x}_\sigma)\bar{\mu}^T f_{x_\sigma}(\bar{\pi}, \bar{w}) dt$$

$$+ \frac{1}{n(\alpha, \beta)} \int_{\Gamma} (\hat{x}_{\alpha\beta} - \bar{x}_{\alpha\beta})\bar{\mu}^T f_{x_{\alpha\beta}}(\bar{\pi}, \bar{w}) dt + \int_{\Gamma} (\hat{c} - \bar{c})\bar{\mu}^T f_c(\bar{\pi}, \bar{w}) dt,$$

which, along with the inequality (8.3.1), gives

$$\int_{\Gamma} (\hat{x} - \bar{x})\bar{\mu}^T f_x(\bar{\pi}, \bar{w}) dt + \int_{\Gamma} (\hat{x}_\sigma - \bar{x}_\sigma)\bar{\mu}^T f_{x_\sigma}(\bar{\pi}, \bar{w}) dt$$

$$+ \frac{1}{n(\alpha, \beta)} \int_{\Gamma} (\hat{x}_{\alpha\beta} - \bar{x}_{\alpha\beta})\bar{\mu}^T f_{x_{\alpha\beta}}(\bar{\pi}, \bar{w}) dt + \int_{\Gamma} (\hat{c} - \bar{c})\bar{\mu}^T f_c(\bar{\pi}, \bar{w}) dt < 0. \qquad (8.3.3)$$

Again, since $\int_{\Gamma} \bar{\nu}^T g(\pi, \bar{u}) dt$ is convex at $(\bar{x}, \bar{c})$, we obtain

$$\int_{\Gamma} \Big\{ \bar{\nu}^T g(\hat{\pi}, \bar{u}) - \bar{\nu}^T g(\bar{\pi}, \bar{u}) \Big\} dt \geq \int_{\Gamma} (\hat{x} - \bar{x})\bar{\nu}^T g_x(\bar{\pi}, \bar{u}) dt + \int_{\Gamma} (\hat{x}_\sigma - \bar{x}_\sigma)\bar{\nu}^T g_{x_\sigma}(\bar{\pi}, \bar{u}) dt$$

$$+ \frac{1}{n(\alpha, \beta)} \int_{\Gamma} (\hat{x}_{\alpha\beta} - \bar{x}_{\alpha\beta})\bar{\nu}^T g_{x_{\alpha\beta}}(\bar{\pi}, \bar{u}) dt + \int_{\Gamma} (\hat{c} - \bar{c})\bar{\nu}^T g_c(\bar{\pi}, \bar{u}) dt.$$

The above inequality together with the robust feasibility of $(\hat{x}, \hat{c})$ in the multi-dimensional multi-objective optimization problem (MOP) and the condition (3), yields

$$\int_{\Gamma} (\hat{x} - \bar{x})\bar{\nu}^T g_x(\bar{\pi}, \bar{u}) dt + \int_{\Gamma} (\hat{x}_\sigma - \bar{x}_\sigma)\bar{\nu}^T g_{x_\sigma}(\bar{\pi}, \bar{u}) dt$$

$$+ \frac{1}{n(\alpha, \beta)} \int_{\Gamma} (\hat{x}_{\alpha\beta} - \bar{x}_{\alpha\beta})\bar{\nu}^T g_{x_{\alpha\beta}}(\bar{\pi}, \bar{u}) dt + \int_{\Gamma} (\hat{c} - \bar{c})\bar{\nu}^T g_c(\bar{\pi}, \bar{u}) dt \leq 0. \qquad (8.3.4)$$

Further, from the assumption $\int_{\Gamma} \bar{\gamma}^T h(\pi, \bar{\upsilon}) dt$ is convex at $(\bar{x}, \bar{c})$ and the robust feasibility of $(\hat{x}, \hat{c})$ in the multi-dimensional multi-objective optimization problem (MOP), we obtain

$$\int_{\Gamma}(\hat{x}-\bar{x})\bar{\gamma}^{T}h_{x}(\bar{\pi},\bar{v})dt+\int_{\Gamma}(\hat{x}_{\sigma}-\bar{x}_{\sigma})\bar{\gamma}^{T}h_{x_{\sigma}}(\bar{\pi},\bar{v})dt$$
$$+\frac{1}{n(\alpha,\beta)}\int_{\Gamma}(\hat{x}_{\alpha\beta}-\bar{x}_{\alpha\beta})\bar{\gamma}^{T}h_{x_{\alpha\beta}}(\bar{\pi},\bar{v})dt+\int_{\Gamma}(\hat{c}-\bar{c})\bar{\gamma}^{T}h_{c}(\bar{\pi},\bar{v})dt\leq 0. \tag{8.3.5}$$

On adding the inequalities (8.3.3), (8.3.4) and (8.3.5), we have

$$\int_{\Gamma}\Big[(\hat{x}-\bar{x})\{\bar{\mu}^{T}f_{x}(\bar{\pi},\bar{w})+\bar{\upsilon}^{T}g_{x}(\bar{\pi},\bar{u})+\bar{\gamma}^{T}h_{x}(\bar{\pi},\bar{v})\}$$
$$+(\hat{x}_{\sigma}-\bar{x}_{\sigma})\{\bar{\mu}^{T}f_{x_{\sigma}}(\bar{\pi},\bar{w})+\bar{\upsilon}^{T}g_{x_{\sigma}}(\bar{\pi},\bar{u})+\bar{\gamma}^{T}h_{x_{\sigma}}(\bar{\pi},\bar{v})\}$$
$$+\frac{1}{n(\alpha,\beta)}(\hat{x}_{\alpha\beta}-\bar{x}_{\alpha\beta})\{\bar{\mu}^{T}f_{x_{\alpha\beta}}(\bar{\pi},\bar{w})+\bar{\upsilon}^{T}g_{x_{\alpha\beta}}(\bar{\pi},\bar{u})+\bar{\gamma}^{T}h_{x_{\alpha\beta}}(\bar{\pi},\bar{v})\}\Big]dt$$
$$+\int_{\Gamma}(\hat{c}-\bar{c})\{\bar{\mu}^{T}f_{c}(\bar{\pi},\bar{w})+\bar{\upsilon}^{T}g_{c}(\bar{\pi},\bar{u})+\bar{\gamma}^{T}h_{c}(\bar{\pi},\bar{v})\}dt<0,$$

which contradicts the relation (8.3.2) and the proof is complete. □

8.4 The Associated Modified Optimization Problem

In this section, we use the modified objective function method to reduce the complexity associated with the considered multi-dimensional multi-objective optimization problem (MOP). In this regard, let $(\bar{x},\bar{c})$ be an arbitrary given robust feasible solution to the multi-dimensional multi-objective optimization problem (MOP). The modified multi-dimensional multi-objective optimization problem associated with the original variational control problem (MOP) is defined as:

$$(\text{MOP})_{(\bar{x},\bar{c})}\quad \min_{(x(\cdot),c(\cdot))}\int_{\Gamma}\Big\{(x(t)-\bar{x}(t))f_{x}(\bar{\pi},w)+(x_{\sigma}(t)-\bar{x}_{\sigma}(t))f_{x_{\sigma}}(\bar{\pi},w)$$
$$+\frac{1}{n(\alpha,\beta)}(x_{\alpha\beta}(t)-\bar{x}_{\alpha\beta}(t))f_{x_{\alpha\beta}}(\bar{\pi},w)+(c(t)-\bar{c}(t))f_{c}(\bar{\pi},w)\Big\}dt$$
$$\text{subject to}\quad g(\pi,u)\leqq 0,\ t\in\Gamma,$$
$$h(\pi,v)=0,\ t\in\Gamma,$$
$$x(t_0)=x_0,\ x(t_1)=x_1,\ x_{\sigma}(t_0)=x_{\sigma 0},\ x_{\sigma}(t_1)=x_{\sigma 1},$$
$$\text{or}\quad x(t)|_{\partial\Gamma}=\text{given},\quad x_{\sigma}(t)|_{\partial\Gamma}=\text{given},$$

where the functionals f, g and h are given as in the original variational control problem (MOP).

The associated robust counterpart of the multi-dimensional multi-objective optimization problem $(\text{MOP})_{(\bar{x},\bar{c})}$ is defined as:

$$
\begin{aligned}
(\text{RMOP})_{(\bar{x},\bar{c})} \quad & \min_{(x(\cdot),c(\cdot))} \int_{\Gamma} \max_{w\in\mathcal{W}} \Big\{(x(t)-\bar{x}(t))f_x(\bar{\pi},w)+(x_\sigma(t)-\bar{x}_\sigma(t))f_{x_\sigma}(\bar{\pi},w) \\
& \quad + \frac{1}{n(\alpha,\beta)}(x_{\alpha\beta}(t)-\bar{x}_{\alpha\beta}(t))f_{x_{\alpha\beta}}(\bar{\pi},w)+(c(t)-\bar{c}(t))f_c(\bar{\pi},w)\Big\}dt \\
\text{subject to} \quad & g(\pi,u)\leqq 0,\ t\in\Gamma,\ \forall u\in\mathcal{U}, \\
& h(\pi,v)=0,\ t\in\Gamma,\ \forall v\in\mathcal{V}, \\
& x(t_0)=x_0,\ x(t_1)=x_1,\ x_\sigma(t_0)=x_{\sigma 0},\ x_\sigma(t_1)=x_{\sigma 1}, \\
\text{or} \quad & x(t)|_{\partial\Gamma}=\text{given},\quad x_\sigma(t)|_{\partial\Gamma}=\text{given},
\end{aligned}
$$

Remark 8.4.0.1 Let us remark that the robust feasible solution set to the problem $(\text{MOP})_{(\bar{x},\bar{c})}$ is the same as in the problem (MOP), and, in consequence, it is also denoted by $\mathcal{D}$.

Definition 8.4.0.1 A point $(\hat{x},\hat{c})\in\mathcal{D}$ is said to be a *weak robust efficient solution* to the modified multi-dimensional multi-objective optimization problem $(\text{MOP})_{(\bar{x},\bar{c})}$ if there does not exist another point $(x,c)\in\mathcal{D}$ such that

$$
\begin{aligned}
& \int_{\Gamma} \max_{w\in\mathcal{W}} \Big[(x-\bar{x})f_x(\bar{\pi},w)+(x_\sigma-\bar{x}_\sigma)f_{x_\sigma}(\bar{\pi},w) \\
& \quad + \frac{1}{n(\alpha,\beta)}(x_{\alpha\beta}-\bar{x}_{\alpha\beta})f_{x_{\alpha\beta}}(\bar{\pi},w)+(c-\bar{c})f_c(\bar{\pi},w)\Big]dt \\
& < \int_{\Gamma} \max_{w\in\mathcal{W}} \Big[(\hat{x}-\bar{x})f_x(\bar{\pi},w)+(\hat{x}_\sigma-\bar{x}_\sigma)f_{x_\sigma}(\bar{\pi},w) \\
& \quad + \frac{1}{n(\alpha,\beta)}(\hat{x}_{\alpha\beta}-\bar{x}_{\alpha\beta})f_{x_{\alpha\beta}}(\bar{\pi},w)+(\hat{c}-\bar{c})f_c(\bar{\pi},w)\Big]dt.
\end{aligned}
$$

In the following, we establish some equivalence results between the problems (MOP) and $(\text{MOP})_{(\bar{x},\bar{c})}$, under appropriate convexity assumptions.

Theorem 8.4.0.1 *Let $(\bar{x},\bar{c})$ be a normal weak robust efficient solution to the multi-objective optimization problem* (MOP). *Further, assume that $\int_{\Gamma}\bar{\upsilon}^T g(\pi,\bar{u})dt$, $\int_{\Gamma}\bar{\gamma}^T h(\pi,\bar{v})dt$ are convex at $(\bar{x},\bar{c})$. Then, $(\bar{x},\bar{c})$ is also a weak robust efficient solution to the modified multi-dimensional multi-objective optimization problem* $(\text{MOP})_{(\bar{x},\bar{c})}$.

Proof By hypothesis, the relations (8.2.1)–(8.2.4), with $\bar{\mu}=e$, are fulfilled for all $t\in\Gamma$, except at discontinuities. Hence, the conditions (8.2.1) and (8.2.2) yield

$$
\begin{aligned}
& \int_{\Gamma} (\tilde{x}-\bar{x})\{e^T f_x(\bar{\pi},\bar{w})+\bar{\upsilon}^T g_x(\bar{\pi},\bar{u})+\bar{\gamma}^T h_x(\bar{\pi},\bar{v}) \\
& \quad - D_\sigma\Big[e^T f_{x_\sigma}(\bar{\pi},\bar{w})+\bar{\upsilon}^T g_{x_\sigma}(\bar{\pi},\bar{u})+\bar{\gamma}^T h_{x_\sigma}(\bar{\pi},\bar{v})\Big] \\
& \quad + \frac{1}{n(\alpha,\beta)}D^2_{\alpha\beta}\Big[e^T f_{x_{\alpha\beta}}(\bar{\pi},\bar{w})+\bar{\upsilon}^T g_{x_{\alpha\beta}}(\bar{\pi},\bar{u})+\bar{\gamma}^T h_{x_{\alpha\beta}}(\bar{\pi},\bar{v})\Big]\}dt
\end{aligned}
$$

$$+\int_{\Gamma}(\tilde{c}-\bar{c})\{e^T f_c(\bar{\pi},\bar{w})+\bar{v}^T g_c(\bar{\pi},\bar{u})+\bar{\gamma}^T h_c(\bar{\pi},\bar{v})\}dt$$
$$=\int_{\Gamma}\Big[(\tilde{x}-\bar{x})\{e^T f_x(\bar{\pi},\bar{w})+\bar{v}^T g_x(\bar{\pi},\bar{u})+\bar{\gamma}^T h_x(\bar{\pi},\bar{v})\}$$
$$+(\tilde{x}_\sigma-\bar{x}_\sigma)\{e^T f_{x_\sigma}(\bar{\pi},\bar{w})+\bar{v}^T g_{x_\sigma}(\bar{\pi},\bar{u})+\bar{\gamma}^T h_{x_\sigma}(\bar{\pi},\bar{v})\}$$
$$+\frac{1}{n(\alpha,\beta)}(\tilde{x}_{\alpha\beta}-\bar{x}_{\alpha\beta})\{e^T f_{x_{\alpha\beta}}(\bar{\pi},\bar{w})+\bar{v}^T g_{x_{\alpha\beta}}(\bar{\pi},\bar{u})+\bar{\gamma}^T h_{x_{\alpha\beta}}(\bar{\pi},\bar{v})\}\Big]dt$$
$$+\int_{\Gamma}(\tilde{c}-\bar{c})\{e^T f_c(\bar{\pi},\bar{w})+\bar{v}^T g_c(\bar{\pi},\bar{u})+\bar{\gamma}^T h_c(\bar{\pi},\bar{v})\}dt=0, \qquad (8.4.1)$$

where we used the formula of integration by parts, the divergence formula and the boundary conditions formulated in the considered problem.

Now, let us proceed by contradiction and consider $(\bar{x},\bar{c})$ is not a weak robust efficient solution to the modified multi-dimensional multi-objective optimization problem $(\text{MOP})_{(\bar{x},\bar{c})}$. Then, there exists $(\tilde{x},\tilde{c})\in\mathcal{D}$ such that

$$\int_{\Gamma}\max_{w\in\mathcal{W}}\Big[(\tilde{x}-\bar{x})f_x(\bar{\pi},w)+(\tilde{x}_\sigma-\bar{x}_\sigma)f_{x_\sigma}(\bar{\pi},w)$$
$$+\frac{1}{n(\alpha,\beta)}(\tilde{x}_{\alpha\beta}-\bar{x}_{\alpha\beta})f_{x_{\alpha\beta}}(\bar{\pi},w)+(\tilde{c}-\bar{c})f_c(\bar{\pi},w)\Big]dt$$
$$<\int_{\Gamma}\max_{w\in\mathcal{W}}\Big[(\bar{x}-\bar{x})f_x(\bar{\pi},w)+(\bar{x}_\sigma-\bar{x}_\sigma)f_{x_\sigma}(\bar{\pi},w)$$
$$+\frac{1}{n(\alpha,\beta)}(\bar{x}_{\alpha\beta}-\bar{x}_{\alpha\beta})f_{x_{\alpha\beta}}(\bar{\pi},w)+(\bar{c}-\bar{c})f_c(\bar{\pi},w)\Big]dt.$$

Since $\max_{w\in\mathcal{W}}f(\pi,w)=f(\pi,\bar{w})$, we get

$$\int_{\Gamma}\Big[(\tilde{x}-\bar{x})f_x(\bar{\pi},\bar{w})+(\tilde{x}_\sigma-\bar{x}_\sigma)f_{x_\sigma}(\bar{\pi},\bar{w})$$
$$+\frac{1}{n(\alpha,\beta)}(\tilde{x}_{\alpha\beta}-\bar{x}_{\alpha\beta})f_{x_{\alpha\beta}}(\bar{\pi},\bar{w})+(\tilde{c}-\bar{c})f_c(\bar{\pi},\bar{w})\Big]dt$$
$$<\int_{\Gamma}\Big[(\bar{x}-\bar{x})f_x(\bar{\pi},\bar{w})+(\bar{x}_\sigma-\bar{x}_\sigma)f_{x_\sigma}(\bar{\pi},\bar{w})$$
$$+\frac{1}{n(\alpha,\beta)}(\bar{x}_{\alpha\beta}-\bar{x}_{\alpha\beta})f_{x_{\alpha\beta}}(\bar{\pi},\bar{w})+(\bar{c}-\bar{c})f_c(\bar{\pi},\bar{w})\Big]dt,$$

equivalent with

$$\int_{\Gamma}\Big[(\tilde{x}-\bar{x})f_x(\bar{\pi},\bar{w})+(\tilde{x}_\sigma-\bar{x}_\sigma)f_{x_\sigma}(\bar{\pi},\bar{w})$$
$$+\frac{1}{n(\alpha,\beta)}(\tilde{x}_{\alpha\beta}-\bar{x}_{\alpha\beta})f_{x_{\alpha\beta}}(\bar{\pi},\bar{w})+(\tilde{c}-\bar{c})f_c(\bar{\pi},\bar{w})\Big]dt<0. \qquad (8.4.2)$$

The robust feasibility of $(\tilde{x}, \tilde{c})$ in the problem (MOP) and the robust necessary efficiency condition (3) yield

$$\int_{\Gamma} \bar{\nu}^T g(\tilde{\pi}, \bar{u})dt - \int_{\Gamma} \bar{\nu}^T g(\bar{\pi}, \bar{u})dt \leq 0. \tag{8.4.3}$$

By hypothesis, since $\int_{\Gamma} \bar{\nu}^T g(\pi, \bar{u})dt$ is convex at $(\bar{x}, \bar{c})$, we obtain

$$\int_{\Gamma} \left\{\bar{\nu}^T g(\tilde{\pi}, \bar{u}) - \bar{\nu}^T g(\bar{\pi}, \bar{u})\right\}dt \geq \int_{\Gamma} (\tilde{x} - \bar{x})\bar{\nu}^T g_x(\bar{\pi}, \bar{u})dt + \int_{\Gamma} (\tilde{x}_\sigma - \bar{x}_\sigma)\bar{\nu}^T g_{x_\sigma}(\bar{\pi}, \bar{u})dt$$
$$+ \frac{1}{n(\alpha, \beta)} \int_{\Gamma} (\tilde{x}_{\alpha\beta} - \bar{x}_{\alpha\beta})\bar{\nu}^T g_{x_{\alpha\beta}}(\bar{\pi}, \bar{u})dt + \int_{\Gamma} (\tilde{c} - \bar{c})\bar{\nu}^T g_c(\bar{\pi}, \bar{u})dt.$$

The above inequality together with relation (8.4.3) yield

$$\int_{\Gamma} (\tilde{x} - \bar{x})\bar{\nu}^T g_x(\bar{\pi}, \bar{u})dt + \int_{\Gamma} (\tilde{x}_\sigma - \bar{x}_\sigma)\bar{\nu}^T g_{x_\sigma}(\bar{\pi}, \bar{u})dt$$
$$+ \frac{1}{n(\alpha, \beta)} \int_{\Gamma} (\tilde{x}_{\alpha\beta} - \bar{x}_{\alpha\beta})\bar{\nu}^T g_{x_{\alpha\beta}}(\bar{\pi}, \bar{u})dt + \int_{\Gamma} (\tilde{c} - \bar{c})\bar{\nu}^T g_c(\bar{\pi}, \bar{u})dt \leq 0. \tag{8.4.4}$$

Further, from the assumption $\int_{\Gamma} \bar{\gamma}^T h(\pi, \bar{v})dt$ is convex at $(\bar{x}, \bar{c})$ and the robust feasibility of $(\tilde{x}, \tilde{c})$ and $(\bar{x}, \bar{c})$ in the multi-dimensional multi-objective optimization problem (MOP), we obtain

$$\int_{\Gamma} (\tilde{x} - \bar{x})\bar{\gamma}^T h_x(\bar{\pi}, \bar{v})dt + \int_{\Gamma} (\tilde{x}_\sigma - \bar{x}_\sigma)\bar{\gamma}^T h_{x_\sigma}(\bar{\pi}, \bar{v})dt$$
$$+ \frac{1}{n(\alpha, \beta)} \int_{\Gamma} (\tilde{x}_{\alpha\beta} - \bar{x}_{\alpha\beta})\bar{\gamma}^T h_{x_{\alpha\beta}}(\bar{\pi}, \bar{v})dt + \int_{\Gamma} (\tilde{c} - \bar{c})\bar{\gamma}^T h_c(\bar{\pi}, \bar{v})dt \leq 0. \tag{8.4.5}$$

On combining the inequalities (8.4.2), (8.4.4) and (8.4.5), we get

$$\int_{\Gamma} \Big[(\tilde{x} - \bar{x})\{e^T f_x(\bar{\pi}, \bar{w}) + \bar{\nu}^T g_x(\bar{\pi}, \bar{u}) + \bar{\gamma}^T h_x(\bar{\pi}, \bar{v})\}$$
$$+ (\tilde{x}_\sigma - \bar{x}_\sigma)\{e^T f_{x_\sigma}(\bar{\pi}, \bar{w}) + \bar{\nu}^T g_{x_\sigma}(\bar{\pi}, \bar{u}) + \bar{\gamma}^T h_{x_\sigma}(\bar{\pi}, \bar{v})\}$$
$$+ \frac{1}{n(\alpha, \beta)}(\tilde{x}_{\alpha\beta} - \bar{x}_{\alpha\beta})\{e^T f_{x_{\alpha\beta}}(\bar{\pi}, \bar{w}) + \bar{\nu}^T g_{x_{\alpha\beta}}(\bar{\pi}, \bar{u}) + \bar{\gamma}^T h_{x_{\alpha\beta}}(\bar{\pi}, \bar{v})\}\Big]dt$$
$$+ \int_{\Gamma} (\tilde{c} - \bar{c})\{e^T f_c(\bar{\pi}, \bar{w}) + \bar{\nu}^T g_c(\bar{\pi}, \bar{u}) + \bar{\gamma}^T h_c(\bar{\pi}, \bar{v})\}dt < 0,$$

which contradicts the relation (8.4.1). Consequently, $(\bar{x}, \bar{c})$ is a weak robust efficient solution to the problem $(\text{MOP})_{(\bar{x}, \bar{c})}$ and this completes the proof. □

Theorem 8.4.0.2 *Let $(\bar{x}, \bar{c})$ be a weak robust efficient solution to the modified multi-objective optimization problem* $(\text{MOP})_{(\bar{x},\bar{c})}$. *Further, assume that* $\max_{w\in\mathcal{W}} f(\pi, w) = f(\pi, \bar{w})$ *and the functional* $\int_\Gamma f(\pi, \bar{w})dt$ *is convex at $(\bar{x}, \bar{c})$. Then, $(\bar{x}, \bar{c})$ is a weak robust efficient solution to the multi-dimensional multi-objective optimization problem* (MOP).

Proof Contrary to the result, we assume that $(\bar{x}, \bar{c})$ is not a weak robust efficient solution to multi-dimensional multi-objective optimization problem (MOP). Then, there exists another point $(\tilde{x}, \tilde{c}) \in \mathcal{D}$ such that

$$\int_\Gamma \max_{w\in\mathcal{W}} f(\tilde{\pi}, w)dt < \int_\Gamma \max_{w\in\mathcal{W}} f(\bar{\pi}, w)dt.$$

Since $\max_{w\in\mathcal{W}} f(\pi, w) = f(\pi, \bar{w})$, we get

$$\int_\Gamma f(\tilde{\pi}, \bar{w})dt < \int_\Gamma f(\bar{\pi}, \bar{w})dt. \tag{8.4.6}$$

By assumption, since $\int_\Gamma f(\pi, \bar{w})dt$ is convex functional at $(\bar{x}, \bar{c})$, we have

$$\begin{aligned}\int_\Gamma \Big\{f(\tilde{\pi}, \bar{w}) - f(\bar{\pi}, \bar{w})\Big\}dt &\geqq \int_\Gamma (\tilde{x} - \bar{x})f_x(\bar{\pi}, \bar{w})dt + \int_\Gamma (\tilde{x}_\sigma - \bar{x}_\sigma)f_{x_\sigma}(\bar{\pi}, \bar{w})dt\\ &+ \frac{1}{n(\alpha, \beta)}\int_\Gamma (\tilde{x}_{\alpha\beta} - \bar{x}_{\alpha\beta})f_{x_{\alpha\beta}}(\bar{\pi}, \bar{w})dt + \int_\Gamma (\tilde{c} - \bar{c})f_c(\bar{\pi}, \bar{w})dt,\end{aligned}$$

which along with the inequality (8.4.6), yield

$$\begin{aligned}&\int_\Gamma (\tilde{x} - \bar{x})f_x(\bar{\pi}, \bar{w})dt + \int_\Gamma (\tilde{x}_\sigma - \bar{x}_\sigma)f_{x_\sigma}(\bar{\pi}, \bar{w})dt\\ &\quad+ \frac{1}{n(\alpha, \beta)}\int_\Gamma (\tilde{x}_{\alpha\beta} - \bar{x}_{\alpha\beta})f_{x_{\alpha\beta}}(\bar{\pi}, \bar{w})dt + \int_\Gamma (\tilde{c} - \bar{c})f_c(\bar{\pi}, \bar{w})dt < 0,\end{aligned}$$

equivalently with

$$\begin{aligned}&\int_\Gamma (\tilde{x} - \bar{x})f_x(\bar{\pi}, \bar{w})dt + \int_\Gamma (\tilde{x}_\sigma - \bar{x}_\sigma)f_{x_\sigma}(\bar{\pi}, \bar{w})dt\\ &\quad+ \frac{1}{n(\alpha, \beta)}\int_\Gamma (\tilde{x}_{\alpha\beta} - \bar{x}_{\alpha\beta})f_{x_{\alpha\beta}}(\bar{\pi}, \bar{w})dt + \int_\Gamma (\tilde{c} - \bar{c})f_c(\bar{\pi}, \bar{w})dt\\ &\quad< \int_\Gamma (\bar{x} - \bar{x})f_x(\bar{\pi}, \bar{w})dt + \int_\Gamma (\bar{x}_\sigma - \bar{x}_\sigma)f_{x_\sigma}(\bar{\pi}, \bar{w})dt\\ &\quad+ \frac{1}{n(\alpha, \beta)}\int_\Gamma (\bar{x}_{\alpha\beta} - \bar{x}_{\alpha\beta})f_{x_{\alpha\beta}}(\bar{\pi}, \bar{w})dt + \int_\Gamma (\bar{c} - \bar{c})f_c(\bar{\pi}, \bar{w})dt.\end{aligned}$$

The above inequality can be rewritten as

$$\int_{\Gamma} \max_{w \in \mathcal{W}} \Big[(\tilde{x} - \bar{x}) f_x(\bar{\pi}, w) + (\tilde{x}_\sigma - \bar{x}_\sigma) f_{x_\sigma}(\bar{\pi}, w) \\ + \frac{1}{n(\alpha, \beta)} (\tilde{x}_{\alpha\beta} - \bar{x}_{\alpha\beta}) f_{x_{\alpha\beta}}(\bar{\pi}, w) + (\tilde{c} - \bar{c}) f_c(\bar{\pi}, w) \Big] dt \\ < \int_{\Gamma} \max_{w \in \mathcal{W}} \Big[(\bar{x} - \bar{x}) f_x(\bar{\pi}, w) + (\bar{x}_\sigma - \bar{x}_\sigma) f_{x_\sigma}(\bar{\pi}, w) \\ + \frac{1}{n(\alpha, \beta)} (\bar{x}_{\alpha\beta} - \bar{x}_{\alpha\beta}) f_{x_{\alpha\beta}}(\bar{\pi}, w) + (\bar{c} - \bar{c}) f_c(\bar{\pi}, w) \Big] dt,$$

which contradicts our assumption that $(\bar{x}, \bar{c})$ is a weak robust efficient solution to the modified multi-dimensional multi-objective optimization problem $(\mathrm{MOP})_{(\bar{x},\bar{c})}$. Hence, the proof is complete. □

Remark 8.4.0.2 As can be easily seen, to derive the equivalence between the original problem (MOP) and its associated modified problem $(\mathrm{MOP})_{(\bar{x},\bar{c})}$, we imposed the convexity assumption only on the constraint functionals. Whereas, to establish the converse of Theorem 8.4.0.1, it is sufficient to consider only the convexity hypothesis of the objective functional.

8.5 The Associated Saddle-Point Efficiency Criterion

In this section, under some convexity assumptions, we establish a connection between a weak robust efficient solution of (MOP) and a *robust saddle-point* associated with a Lagrange functional (Lagrangian) corresponding to the modified multi-dimensional variational control problem $(\mathrm{MOP})_{(\bar{x},\bar{c})}$. In this regard, in accordance to Treanţă [7, 9], Jayswal and Preeti [4], Preeti et al. [10], we formulate the next definitions.

Definition 8.5.0.1 The Lagrange functional $L((x, c), \nu, \gamma, w, u, v) : \mathcal{X} \times \mathcal{C} \times \mathbb{R}^m_+ \times \mathbb{R}^n \times \mathcal{W} \times \mathcal{U} \times \mathcal{V} \to \mathbb{R}^s$ associated with the multi-dimensional multi-objective optimization problem $(\mathrm{MOP})_{(\bar{x},\bar{c})}$ is defined as

$$L((x, c), \nu, \gamma, w, u, v) = \int_{\Gamma} \Big\{ \max_{w \in \mathcal{W}} \Big[(x(t) - \bar{x}(t)) f_x(\bar{\pi}, w) + (x_\sigma(t) - \bar{x}_\sigma(t)) f_{x_\sigma}(\bar{\pi}, w) \\ + \frac{1}{n(\alpha, \beta)} (x_{\alpha\beta}(t) - \bar{x}_{\alpha\beta}(t)) f_{x_{\alpha\beta}}(\bar{\pi}, w) + (c(t) - \bar{c}(t)) f_c(\bar{\pi}, w) \Big] \\ + \nu^T(t) g(\pi, u) e + \gamma^T(t) h(\pi, v) e \Big\} dt.$$

Definition 8.5.0.2 A point $((\bar{x}, \bar{c}), \bar{\nu}, \bar{\gamma}, \bar{w}, \bar{u}, \bar{v}) \in \mathcal{D} \times \mathbb{R}^m_+ \times \mathbb{R}^n \times \mathcal{W} \times \mathcal{U} \times \mathcal{V}$ is said to be a *robust saddle-point* for the Lagrange functional $L((x, c), \nu, \gamma, w, u, v)$ associated with the modified multi-dimensional variational control problem $(\mathrm{MOP})_{(\bar{x},\bar{c})}$ if the following relations are fulfilled:

(i) $L((\bar{x},\bar{c}),\nu,\gamma,w,u,v) \leqq L((\bar{x},\bar{c}),\bar{\nu},\bar{\gamma},w,\bar{u},\bar{v}),\ \forall \nu \in \mathbb{R}^m_+,\ \forall \gamma \in \mathbb{R}^n,$
$\forall u \in \mathcal{U},\ \forall v \in \mathcal{V},$
(ii) $L((x,c),\bar{\nu},\bar{\gamma},w,\bar{u},\bar{v}) \geqq L((\bar{x},\bar{c}),\bar{\nu},\bar{\gamma},w,\bar{u},\bar{v}),\ \forall (x,c) \in \mathscr{X} \times \mathscr{C}.$

Further, taking into account the aforementioned definitions, we formulate the following two results.

Theorem 8.5.0.1 *Let* $(\bar{x},\bar{c})$ *be a robust feasible solution to the multi-dimensional multi-objective optimization problem* (MOP). *Assume that* $\max_{w\in\mathcal{W}} f(\pi,w) = f(\pi,\bar{w})$ *and the objective functional* $\int_\Gamma f(\pi,\bar{w})dt$ *is convex at* $(\bar{x},\bar{c})$. *If* $((\bar{x},\bar{c}),\bar{\nu},\bar{\gamma},\bar{w},\bar{u},\bar{v})$ *is a robust saddle-point for the Lagrange functional* $L((x,c),\nu,\gamma,w,u,v)$ *associated with the modified multi-dimensional variational control problem* $(\mathrm{MOP})_{(\bar{x},\bar{c})}$, *then* $(\bar{x},\bar{c})$ *is a weak robust efficient solution to the multi-dimensional multi-objective optimization problem* (MOP).

Proof Contrary to the result, let us assume that $(\bar{x},\bar{c})$ is not a weak robust efficient solution to the multi-dimensional multi-objective optimization problem (MOP). Proceeding on the line of Theorem 8.4.0.2, we get

$$\begin{aligned}
&\int_\Gamma \max_{w\in\mathcal{W}}\Big[(\tilde{x}-\bar{x})f_x(\tilde{\pi},w)+(\tilde{x}_\sigma-\bar{x}_\sigma)f_{x_\sigma}(\tilde{\pi},w)\\
&\qquad+\frac{1}{n(\alpha,\beta)}(\tilde{x}_{\alpha\beta}-\bar{x}_{\alpha\beta})f_{x_{\alpha\beta}}(\tilde{\pi},w)+(\tilde{c}-\bar{c})f_c(\tilde{\pi},w)\Big]dt\\
&\qquad<\int_\Gamma \max_{w\in\mathcal{W}}\Big[(\bar{x}-\bar{x})f_x(\bar{\pi},w)+(\bar{x}_\sigma-\bar{x}_\sigma)f_{x_\sigma}(\bar{\pi},w)\\
&\qquad+\frac{1}{n(\alpha,\beta)}(\bar{x}_{\alpha\beta}-\bar{x}_{\alpha\beta})f_{x_{\alpha\beta}}(\bar{\pi},w)+(\bar{c}-\bar{c})f_c(\bar{\pi},w)\Big]dt,
\end{aligned}\tag{8.5.1}$$

for some $(\tilde{x},\tilde{c}) \in \mathcal{D}$.

From the feasibility of $(\tilde{x},\tilde{c})$ to the problem (MOP) and $\bar{\nu} \in \mathbb{R}^m_+$, we get

$$\int_\Gamma \bar{\nu}^T g(\tilde{\pi},\bar{u})dt \leq 0. \tag{8.5.2}$$

On the other hand, since $((\bar{x},\bar{c}),\bar{\nu},\bar{\gamma},\bar{w},\bar{u},\bar{v})$ is a robust saddle-point for the Lagrange functional $L((x,c),\nu,\gamma,w,u,v)$ associated with the modified multi-dimensional variational control problem $(\mathrm{MOP})_{(\bar{x},\bar{c})}$, by using Definition 8.5.0.2 (i), we have

$$L((\bar{x},\bar{c}),\nu,\gamma,w,u,v) \leqq L((\bar{x},\bar{c}),\bar{\nu},\bar{\gamma},w,\bar{u},\bar{v}),\ \forall \nu \in \mathbb{R}^m_+,\ \forall \gamma \in \mathbb{R}^n, \forall u \in \mathcal{U},\ \forall v \in \mathcal{V},$$

which, in virtue of the definition of Lagrange functional, can be rewritten as

$$\int_{\Gamma}\Big\{\max_{w\in\mathcal{W}}\Big[(\bar{x}(t)-\bar{x}(t))f_x(\bar{\pi},w)+(\bar{x}_\sigma(t)-\bar{x}_\sigma(t))f_{x_\sigma}(\bar{\pi},w)$$
$$+\frac{1}{n(\alpha,\beta)}(\bar{x}_{\alpha\beta}(t)-\bar{x}_{\alpha\beta}(t))f_{x_{\alpha\beta}}(\bar{\pi},w)+(\bar{c}(t)-\bar{c}(t))f_c(\bar{\pi},w)\Big]$$
$$+\nu^T(t)g(\bar{\pi},u)e+\gamma^T(t)h(\bar{\pi},v)e\Big\}dt$$
$$\leqq\int_{\Gamma}\Big\{\max_{w\in\mathcal{W}}\Big[(\bar{x}(t)-\bar{x}(t))f_x(\bar{\pi},w)+(\bar{x}_\sigma(t)-\bar{x}_\sigma(t))f_{x_\sigma}(\bar{\pi},w)$$
$$+\frac{1}{n(\alpha,\beta)}(\bar{x}_{\alpha\beta}(t)-\bar{x}_{\alpha\beta}(t))f_{x_{\alpha\beta}}(\bar{\pi},w)+(\bar{c}(t)-\bar{c}(t))f_c(\bar{\pi},w)\Big]$$
$$+\bar{\nu}^T(t)g(\bar{\pi},\bar{u})e+\bar{\gamma}^T(t)h(\bar{\pi},\bar{v})e\Big\}dt.$$

Since $\max_{w\in\mathcal{W}} f(\pi,w)=f(\pi,\bar{w})$, it follows that

$$\int_{\Gamma}\Big\{(\bar{x}(t)-\bar{x}(t))f_x(\bar{\pi},\bar{w})+(\bar{x}_\sigma(t)-\bar{x}_\sigma(t))f_{x_\sigma}(\bar{\pi},\bar{w})$$
$$+\frac{1}{n(\alpha,\beta)}(\bar{x}_{\alpha\beta}(t)-\bar{x}_{\alpha\beta}(t))f_{x_{\alpha\beta}}(\bar{\pi},\bar{w})+(\bar{c}(t)-\bar{c}(t))f_c(\bar{\pi},\bar{w})$$
$$+\nu^T(t)g(\bar{\pi},u)+\gamma^T(t)h(\bar{\pi},v)\Big\}dt$$
$$\leqq\int_{\Gamma}\Big\{(\bar{x}(t)-\bar{x}(t))f_x(\bar{\pi},\bar{w})+(\bar{x}_\sigma(t)-\bar{x}_\sigma(t))f_{x_\sigma}(\bar{\pi},\bar{w})$$
$$+\frac{1}{n(\alpha,\beta)}(\bar{x}_{\alpha\beta}(t)-\bar{x}_{\alpha\beta}(t))f_{x_{\alpha\beta}}(\bar{\pi},\bar{w})+(\bar{c}(t)-\bar{c}(t))f_c(\bar{\pi},\bar{w})$$
$$+\bar{\nu}^T(t)g(\bar{\pi},\bar{u})e+\bar{\gamma}^T(t)h(\bar{\pi},\bar{v})e\Big\}dt.$$

If we set $\nu=0$ and $\gamma=0$ in the above inequality, we obtain

$$\int_{\Gamma}\bar{\nu}^T g(\bar{\pi},\bar{u})dt\geq 0. \tag{8.5.3}$$

From (8.5.2) and (8.5.3), it follows that

$$\int_{\Gamma}\bar{\nu}^T g(\tilde{\pi},\bar{u})dt\leq\int_{\Gamma}\bar{\nu}^T g(\bar{\pi},\bar{u})dt,$$

which, along with the inequality (8.5.1), give

$$\int_{\Gamma}\Big\{\max_{w\in\mathcal{W}}\Big[(\tilde{x}-\bar{x})f_x(\bar{\pi},w)+(\tilde{x}_\sigma-\bar{x}_\sigma)f_{x_\sigma}(\bar{\pi},w)$$
$$+\frac{1}{n(\alpha,\beta)}(\tilde{x}_{\alpha\beta}-\bar{x}_{\alpha\beta})f_{x_{\alpha\beta}}(\bar{\pi},w)+(\tilde{c}-\bar{c})f_c(\bar{\pi},w)\Big]+\bar{\nu}^T g(\tilde{\pi},\bar{u})e\Big\}dt$$

$$< \int_{\Gamma} \Big\{ \max_{w \in \mathcal{W}} \Big[(\bar{x} - \bar{x}) f_x(\bar{\pi}, w) + (\bar{x}_\sigma - \bar{x}_\sigma) f_{x_\sigma}(\bar{\pi}, w) + \frac{1}{n(\alpha, \beta)} (\bar{x}_{\alpha\beta} - \bar{x}_{\alpha\beta}) f_{x_{\alpha\beta}}(\bar{\pi}, w) + (\bar{c} - \bar{c}) f_c(\bar{\pi}, w) \Big] + \bar{\nu}^T g(\bar{\pi}, \bar{u}) e \Big\} dt,$$

equivalently with

$$\int_{\Gamma} \Big\{ \max_{w \in \mathcal{W}} \Big[(\tilde{x} - \bar{x}) f_x(\tilde{\pi}, w) + (\tilde{x}_\sigma - \bar{x}_\sigma) f_{x_\sigma}(\tilde{\pi}, w) + \frac{1}{n(\alpha, \beta)} (\tilde{x}_{\alpha\beta} - \bar{x}_{\alpha\beta}) f_{x_{\alpha\beta}}(\tilde{\pi}, w) + (\tilde{c} - \bar{c}) f_c(\tilde{\pi}, w) \Big] + \bar{\nu}^T g(\tilde{\pi}, \bar{u}) e + \bar{\gamma}^T h(\tilde{\pi}, \bar{v}) e \Big\} dt$$

$$< \int_{\Gamma} \Big\{ \max_{w \in \mathcal{W}} \Big[(\bar{x} - \bar{x}) f_x(\bar{\pi}, w) + (\bar{x}_\sigma - \bar{x}_\sigma) f_{x_\sigma}(\bar{\pi}, w) + \frac{1}{n(\alpha, \beta)} (\bar{x}_{\alpha\beta} - \bar{x}_{\alpha\beta}) f_{x_{\alpha\beta}}(\bar{\pi}, w) + (\bar{c} - \bar{c}) f_c(\bar{\pi}, w) \Big] + \bar{\nu}^T g(\bar{\pi}, \bar{u}) e + \bar{\gamma}^T h(\bar{\pi}, \bar{v}) e \Big\} dt,$$

or

$$L((\tilde{x}, \tilde{c}), \bar{\nu}, \bar{\gamma}, \bar{w}, \bar{u}, \bar{v}) < L((\bar{x}, \bar{c}), \bar{\nu}, \bar{\gamma}, \bar{w}, \bar{u}, \bar{v}), \ (\tilde{x}, \tilde{c}) \in \mathscr{X} \times \mathscr{C},$$

which contradicts Definition 8.5.0.2. The proof is complete. □

Theorem 8.5.0.2 *Let $(\bar{x}, \bar{c})$ be a normal weak robust efficient solution to the multi-objective optimization problem* (MOP). *Assume that* $\max_{w \in \mathcal{W}} f(\pi, w) = f(\pi, \bar{w})$ *and the constraint functionals $\int_{\Gamma} \bar{\nu}^T g(\pi, \bar{u}) dt$, $\int_{\Gamma} \bar{\gamma}^T h(\pi, \bar{v}) dt$ are convex at $(\bar{x}, \bar{c})$. Then, the point $((\bar{x}, \bar{c}), \bar{\nu}, \bar{\gamma}, \bar{w}, \bar{u}, \bar{v})$ is a robust saddle-point for the Lagrange functional $L((x, c), \nu, \gamma, w, u, v)$ associated with the modified multi-dimensional variational control problem* $(\text{MOP})_{(\bar{x}, \bar{c})}$.

Proof By hypothesis, the relations (8.2.1)–(8.2.4), with $\bar{\mu} = e$, are fulfilled for all $t \in \Gamma$, except at discontinuities. Hence, the conditions (8.2.1) and (8.2.2) yield

$$\int_{\Gamma} (x - \bar{x}) \{ e^T f_x(\bar{\pi}, \bar{w}) + \bar{\nu}^T g_x(\bar{\pi}, \bar{u}) + \bar{\gamma}^T h_x(\bar{\pi}, \bar{v}) - D_\sigma \Big[e^T f_{x_\sigma}(\bar{\pi}, \bar{w}) + \bar{\nu}^T g_{x_\sigma}(\bar{\pi}, \bar{u}) + \bar{\gamma}^T h_{x_\sigma}(\bar{\pi}, \bar{v}) \Big] + \frac{1}{n(\alpha, \beta)} D^2_{\alpha\beta} \Big[e^T f_{x_{\alpha\beta}}(\bar{\pi}, \bar{w}) + \bar{\nu}^T g_{x_{\alpha\beta}}(\bar{\pi}, \bar{u}) + \bar{\gamma}^T h_{x_{\alpha\beta}}(\bar{\pi}, \bar{v}) \Big] \} dt$$

$$+ \int_{\Gamma} (c - \bar{c}) \{ e^T f_c(\bar{\pi}, \bar{w}) + \bar{\nu}^T g_c(\bar{\pi}, \bar{u}) + \bar{\gamma}^T h_c(\bar{\pi}, \bar{v}) \} dt$$

$$= \int_{\Gamma} \Big[(x - \bar{x}) \{ e^T f_x(\bar{\pi}, \bar{w}) + \bar{\nu}^T g_x(\bar{\pi}, \bar{u}) + \bar{\gamma}^T h_x(\bar{\pi}, \bar{v}) \}$$

$$+ (x_\sigma - \bar{x}_\sigma)\{e^T f_{x_\sigma}(\bar{\pi}, \bar{w}) + \bar{v}^T g_{x_\sigma}(\bar{\pi}, \bar{u}) + \bar{\gamma}^T h_{x_\sigma}(\bar{\pi}, \bar{v})\}$$

$$+ \frac{1}{n(\alpha, \beta)}(x_{\alpha\beta} - \bar{x}_{\alpha\beta})\{e^T f_{x_{\alpha\beta}}(\bar{\pi}, \bar{w}) + \bar{v}^T g_{x_{\alpha\beta}}(\bar{\pi}, \bar{u}) + \bar{\gamma}^T h_{x_{\alpha\beta}}(\bar{\pi}, \bar{v})\}\Big]dt$$

$$+ \int_\Gamma (c - \bar{c})\{e^T f_c(\bar{\pi}, \bar{w}) + \bar{v}^T g_c(\bar{\pi}, \bar{u}) + \bar{\gamma}^T h_c(\bar{\pi}, \bar{v})\}dt = 0, \tag{8.5.4}$$

where we used the formula of integration by parts, the divergence formula and the boundary conditions formulated in the considered problem.

Further, taking into account the assumption of convexity for the following multiple integral functionals $\int_\Gamma \bar{v}^T g(\pi, \bar{u})dt$, $\int_\Gamma \bar{\gamma}^T h(\pi, \bar{v})dt$ at $(\bar{x}, \bar{u})$, we obtain

$$\int_\Gamma \Big\{\bar{v}^T g(\pi, \bar{u}) - \bar{v}^T g(\bar{\pi}, \bar{u})\Big\}dt \geq \int_\Gamma (x - \bar{x})\bar{v}^T g_x(\bar{\pi}, \bar{u})dt$$

$$+ \int_\Gamma (x_\sigma - \bar{x}_\sigma)\bar{v}^T g_{x_\sigma}(\bar{\pi}, \bar{u})dt + \frac{1}{n(\alpha, \beta)} \int_\Gamma (x_{\alpha\beta} - \bar{x}_{\alpha\beta})\bar{v}^T g_{x_{\alpha\beta}}(\bar{\pi}, \bar{u})dt$$

$$+ \int_\Gamma (c - \bar{c})\bar{v}^T g_c(\bar{\pi}, \bar{u})dt, \quad \int_\Gamma \Big\{\bar{\gamma}^T h(\pi, \bar{v}) - \bar{\gamma}^T h(\bar{\pi}, \bar{v})\Big\}dt$$

$$\geq \int_\Gamma (x - \bar{x})\bar{\gamma}^T h_x(\bar{\pi}, \bar{v})dt + \int_\Gamma (x_\sigma - \bar{x}_\sigma)\bar{\gamma}^T h_{x_\sigma}(\bar{\pi}, \bar{v})dt$$

$$+ \frac{1}{n(\alpha, \beta)} \int_\Gamma (x_{\alpha\beta} - \bar{x}_{\alpha\beta})\bar{\gamma}^T h_{x_{\alpha\beta}}(\bar{\pi}, \bar{v})dt + \int_\Gamma (c - \bar{c})\bar{\gamma}^T h_c(\bar{\pi}, \bar{v})dt,$$

implying

$$\int_\Gamma \Big\{\bar{v}^T g(\pi, \bar{u}) + \bar{\gamma}^T h(\pi, \bar{v})\Big\}dt - \int_\Gamma \Big\{\bar{v}^T g(\bar{\pi}, \bar{u}) + \bar{\gamma}^T h(\bar{\pi}, \bar{v})\Big\}dt$$

$$\geq \int_\Gamma (x - \bar{x})\Big[\bar{v}^T g_x(\bar{\pi}, \bar{u}) + \gamma^T h_x(\bar{\pi}, \bar{v})\Big]dt$$

$$+ \int_\Gamma (x_\sigma - \bar{x}_\sigma)\Big[\bar{v}^T g_{x_\sigma}(\bar{\pi}, \bar{u}) + \bar{\gamma}^T h_{x_\sigma}(\bar{\pi}, \bar{v})\Big]dt$$

$$+ \frac{1}{n(\alpha, \beta)} \int_\Gamma (x_{\alpha\beta} - \bar{x}_{\alpha\beta})\Big[\bar{v}^T g_{x_{\alpha\beta}}(\bar{\pi}, \bar{u}) + \bar{\gamma}^T h_{x_{\alpha\beta}}(\bar{\pi}, \bar{v})\Big]dt$$

$$+ \int_\Gamma (c - \bar{c})\Big[\bar{v}^T g_c(\bar{\pi}, \bar{u}) + \bar{\gamma}^T h_c(\bar{\pi}, \bar{v})\Big]dt$$

$$= - \int_\Gamma (x - \bar{x})e^T f_x(\bar{\pi}, \bar{w})dt - \int_\Gamma (x_\sigma - \bar{x}_\sigma)e^T f_{x_\sigma}(\bar{\pi}, \bar{w})dt$$

$$- \frac{1}{n(\alpha, \beta)} \int_\Gamma (x_{\alpha\beta} - \bar{x}_{\alpha\beta})e^T f_{x_{\alpha\beta}}(\bar{\pi}, \bar{w})dt - \int_\Gamma (c - \bar{c})e^T f_c(\bar{\pi}, \bar{w})dt,$$

by considering (8.5.4). The previous inequality can be formulated as follows

$$\begin{aligned}
&\int_{\Gamma} (x - \bar{x}) e^T f_x(\bar{\pi}, \bar{w}) dt + \int_{\Gamma} (x_\sigma - \bar{x}_\sigma) e^T f_{x_\sigma}(\bar{\pi}, \bar{w}) dt \\
&\quad + \frac{1}{n(\alpha, \beta)} \int_{\Gamma} (x_{\alpha\beta} - \bar{x}_{\alpha\beta}) e^T f_{x_{\alpha\beta}}(\bar{\pi}, \bar{w}) dt + \int_{\Gamma} (c - \bar{c}) f_c(\bar{\pi}, \bar{w}) dt \\
&\quad + \int_{\Gamma} \left\{ \bar{\nu}^T g(\pi, \bar{u}) + \bar{\gamma}^T h(\pi, \bar{v}) \right\} dt \\
&\quad \geq \int_{\Gamma} (\bar{x} - \bar{x}) e^T f_x(\bar{\pi}, \bar{w}) dt + \int_{\Gamma} (\bar{x}_\sigma - \bar{x}_\sigma) e^T f_{x_\sigma}(\bar{\pi}, \bar{w}) dt \\
&\quad + \frac{1}{n(\alpha, \beta)} \int_{\Gamma} (\bar{x}_{\alpha\beta} - \bar{x}_{\alpha\beta}) e^T f_{x_{\alpha\beta}}(\bar{\pi}, \bar{w}) dt + \int_{\Gamma} (\bar{c} - \bar{c}) e^T f_c(\bar{\pi}, \bar{w}) dt \\
&\quad + \int_{\Gamma} \left\{ \bar{\nu}^T g(\bar{\pi}, \bar{u}) + \bar{\gamma}^T h(\bar{\pi}, \bar{v}) \right\} dt,
\end{aligned}$$

which involves the inequality

$$L((x, c), \bar{\nu}, \bar{\gamma}, w, \bar{u}, \bar{v}) \geqq L((\bar{x}, \bar{c}), \bar{\nu}, \bar{\gamma}, w, \bar{u}, \bar{v}), \ \forall (x, c) \in \mathscr{X} \times \mathscr{C}. \quad (8.5.5)$$

Also, the following inequality is satisfied

$$\int_{\Gamma} \nu^T g(\bar{\pi}, u) dt + \int_{\Gamma} \gamma^T h(\bar{\pi}, v) dt \leq 0$$

for all $(\nu, \gamma) \in \mathbb{R}_+^m \times \mathbb{R}^n$, $(u, v) \in \mathcal{U} \times \mathcal{V}$ and, using the feasibility of $(\bar{x}, \bar{u})$, we obtain

$$\int_{\Gamma} \nu^T g(\bar{\pi}, u) dt + \int_{\Gamma} \gamma^T h(\bar{\pi}, v) dt \leq \int_{\Gamma} \bar{\nu}^T g(\bar{\pi}, \bar{u}) dt + \int_{\Gamma} \bar{\gamma}^T h(\bar{\pi}, \bar{v}) dt,$$

or, equivalently,

$$\begin{aligned}
&\int_{\Gamma} (\bar{x} - \bar{x}) e^T f_x(\bar{\pi}, \bar{w}) dt + \int_{\Gamma} (\bar{x}_\sigma - \bar{x}_\sigma) e^T f_{x_\sigma}(\bar{\pi}, \bar{w}) dt \\
&\quad + \frac{1}{n(\alpha, \beta)} \int_{\Gamma} (\bar{x}_{\alpha\beta} - \bar{x}_{\alpha\beta}) e^T f_{x_{\alpha\beta}}(\bar{\pi}, \bar{w}) dt + \int_{\Gamma} (\bar{c} - \bar{c}) e^T f_c(\bar{\pi}, \bar{w}) dt \\
&\quad + \int_{\Gamma} \nu^T g(\bar{\pi}, u) dt + \int_{\Gamma} \gamma^T h(\bar{\pi}, v) dt \\
&\quad \leq \int_{\Gamma} (\bar{x} - \bar{x}) e^T f_x(\bar{\pi}, \bar{w}) dt + \int_{\Gamma} (\bar{x}_\sigma - \bar{x}_\sigma) e^T f_{x_\sigma}(\bar{\pi}, \bar{w}) dt \\
&\quad + \frac{1}{n(\alpha, \beta)} \int_{\Gamma} (\bar{x}_{\alpha\beta} - \bar{x}_{\alpha\beta}) e^T f_{x_{\alpha\beta}}(\bar{\pi}, \bar{w}) dt + \int_{\Gamma} (\bar{c} - \bar{c}) e^T f_c(\bar{\pi}, \bar{w}) dt \\
&\quad + \int_{\Gamma} \bar{\nu}^T g(\bar{\pi}, \bar{u}) dt + \int_{\Gamma} \bar{\gamma}^T h(\bar{\pi}, \bar{v}) dt,
\end{aligned}$$

involving

$$L((\bar{x},\bar{c}),\nu,\gamma,w,u,v) \leqq L((\bar{x},\bar{c}),\bar{\nu},\bar{\gamma},w,\bar{u},\bar{v}),\ \forall\nu\in\mathbb{R}^m_+,\ \forall\gamma\in\mathbb{R}^n,\ \forall u\in\mathcal{U},\ \forall v\in\mathcal{V}. \quad (8.5.6)$$

In consequence, by (8.5.5) and (8.5.6), we conclude that $((\bar{x},\bar{c}),\bar{\nu},\bar{\gamma},\bar{w},\bar{u},\bar{v})$ is a robust saddle-point for the Lagrange functional $L((x,c),\nu,\gamma,w,u,v)$ associated with the modified multi-dimensional variational control problem $(\text{MOP})_{(\bar{x},\bar{c})}$ and the proof is complete. □

References

1. A. Jayswal, T. Antczak, S. Jha, Modified objective function approach for multitime variational problems. Turkish J. Math. **42**, 1111–1129 (2018)
2. A. Jayswal, T. Antczak, S. Jha, On equivalence between a variational problem and its modified variational problem with the η objective function under invexity. Int. Trans. Op. Res. **26**, 2053–2070 (2019)
3. T. Antczak, A. Jayswal, S. Jha, The modified objective function method for univex multiobjective variational problems. Bull. Iranian Math. Soc. **45**, 267–282 (2019)
4. A.J. Preeti, Saddle point criteria for multi-dimensional control optimisation problem involving first-order PDE constraints. Internat. J. Control **94**, 1567–1576 (2021)
5. X.K. Sun, K.L. Teo, J. Zeng, X.L. Guo, On approximate solutions and saddle point theorems for robust convex optimization. Optim. Lett. **14**, 1711–1730 (2020)
6. S. Treanţă, On modified interval-valued variational control problems with first-order PDE constraints. Symmetry-Basel **12**, 472 (2020)
7. S. Treanţă, On a modified optimal control problem with first-order PDE constraints and the associated saddle-point optimality criterion. Eur. J. Control. **51**, 1–9 (2020)
8. S. Treanţă, Saddle-point optimality criteria in modified variational control problems with PDE constraints. Opt. Control Appl. Meth. **41**, 1160–1175 (2020)
9. S. Treanţă, Saddle-point optimality criteria involving (ρ, b, d)-invexity and (ρ, b, d)-pseudoinvexity in interval-valued optimization problems. Int. J. Control (2021). https://doi.org/10.1080/00207179.2020.1837960
10. A.J. Preeti, M. Arana-Jiménez, Robust saddle-point criteria for multi-dimensional optimization problems with data uncertainty. Int. J. Control (2020). https://doi.org/10.1080/00207179.2020.1759827
11. X.W. Liu, Y.X. Yuan, A robust algorithm for optimization with general equality and inequality constraints. SIAM J. Sci. Comput. **22**, 517–534 (2000)
12. V. Jeyakumar, J.H. Wang, G. Li, Lagrange multiplier characterizations of robust best approximations under constraint data uncertainty. J. Math. Anal. Appl. **393**, 285–297 (2012)
13. H.Z. Wei, C.R. Chen, S.J. Li, Characterizations for optimality conditions of general robust optimization problems. J. Optim. Theory Appl. **177**, 835–856 (2018)
14. S. Treanţă, Efficiency in uncertain variational control problems. Neural Comput. Appl. **33**, 5719–5732 (2021)
15. D.J. Saunders, *The Geometry of Jet Bundles*, London Math. Soc. Lecture Notes Series, vol. 142 (Cambridge University Press, Cambridge, 1989)
16. S. Treanţă, M. Arana-Jiménez, KT-pseudoinvex multidimensional control problem. Optim. Control Appl. Meth. **39**, 1291–1300 (2018)
17. Şt. Mititelu, Optimality and duality for invex multi-dimensional control problems with mixed constraints. J. Adv. Math. Stud. **2**, 25–34 (2009)

18. S. Treanţă, Multiobjective fractional variational problem on higher-order jet bundles. Commun. Math. Stat. **4**, 323–340 (2016)
19. Şt. Mititelu, S. Treanţă, Efficiency conditions in vector control problems governed by multiple integrals. J. Appl. Math. Comput. **57**, 647–665 (2018)
20. I. Ahmad, T.R. Gulati, Mixed type duality for multiobjective variational problems with generalized (F; ρ)-convexity. J. Math. Anal. Appl. **306**, 669–683 (2005)
21. T.R. Gulati, G. Mehndiratta, K. Verma, Symmetric duality for second-order fractional programs. Optim. Lett. **7**, 1341–1352 (2013)
22. C. Udrişte and I. Ţevy, Multitime dynamic programming for curvilinear integral actions. J. Optim. Theory Appl. **146**, 189–207 (2010)
23. C. Udrişte, A. Bejenaru, Multitime optimal control with area integral costs on boundary. Balkan J. Geom. Appl. **16**, 138–154 (2011)

The manufacturer's authorised representative in the EU is Springer Nature Customer Service Centre GmbH, Europaplatz 3, 69115 Heidelberg, Germany. If you have any concerns regarding our products, please contact ProductSafety@springernature.com

Printed and bound by CPI Group (UK) Ltd, Croydon, CR0 4YY
30/09/2026
02225237-0001